DEPARTMENT OF THE INTERIOR

UNITED STATES GEOLOGICAL SURVEY

GEORGE OTIS SMITH, DIRECTOR

WATER-SUPPLY PAPER 269

SURFACE WATER SUPPLY OF THE UNITED STATES

1909

PART IX. COLORADO RIVER BASIN

PREPARED UNDER THE DIRECTION OF M. O. LEIGHTON

BY

W. B. FREEMAN AND R. H. BOLSTER

WASHINGTON

GOVERNMENT PRINTING OFFICE

1911

CONTENTS.

ILLUSTRATIONS.

6

SURFACE WATER SUPPLY OF THE COLORADO RIVER BASIN, 1909.

By W. B. FREEMAN and R. H. BOLSTER.

INTRODUCTION.

AUTHORITY FOR INVESTIGATIONS.

This volume contains results of flow measurements made on certain streams in the United States. The work was performed by the water-resources branch of the United States Geological Survey, either independently or in cooperation with organizations mentioned herein. These investigations are authorized by the organic law of the Geological Survey (Stat. L., vol. 20, p. 394), which provides, among other things, as follows:

Provided that this officer [the Director] shall have the direction of the Geological Survey and the classification of public lands and examination of the geological structure, mineral resources, and products of the national domain.

Inasmuch as water is the most abundant and. most valuable mineral in nature, the investigation of water resources is included under the above provision for investigating mineral resources. The work has been supported since the fiscal year ending June 30, 1895, by appropriations in successive sundry civil bills passed by Congress under the following item:

For gaging the streams and determining the water supply of the United States, and for the investigation of underground currents and artesian wells, and for the preparation of reports upon the best methods of utilizing the water resources.

The various appropriations that have been made for this purpose are as follows:

Annual appropriations for the fiscal year ending June 30—

1895	$12,500
1896	20,000
1897 to 1900, inclusive	50,000
1901 to 1902, inclusive	100,000
1903 to 1906, inclusive	200,000
1907	150,000
1908 to 1910, inclusive	100,000
1911	150,000

SCOPE OF INVESTIGATIONS.

These investigations of stream flow are not complete nor do they include all the river systems or parts thereof that might purposefully be studied. The scope of the work is limited by the appropriations available. The field covered is the widest and the character of the work is believed to be the best possible under the controlling conditions. The work would undoubtedly have greater scientific importance and ultimately be of more practical value if the money now expended for wide areas were concentrated on a few small drainage basins; but such a course is impossible because general appropriations made by Congress are applicable to all parts of the country. Each part demands its proportionate share of the benefits.

It is essential that records of stream flow shall be kept during a period of years long enough to determine within reasonable limits the entire range of flow from the absolute maximum to the absolute minimum. The length of such a period manifestly differs for different streams. Experience has shown that the records for some streams should cover from 5 to 10 years, and for other streams 20 years or even more, the limit being determined by the relative importance of the stream and the interdependence of the results with other long-time records on adjacent streams.

In the performance of this work an effort is made to reach the highest degree of precision possible with a rational expenditure of time and a judicious expenditure of a small amount of money. In all engineering work there is a point beyond which refinement is needless and wasteful, and this statement applies with especial force to stream-flow measurements. It is confidently believed that the stream-flow data presented in the publications of the Survey are in general sufficiently accurate for all practical purposes. Many of the records are, however, of insufficient length, owing to the unforeseen reduction of appropriations and consequent abandonment of stations. All persons are cautioned to exercise the greatest care in using such incomplete records.

Records have been obtained at more than 1,550 different points in the United States, and in addition the surface water supply of small areas in Seward Peninsula and the Yukon-Tanana region, Alaska, has been investigated. During 1909 regular gaging stations were maintained by the Survey and cooperating organizations at about 850 points in the United States, and many miscellaneous measurements were made at other points. Data were also obtained in regard to precipitation, evaporation, storage reservoirs, river profiles, and water power in many sections of the country and will be made available in the regular surface water-supply papers and in special papers from time to time.

PURPOSES OF THE WORK.

The results contained in this volume are requisite to meet the immediate demands of many public interests, including navigation, irrigation, domestic water supply, water power, swamp and overflow land drainage, and flood prevention.

Navigation.—The Federal Government has expended more than $250,000,000 for the improvement of inland navigation, and prospective expenditures will approximate several times this amount. It is obvious that the determination of stream flow is necessary to the intelligent solution of the many problems involved.

Irrigation.—The United States is now expending $51,000,000 on Federal irrigation systems, and this amount is far exceeded by the private expenditures of this nature in the arid West. The integrity of any irrigation system depends absolutely on the amount of water available. Therefore investigations of stream flow in that portion of the country are not only of first importance in the redemption of the lands, but constitute an insurance of Federal and private investments.

Domestic water supply.—The highest use of water is for domestic supply, and although this branch of the subject is of less direct Federal interest than the branches already named, it nevertheless has so broad a significance with respect to the general welfare that the Federal Government is ultimately and intimately concerned.

Water power.—The development of the water power of the country is an economic necessity. Our stock of coal is being rapidly depleted, and the cost of steam power is increasing accordingly. Industrial growth and, as a consequence, the progress of the United States as a nation will cease if cheap power is not available. Water power affords the only avenue now open. When the electric transmission of power was accomplished, the relation of our water powers to national economy changed entirely. Before the day of electric transmission water power was important only at the locality at which it was generated, but it has now become a public utility in which the individual citizen is vitally interested. Inasmuch as the amount of water power that may be made available depends on the flow of rivers, the investigation of flow becomes a prerequisite in the judicious management of this source of energy.

Drainage of swamp and overflowed lands.—More than 70,000,000 acres of the richest land in this country are now practically worthless or of precarious value by reason of overflow and swamp conditions. When this land is drained, it becomes exceedingly productive, and its value increases many fold. Such reclamation would add to the national assets at least $700,000,000. The study of run-off is the first consideration in connection with drainage projects. If by the drainage of a large area into any particular channel that channel becomes so gorged with water which it had not hitherto been called upon to convey that overflow is caused in places where previously the land was not subject to inundation, then drainage results merely in an exchange of land values. This is not the purpose of drainage improvement.

Flood prevention.—The damage from floods in the United States probably exceeds on the average $100,000,000 annually, and in the year 1908, according to estimates based on reliable data, the aggregate damage approximated $250,000,000. Such an annual tax on the property of great regions should be reduced in the orderly progress of government. It goes without saying that any consideration of flood prevention must be based on a thorough knowledge of stream flow, both in the contributing areas which furnish the water and along the great lowland rivers.

PUBLICATIONS.

The data on stream flow collected by the United States Geological Survey since its inception have appeared in the annual reports, bulletins, and water-supply papers. Owing to natural processes of evolution and to changes in governmental requirements, the character of the work and the territory covered by these different publications has varied greatly. For the purpose of uniformity in the presentation of reports a general plan has been agreed upon by the United States Reclamation Service, the United States Forest Service, the United States Weather Bureau, and the United States Geological Survey, according to which the area of the United States has been divided into 12 parts, whose boundaries coincide with certain natural drainage lines. The areas so described are indicated by the following list of papers on surface water supply for 1909. The dividing line between the North Atlantic and South Atlantic drainage areas lies between York and James rivers.

Papers on surface water supply of the United States, 1909.

Part.	No.	Title.	Part.	No.	Title.
I	261	North Atlantic coast.	VI	266	Missouri River basin.
II	262	South Atlantic coast and eastern Gulf of Mexico.	VII	267	Lower Mississippi River basin.
			VIII	268	Western Gulf of Mexico.
III	263	Ohio River basin.	IX	269	Colorado River basin.
IV	264	St. Lawrence River basin.	X	270	Great Basin.
V	265	Upper Mississippi River and Hudson Bay basin.	XI	271	California.
			XII	272	North Pacific coast.

The following table gives the character of data regarding stream flow at regular stations to be found in the various publications of the United States Geological Survey, exclusive of all special papers.

Numbers of reports are inclusive, and dates also are inclusive, so far as the data are available:

Stream-flow data in reports of the United States Geological Survey.

[Ann.=Annual Report; B.=Bulletin; W. S.=Water-Supply Paper.]

Report.	Character of data.	Year.
10th Ann., pt. 2........	Descriptive information only..................................	
11th Ann., pt. 2........	Monthly discharge..	1884 to Sept., 1890.
12th Ann., pt. 2........	.....do..	1884 to June 30, 1891.
13th Ann., pt. 3........	Mean discharge in second-feet...............................	1884 to Dec. 31, 1892.
14th Ann., pt. 2........	Monthly discharge (long-time records, 1871 to 1893)...........	1888 to Dec. 31, 1893.
B. 131.................	Descriptions, measurements, gage heights, and ratings........	1893 and 1894.
16th Ann., pt. 2........	Descriptive information only.................................	
B. 140.................	Descriptions, measurements, gage heights, ratings, and monthly discharge (also many data covering earlier years).	1895.
W. S. 11..............	Gage heights (also gage heights for earlier years)..............	1896.
18th Ann., pt. 4........	Descriptions, measurements, ratings, and monthly discharge (also similar data for earlier years.)	1895 and 1896.
W. S. 15..............	Descriptions, measurements, and gage heights, eastern United States, eastern Mississippi River, and Missouri River above junction with Kansas.	1897.
W. S. 16..............	Descriptions, measurements, and gage heights, western Mississippi River below junction of Missouri and Platte, and western United States.	1897.
19th Ann., pt. 4........	Descriptions, measurements, ratings, and monthly discharge (also some long-time records).	1897.
W. S. 27..............	Measurements, ratings, and gage heights, eastern United States, eastern Mississippi River, and Missouri River.	1898.
W. S. 28..............	Measurements, ratings, and gage heights, Arkansas River and western United States.	1898.
20th Ann., pt. 4........	Monthly discharge (also for many earlier years)...............	1898.
W. S. 35 to 39.........	Descriptions, measurements, gage heights, and ratings........	1899.
21st Ann., pt. 4........	Monthly discharge...	1899.
W. S. 47 to 52.........	Descriptions, measurements, gage heights, and ratings........	1900.
22d Ann., pt. 4........	Monthly discharge...	1900.
W. S. 65, 66...........	Descriptions, measurements, gage heights, and ratings........	1901.
W. S. 75..............	Monthly discharge...	1901.
W. S. 82 to 85.........	Complete data..	1902.
W. S. 97 to 100........	.....do..	1903.
W. S. 124 to 135.......	.....do..	1904.
W. S. 165 to 178.......	.....do..	1905.
W. S. 201 to 214.......	Complete data, except descriptions...........................	1906.
W. S. 241 to 252.......	Complete data..	1907–8.
W. S. 261 to 272.......	.....do..	1909.

NOTE.—No data regarding stream flow are given in the 15th and 17th annual reports.

The records at most of the stations discussed in these reports extend over a series of years. An index of the reports containing records prior to 1904 has been published in Water-Supply Paper 119. The first table which follows gives, by years and drainage basins, the numbers of the papers on surface water supply published from 1899 to 1909. Wherever the data for a drainage basin appear in two papers the number of one is placed in parentheses and the portion of the basin covered by that paper is indicated in the second table. For example, in 1904 the data for Missouri River were published in Water-Supply Papers 130 and 131, and the portion of the records contained in Water-Supply Paper 131, as indicated by the second table, is that relating to Platte and Kansas rivers.

Numbers of water-supply papers containing results of stream measurements, 1899–1909.

	1899 a	1900 b	1901	1902	1903	1904	1905	1906	1907–8	1909
Atlantic coast and eastern Gulf of Mexico:										
New England rivers..	35	47	65,75	82	97	124	165	201	241	261
Hudson River to Delaware River, inclusive..........	35	47,(48)	65,75	82	97	125	166	202	241	261
Susquehanna River to York River, inclusive.............	35	48	65,75	82	97	126	167	203	241	261
James River to Yadkin River, inclusive...............	(35),36	48	65,75	(82),83	(97),98	126	167	203	242	262
Santee River to Pearl River, inclusive....	36	48	65,75	83	98	127	168	204	242	262
St. Lawrence River......	36	49	65,75	(82),83	97	129	170	206	244	264
Hudson Bay.............			66,75	85	100	130	171	207	245	265
Mississippi River:										
Ohio River..........	36	48,(49)	65,75	83	98	128	169	205	243	263
Upper Mississippi River..............	36	49	65,75	83	98,(99)	128, (130)	171	207	245	265
Missouri River.......	(36),37	49,(50)	66,75	84	99	130, (131)	172	208	246	266
Lower Mississippi River..............	37	50	(65), 65,75	(83),84	(98),99	(128), 131	(169), 173	(205), 209	247	267
Western Gulf of Mexico..	37	50	66,75	84	99	132	174	210	248	268
Pacific coast and Great Basin:										
Colorado River.......	(37),38	50	66,75	85	100	133, (134)	175, (177)	211, (213)	249, (251)	269, (271)
Great Basin..........	38,(39)	51	66,75	85	100	133, (134)	176, (177)	212, (213)	250, (251)	270, (271)
South Pacific coast to Klamath River, inclusive..........	(38),39	51	66,75	85	100	134	177	213	251	271
North Pacific coast...	38	51	66,75	85	100	135	(177), 178	214	252	272

a Rating tables and index to Water-Supply Papers 35–39 contained in Water-Supply Paper 39.
b Rating tables and index to Water-Supply Papers 47–52 and data on precipitation, wells, and irrigation in California and Utah contained in Water-Supply Paper 52.

Numbers of water-supply papers containing data covering portions of drainage basins.

No.	River basin.	Tributaries included.
35	James............................	
36	Missouri.........................	Gallatin.
37	Colorado.........................	Green, Gunnison, Grand above junction with Gunnison.
38	Sacramento.......................	Except Kings and Kern.
39	Great Basin......................	Mohave.
48	Delaware.........................	Wissahickon and Schuylkill.
49	Ohio.............................	Scioto.
50	Missouri.........................	Loup and Platte near Columbus, Nebr. All tributaries below junction with Platte.
65	Lower Mississippi.................	Yazoo.
82	James............................ St. Lawrence.....................	Lake Ontario, tributaries to St. Lawrence River proper.
83	Lower Mississippi.................	Yazoo.
97	James............................	
98	Lower Mississippi.................	Do.
99	Upper Mississippi.................	Tributaries from the west.
128	Lower Mississippi.................	Yazoo.
130	Upper Mississippi.................	Tributaries from the west.
131	Missouri.........................	Platte, Kansas.
134	Colorado......................... Great Basin......................	Data near Yuma, Ariz., repeated. Susan, Owens, Mohave.
169	Lower Mississippi.................	Yazoo.
177	Colorado......................... Great Basin...................... North Pacific coast..............	Below junction with Gila. Susan repeated, Owens, Mohave. Rogue, Umpqua, Siletz.
205	Lower Mississippi.................	Yazoo, Homochitto.
213	Colorado......................... Great Basin......................	Data at Hardyville repeated; at Yuma, Salton Sea. Owens, Mohave.
251	Colorado.........................	Yuma and Salton Sea stations repeated.
271	Great Basin......................	Owens River Basin.

The order of treatment of stations in any basin in these papers is downstream. The main stem of any river is determined on the basis of drainage area, local changes in name and lake surface being disregarded. After all stations from the source to the mouth of the main stem of the river have been given, the tributaries are taken up in regular order from source to mouth. The tributaries are treated the same as the main stream, all stations in each tributary basin being given before taking up the next one below.

The exceptions to this rule occur in the records for Mississippi River, which are given in four parts, as indicated above, and in the records for large lakes, where it is often clearer to take up the streams in regular order around the rim of the lake than to cross back and forth over the lake surface.

DEFINITION OF TERMS.

The volume of water flowing in a stream—the "run-off" or "discharge"—is expressed in various terms, each of which has become associated with a certain class of work. These terms may be divided into two groups: (1) Those which represent a rate of flow, as second-feet, gallons per minute, miner's inches, and run-off in second-feet per square mile, and (2) those which represent the actual quantity of water, as run-off in depth in inches and acre-feet. They may be defined as follows:

"Second-foot" is an abbreviation for cubic foot per second and is the rate of discharge of water flowing in a stream 1 foot wide, 1 foot deep, at a rate of 1 foot per second. It is generally used as a fundamental unit from which others are computed by the use of the factors given in the following table of equivalents.

"Gallons per minute" is generally used in connection with pumping and city water supply.

The "miner's inch" is the rate of discharge of water that passes through an orifice 1 inch square under a head which varies locally. It is commonly used by miners and irrigators throughout the West, and is defined by statute in each State in which it is used.

"Second-feet per square mile" is the average number of cubic feet of water flowing per second from each square mile of area drained, on the assumption that the run-off is distributed uniformly both as regards time and area.

"Run-off in depth in inches on drainage area" is the depth to which the drainage area would be covered if all the water flowing from it in a given period were conserved and uniformly distributed on the surface. It is used for comparing run-off with rainfall, which is usually expressed in depth in inches.

"Acre-foot" is equivalent to 43,560 cubic feet, and is the quantity required to cover an acre to the depth of 1 foot. It is commonly used in connection with storage for irrigation work.

CONVENIENT EQUIVALENTS.

The following is a list of convenient equivalents for use in hydraulic computations:

1 second-foot equals 40 California miner's inches (law of March 23, 1901).
1 second-foot equals 38.4 Colorado miner's inches.
1 second-foot equals 40 Arizona miner's inches.
1 second-foot equals 7.48 United States gallons per second; equals 448.8 gallons per minute; equals 646,272 gallons for one day.
1 second-foot equals 6.23 British imperial gallons per second.
1 second-foot for one year covers 1 square mile 1.131 feet or 13.572 inches deep.
1 second-foot for one year equals 31,536,000 cubic feet.
1 second-foot equals about 1 acre-inch per hour.
1 second-foot for one day covers 1 square mile 0.03719 inch deep.
1 second-foot for one 28-day month covers 1 square mile 1.041 inches deep.
1 second-foot for one 29-day month covers 1 square mile 1.079 inches deep.
1 second-foot for one 30-day month covers 1 square mile 1.116 inches deep.
1 second-foot for one 31-day month covers 1 square mile 1.153 inches deep.
1 second-foot for one day equals 1.983 acre-feet.
1 second-foot for one 28-day month equals 55.54 acre-feet.
1 second-foot for one 29-day month equals 57.52 acre-feet.
1 second-foot for one 30-day month equals 59.50 acre-feet.
1 second-foot for one 31-day month equals 61.49 acre-feet.
100 California miner's inches equals 18.7 United States gallons per second.
100 California miner's inches equals 96.0 Colorado miner's inches.
100 California miner's inches for one day equals 4.96 acre-feet.
100 Colorado miner's inches equals 2.60 second-feet.
100 Colorado miner's inches equals 19.5 United States gallons per second.
100 Colorado miner's inches equals 104 California miner's inches.
100 Colorado miner's inches for one day equals 5.17 acre-feet.
100 United States gallons per minute equals 0.223 second-foot.
100 United States gallons per minute for one day equals 0.442 acre-foot.
1,000,000 United States gallons per day equals 1.55 second-feet.
1,000,000 United States gallons equals 3.07 acre-feet.
1,000,000 cubic feet equals 22.95 acre-feet.
1 acre-foot equals 325,850 gallons.
1 inch deep on 1 square mile equals 2,323,200 cubic feet.
1 inch deep on 1 square mile equals 0.0737 second-foot per year.
1 foot equals 0.3048 meter.
1 mile equals 1.60935 kilometers.
1 mile equals 5,280 feet. ·
1 acre equals 0.4047 hectare.
1 acre equals 43,560 square feet.
1 acre equals 209 feet square, nearly.
1 square mile equals 2.59 square kilometers.
1 cubic foot equals 0.0283 cubic meter.
1 cubic foot equals 7.48 gallons.
1 cubic foot of water weighs 62.5 pounds.

1 cubic meter per minute equals 0.5886 second-foot.

1 horsepower equals 550 foot-pounds per second.

1 horsepower equals 76.0 kilogram-meters per second.

1 horsepower equals 746 watts.

1 horsepower equals 1 second-foot falling 8.80 feet.

1⅓ horsepower equals about 1 kilowatt.

To calculate water power quickly: $\dfrac{\text{Sec.-ft.} \times \text{fall in feet}}{11}$ =net horsepower on water wheel realizing 80 per cent of theoretical power.

EXPLANATION OF TABLES.

For each drainage basin there is given a brief general description covering such features as area, source, tributaries, topography, geology, forestation, rainfall, ice conditions, irrigation, storage, power possibilities, and other special features of importance or interest.

For each regular current-meter gaging station are given in general, and so far as available, the following data: Description of station, list of discharge measurements, table of daily gage heights, table of daily discharges, table of monthly and yearly discharges, and run-off. For stations located at weirs or dams the gage-height table is omitted.

In addition to statements regarding the location and installation of current-meter stations the descriptions give information in regard to any conditions which may affect the constancy of the relation of gage height to discharge, covering such points as ice, logging, shifting conditions of flow, and backwater; also information regarding diversions which decrease the total flow at the measuring section. Statements are also made regarding the accuracy and reliability of the data.

The discharge-measurement table gives the results of the discharge measurements made during the year, including the date, name of hydrographer, width and area of cross section, gage height, and discharge in second-feet.

The table of daily gage height gives the daily fluctuations of the surface of the river as found from the mean of the gage readings taken each day. At most stations the gage is read in the morning and in the evening. The gage height given in the table represents the elevation of the surface of the water above the zero of the gage. All gage heights during ice conditions, backwater from obstructions, etc., are published as recorded, with suitable footnotes. The rating is not applicable for such periods unless the proper correction to the gage heights is known and applied. Attention is called to the fact that the zero of the gage is placed at an arbitrary datum and has no relation to zero flow or the bottom of the river. In general, the zero is located somewhat below the lowest known flow, so that negative readings shall not occur.

The discharge measurements and gage heights are the base data from which rating tables, daily discharge tables, and monthly discharge tables are computed.

The rating table gives, either directly or by interpolation, the discharge in second-feet corresponding to every stage of the river recorded during the period for which it is applicable. It is not published in this report, but can be determined from the gage heights and discharges for the purposes of verifying the published results, as follows:

First plot the discharge measurements for the current and earlier years on cross-section paper with gage heights in feet as ordinates and discharges in second-feet as abscissas. Then tabulate a number of gage heights taken from the daily gage-height table for the complete range of stage given and the corresponding discharges for the days selected from the daily discharge table and plot the values on cross-section paper. The last points plotted will define the rating curve used and will lie among the plotted discharge measurements. After drawing the rating curve, a table can be developed by scaling off the discharge in second-feet for each tenth foot of gage height. These values should be so adjusted that the first differences shall always be increasing or constant, except for known backwater conditions.

The table of daily discharge gives the discharges in second-feet corresponding to the observed gage heights as determined from the rating tables.

In the table of monthly discharge the column headed "Maximum" gives the mean flow, as determined from the rating table, for the day when the mean gage height was highest. As the gage height is the mean for the day, it does not indicate correctly the stage when the water surface was at crest height and the corresponding discharge consequently larger than given in the maximum column. Likewise, in the column of "Minimum" the quantity given is the mean flow for the day when the mean gage height was lowest. The column headed "Mean" is the average flow in cubic feet for each second during the month. On this the computations for the remaining columns, which are defined on page 13, are based.

FIELD METHODS OF MEASURING STREAM FLOW.

There are three distinct methods of determining the flow of open-channel streams: (1) By measurements of slope and cross section and the use of Chezy's and Kutter's formulas; (2) by means of a weir or dam; (3) by measurements of the velocity of the current and of the area of the cross section. The method chosen depends on the local physical conditions, the degree of accuracy desired, the funds available, and the length of time that the record is to be continued.

Slope method.—Much information has been collected relative to the coefficients to be used in the Chezy formula, $v = c\sqrt{Rs}$. This has been utilized by Kutter, both in developing his formula for c and in determining the values of the coefficient n which appears therein. The results obtained by the slope method are in general only roughly approximate, owing to the difficulty in obtaining accurate data and the uncertainty of the value for n to be used in Kutter's formula. The most common use of this method is in estimating the flood discharge of a stream when the only data available are the cross section, the slope as shown by marks along the bank, and a knowledge of the general conditions. It is seldom used by the United States Geological Survey.[1]

Weir method.—Relatively few stations are maintained at weirs or dams by the United States Geological Survey. Standard types of sharp-crested and broad-crested weirs within the limits for which accurate coefficients have been experimentally obtained give very accurate records of discharge if properly maintained. At practically all broad-crested weirs, however, there is a diversion of water either through or around the dam, usually for the purpose of development of water power. The flow is often complicated and the records are subject to errors from such sources as leakage through the dam, backwater at high stages, uncertainty regarding coefficient, irregularity of crest, obstructions from logs or ice, use of flashboards, old turbines with imperfect ratings, and many others depending on the type of development and the uses of the diverted water.

In general records of discharge at dams are usually accurate enough for practical use if no others are available. It has been the general experience of the United States Geological Survey, however, that records at current-meter gaging stations under unobstructed channel conditions are more accurate than those collected at dams, and where the conditions are reasonably favorable are practically as good as those obtained at sharp-crested weirs.[2]

Velocity method.—Streams in general present throughout their courses to a greater or less extent all conditions of permanent, semi-permanent, and varying conditions of flow. In accordance with the location of the measuring section with respect to these physical conditions, current-meter gaging stations may in general be divided into four classes—(1) those with permanent conditions of flow;

---

[1] Full information regarding this method is given in the various textbooks on hydraulics.

[2] The determination of discharge over the different types of weirs and dams is treated fully in "Weir experiments, coefficients, and formulas" (Water-Supply Paper 200) and in the various text-books on hydraulics. "Turbine water-wheel tests and power tables" (Water-Supply Paper 180) treats of the discharge through turbines when used as meters. The edition of the latter water-supply paper is nearly exhausted. It can, however, be consulted at most of the larger libraries of the country or it can be obtained from the superintendent of documents, Washington, D. C., at a cost of 20 cents.

(2) those with beds which change only during extreme high water; (3) those with beds which change frequently, but which do not cause a variation of more than about 5 per cent of the discharge curves from year to year; and (4) those with constantly shifting beds. In determining the daily flow different office methods are necessary for each class. The field data on which the determinations are based and the methods of collecting them are, however, in general the same.

Great care is taken in the selection and equipment of gaging stations for determining discharge by velocity measurements, in order that the data may have the required degree of accuracy. They are located, as far as possible, at such points that the relation between gage height and discharge will always remain constant for any given stage. The experience of engineers of the Geological Survey has been that permanency of conditions of flow is the prime requisite of any current-meter gaging station when maintained for several years unless funds are available to cover all changes in conditions of flow. A straight, smooth section without cross currents, backwater, boils, etc., at any stage is highly desirable, but on most streams is not attainable except at the cost of a cable equipment. Rough, permanent sections, if measurements are properly made by experienced engineers, taking measuring points at a distance apart of 5 per cent or less of the total width, will, within reasonable limits, yield better results for a given outlay of money than semi-permanent or shifting sections with smooth, uniform current. So far as possible stations are located where the banks are high and not subject to overflow at high stages and out of the influence of tributary streams, dams, or other artificial obstructions which might affect the relation between gage height and discharge.

A gaging station consists essentially of a gage for determining the daily fluctuations of stage of the river and some structure or apparatus from which discharge measurements are made, usually a bridge or cable.

The two factors required to determine the discharge of a stream past a section perpendicular to the mean direction of the current are the area of the cross section and the mean velocity of flow normal to that section.

In making a measurement with a current meter a number of points, called measuring points, are measured off above and in the plane of the measuring section at which observations of depth and velocity are taken. (See Pl. I, A.) These points are spaced equally for those parts of the section where the flow is uniform and smooth and are spaced unequally for other parts, according to the discretion and judgment of the engineer. In general the points should not be spaced farther apart than 5 per cent of the channel width, nor farther

A. FOR BRIDGE MEASUREMENT.

B. FOR WADING MEASUREMENT.

TYPICAL GAGING STATIONS.

apart than the approximate mean depth of the section at the time of measurement.

The measuring points divide the total cross section into elementary strips at each end of which observations of depth and velocity are made. The discharge of any elementary strip is the product of the average of the depths at the two ends times the width of the strip times the average of the mean velocities at the two ends of the strip. The sum of the discharges of the elementary strips is the total discharge of the stream.[1]

Depths for the determination of the area are usually obtained by sounding with the current meter and cable. In rough sections or swift currents an ordinary weight and cable are used, particular care being taken that all observations shall be in the plane of the cross section.

Two methods of determining the velocity of flow of a stream are in general use—the float method and the current-meter method.

The float method, with its various modifications of surface, subsurface, and tube or rod floats, is now considered obsolete in the ordinary practice of the United States Geological Survey. The use of this method is limited to special conditions where it is impracticable to use the current meter, such as in places where large quantities of ice or débris which may damage the meter are flowing with the current, and for miscellaneous measurements or other work where a high degree of accuracy is not necessary. Tube floats are very satisfactory for use in canals with regular bottoms and even flow of current. Measurements by the float method are made as follows: The velocity of flow of the stream is obtained by observing the time which it takes floats set free at different points across the stream to pass between two range lines about 200 feet apart. The area used is the mean value obtained from several cross sections measured between the two range lines. The chief disadvantages of this method are difficulty in obtaining the correct value of mean area for the course used and uncertainty regarding the proper coefficient to apply to the observed velocity.[2]

The Price current meter is now used almost to the exclusion of other types of meters by the United States Geological Survey in the determination of the velocity of flow of water in open channels, a use for which it is adapted under practically all conditions.[3]

Plate II shows in the center the new type of penta-recording current meter equipped for measurements at bridge and cable stations; on the left the same type of meter is shown equipped for wading meas-

---

[1] For a discussion of methods of computing the discharge of a stream see Engineering News, June 25, 1908.

[2] Further information regarding the float method is given in Water-Supply Paper 95 and the various text-books on stream flow.

[3] See Hoyt, J. C., and others, Use and care of the current meter as practiced by the U. S. Geological Survey: Trans. Am. Soc. Civil Eng., vol. 66, 1910, p. 70.

urements, to record by the acoustic method; on the right the meter
is shown equipped to record electrically. (See Pl. I, *B*.) Briefly,
the meter consists of six cups attached to a vertical shaft which
revolves on a conical hardened steel point when immersed in moving water. The revolutions are indicated electrically. The rating,
or relation between the velocity of moving water and the revolutions
of the wheel, is determined for each meter by drawing it through still
water for a given distance at different speeds and noting the number
of revolutions for each run. From these data a rating table is prepared which gives the velocity per second of moving water for any
number of revolutions in a given time interval. The ratio of revolutions per second to velocity of flow in feet per second is very nearly a
constant for all speeds and is approximately 0.45.

Three classes of methods of measuring velocity with current meters
are in general use—multiple-point, single-point, and integration.

The two principal multiple-point methods in general use are the
vertical velocity curve and 0.2 and 0.8 depth.

In the vertical velocity curve method a series of velocity determinations are made in each vertical at regular intervals, usually about
10 to 20 per cent of the depth apart. By plotting these velocities as
abscissas and their depths as ordinates and drawing a smooth curve
among the resulting points, the vertical velocity curve is developed.
This curve shows graphically the magnitude and changes in velocity
from the surface to the bottom of the stream. The mean velocity
in the vertical is then obtained by dividing the area bounded by
this velocity curve and its axis by the depth. This method of obtaining the mean velocity in the vertical is probably the best known,
but on account of the length of time required to make a complete
measurement its use is largely limited to the determination of coefficients for purposes of comparison and to measurements under ice.

In the second multiple-point method the meter is held successively
at 0.2 and 0.8 depth, and the mean of the velocities at these two points
is taken as the mean velocity for that vertical. (See Pl. I, *A*.) On the
assumption that the vertical velocity curve is a common parabola
with horizontal axis, the mean of the velocities at 0.22 and 0.79 depth
will give (closely) the mean velocity in the vertical. Actual observations under a wide range of conditions show that this multiple-point
method gives the mean velocity very closely for open-water conditions and that a completed measurement seldom varies as much
as 1 per cent from the value given by the vertical velocity curve
method. Moreover, the indications are that it holds nearly as well
for ice-covered rivers. It is very extensively used in the regular
practice of the United States Geological Survey.

The single-point method consists in holding the meter either at
the depth of the thread of mean velocity or at an arbitrary depth

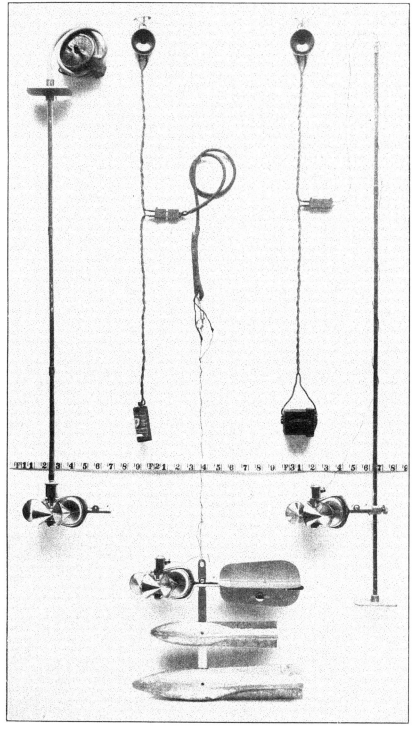

SMALL PRICE CURRENT METERS.

for which the coefficient for reducing to mean velocity has been determined or must be assumed.

Extensive experiments by means of vertical velocity curves show that the thread of mean velocity generally occurs between 0.5 and 0.7 total depth. In general practice the thread of mean velocity is considered to be at 0.6 depth, and at this point the meter is held in most of the measurements made by the single-point method. A large number of vertical velocity curve measurements, taken on many streams and under varying conditions, show that the average coefficient for reducing the velocity obtained at 0.6 depth to mean velocity is practically unity. The variation of the coefficient from unity in individual cases is, however, greater than in the 0.2 and 0.8 method, and the general results are not as satisfactory.

In the other principal single-point method the meter is held near the surface, usually 1 foot below, or low enough to be out of the effect of the wind or other disturbing influences. This is known as the sub-surface method. The coefficient for reducing the velocity taken at the subsurface to the mean has been found to be in general from about 0.85 to 0.95, depending on the stage, velocity, and channel conditions. The higher the stage the larger the coefficient. This method is especially adapted for flood measurements, or when the velocity is so great that the meter can not be kept in the correct position for the other methods.

The vertical integration method consists in moving the meter at a slow, uniform speed from the surface to the bottom and back again to the surface and noting the number of revolutions and the time taken in the operation. This method has the advantage that the velocity at each point of the vertical is measured twice. It is useful as a check on the point methods. In using the Price meter great care should be taken that the vertical movement of the meter is not rapid enough to vitiate the accuracy of the resulting velocity.

The determination of the flow of an ice-covered stream is difficult, owing to diversity and instability of conditions during the winter period and also to lack of definite information in regard to the laws of flow of water under ice. The method now employed is to make frequent discharge measurements during the frozen periods by the 0.2 and 0.8 and the vertical velocity curve methods, and to keep an accurate record of the conditions, such as the gage height to the surface of the water as it rises in a hole cut in the ice and the thickness and character of the ice. From these data an approximate estimate of the daily flow can be made by constructing a rating curve (really a series of curves) similar to that used for open channels, but considering, in addition to gage heights and discharge, the varying thickness of ice.[1]

---

[1] For information in regard to flow under ice cover see Water-Supply Paper U. S. Geol. Survey, No. 187.

OFFICE METHODS OF COMPUTING AND STUDYING DISCHARGE AND
RUN-OFF.

At the end of each year the field or base data for current-meter
gaging stations, consisting of daily gage heights, discharge measurements, and full notes, are assembled. The measurements are plotted
on cross-section paper and rating curves are drawn wherever feasible.
The rating tables prepared from these curves are then applied to
the tables of daily gage heights to obtain the daily discharges, and
from these applications the tables of monthly discharge and run-off
are computed.

Rating curves are drawn and studied with special reference to
the class of channel conditions which they represent. (See p. 17.)
The discharge measurements for all classes of stations when plotted
with gage heights in feet as ordinates and discharges in second-feet
as abscissas define rating curves which are more or less generally
parabolic in form. In many cases curves of area in square feet
and mean velocity in feet per second are also constructed to the
same scale of ordinates as the discharge curve. These are used
mainly to extend the discharge curves beyond the limits of the
plotted discharge measurements, and for checking purposes to avoid
errors in the form of the discharge curve and to determine and eliminate erroneous measurements.

For every rating table the following assumptions are made for
the period of application of the table: (a) That the discharge is a
function of and increases gradually with the stage; (b) that the discharge is the same whenever the stream is at a given stage, and
hence such changes in conditions of flow as may have occurred
during the period of application are either compensating or negligible,
except that the rating is not applicable for known conditions of ice,
log jams, or other similar obstructions; (c) that the increased and
decreased discharge due to change of slope on rising and falling stages
is either negligible or compensating.

As already stated, the gaging stations may be divided into several
classes, as indicated in the following paragraphs:

The stations of class 1 represent the most favorable conditions for
an accurate rating and are also the most economical to maintain.
The bed of the stream is usually composed of rock and is not subject
to the deposit of sediment and loose material. This class includes
also many stations located in a pool below which is a permanent
rocky riffle that controls the flow like a weir. Provided the control
is sufficiently high and close to the gage to prevent cut and fill at the
gaging point from materially affecting the slope of the water surface,
the gage height will for all practical purposes be a true index of the
discharge. Discharge measurements made at such stations usually

plot within 2 or 3 per cent of the mean-discharge curve, and the rating developed from that curve represents a very high degree of accuracy. Stations of this type are found in the north Atlantic coast drainage basins.

Class 2 is confined mainly to stations on rough, mountainous streams with steep slopes. The beds of such streams are as a rule comparatively permanent during low and medium stages and when the flow is sufficiently well defined by an adequate number of discharge measurements before and after each flood the stations of this class give nearly as good results as those of class 1. As it is seldom possible to make measurements covering the time of change at flood stage, the assumption is often made that the curves before and after the flood converged to a common point at the highest gage height recorded during the flood. Hence the only uncertain period occurs during the few days of highest gage heights covering the period of actual change in conditions of flow. Stations of this type are found in the upper Missouri River basin.

Class 3 includes most of the current-meter gaging stations maintained by the United States Geological Survey. If sufficient measurements could be made at stations of this class, results would be obtained nearly equaling those of class 1, but owing to the limited funds at the disposal of the Survey this is manifestly impossible, nor is it necessary for the uses to which discharge data are applied. The critical points are, as a rule, at relatively high or low stages. The percentage error, however, is greater at low stages. No absolute rule can be laid down for stations of this class. Each rating curve must be constructed mainly on the basis of the measurements of the current year, the engineer being guided largely by the past history of the station and the following general law: If all measurements ever made at a station of this class are plotted on cross-section paper they will define a mean curve which may be called a standard curve. It has been found in practice that if after a change caused by high stage a relatively constant condition of flow occurs at medium and low stages, all measurements made after the change will plot on a smooth curve which is practically parallel to the standard curve with respect to their ordinates or gage heights. This law of the parallelism of ratings is the fundamental basis of all ratings and estimates at stations with semipermanent and shifting channels. It is not absolutely correct but, with few exceptions, answers all the practical requirements of estimates made at low and medium stages after a change at a high stage. This law appears to hold equally true whether the change occurs at the measuring section or at some controlling point below. The change is, of course, fundamentally due to change in the channel caused by cut, or fill, or both, at or near the measuring section. For all except small streams the changes in section usually occur at the bottom. The following simple but typical examples illustrate this law:

(a) If 0.5 foot of planking were to be nailed on the bottom of a well-rated wooden flume of rectangular section there would result, other conditions of flow being equal, new curves of discharge, area, and velocity, each plotting 0.5 foot above the original curves when referred to the original gage. In other words, this condition would be analogous to a uniform fill or cut in a river channel which either reduces or increases all three values of discharge, area, and velocity for any gage height. In practice, however, such ideal conditions rarely exist.

(b) In the case of a cut or fill at the measuring section there is a marked tendency toward decrease or increase, respectively, of the velocity. In other words, the velocity has a compensating effect and if the compensation is exact at all stages the discharge at a given stage will be the same under both the new and the old conditions.

(c) In the case of uniform change along the crest of a weir or rocky control, the area curve will remain the same as before the change, and it can be shown that here again the change in velocity curve is such that it will produce a new discharge curve essentially parallel to the original discharge curve with respect to their ordinates.

Of course in actual practice such simple changes of section do not occur. The changes are complicated and lack uniformity, a cut at one place being largely offset by a fill at another and vice versa. If these changes are very radical and involve large percentages of the total area—as, for example, on small streams—there may result a wide departure from the law of parallelism of ratings. In complicated changes of section the corresponding changes in velocity which tend to produce a new parallel discharge curve may interfere with each other materially, causing eddies, boils, backwater, and radical changes in slope. In such extreme conditions, however, the measuring section would more properly fall under class 4 and would require very frequent measurements of discharge. Special stress is laid on the fact that in the lack of other data to the contrary the utilization of this law will yield the most probable results.

Slight changes at low or medium stages of an oscillating character are usually averaged by a mean curve drawn among them parallel to the standard curve, and if the individual measurements do not vary more than 5 per cent from the rating curve the results are considered good for stations of this class. Stations of this type are found in the south Atlantic coast drainage basins.

Class 4 comprises stations that have soft, muddy, or sandy beds. Good results can be obtained from such sections only by frequent discharge measurements, the frequency varying from a measure-

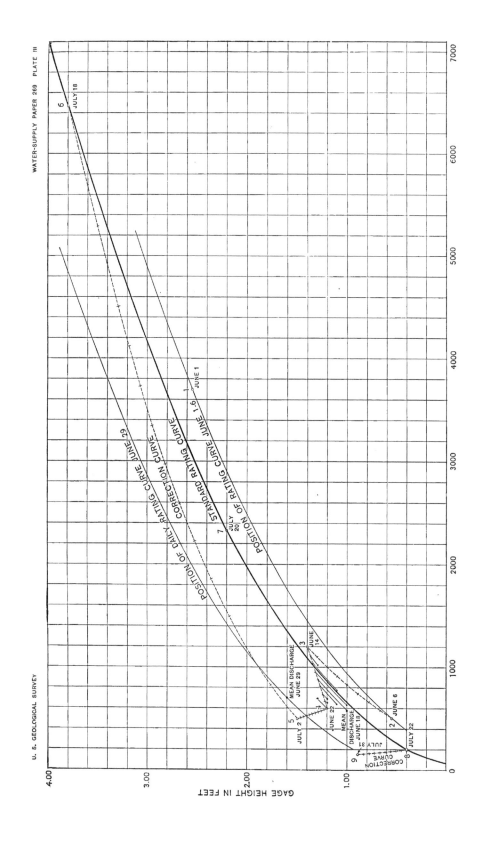

GAGE HEIGHT IN FEET

ment every two or three weeks to a measurement every day, according to the rate of diurnal change in conditions of flow.

The following method of determining the daily discharge of streams of this class is now used by the engineers of the United States Geological Survey almost exclusively, owing to the rapidity with which the necessary computations can be made, the clearness with which all changes in conditions of flow, so far as known, can be followed and the accuracy of the results obtained.

In the graphic method of determining the daily discharge of streams with changeable beds, which was devised by R. H. Bolster, the discharge measurements for the entire year are first plotted with discharges as abscissas and gage heights as ordinates. The points so plotted are considered chronologically and, even though scattered, will usually locate one or more fairly well-defined curves, called standard curves. (See Pl. III.) In general, the number and position of these standard curves are determined by the radical changes in the stream bed due to floods.

When stream beds change very rapidly it is necessary to change the position of the rating curve each day, making a new curve daily. This daily curve is of the same form as the standard curve and is parallel to it with respect to ordinates. For a day when a measurement is made the rating curve passes through such plotted measurement, the discharge for the day being read off from the scale of discharge in second-feet at the point of intersection of the curve and the mean gage height for the day. In order to locate the rating curve for other days, curves are drawn connecting consecutive measurements. They are called correction curves and should have the same curvature as that portion of the standard curve which lies vertically above or below them. They are divided into as many equal parts as there are days intervening between the measurements, on the assumption that the change in conditions of flow between any two consecutive measurements is uniform from day to day. The daily rating curve will then pass through these points of division, and the discharge is read directly from these curves at the point of intersection with the observed daily gage height.

In order to facilitate the use of this method and obviate the drawing of daily rating curves, dividers are employed. Care is exercised to keep both points in the same vertical line of discharge, then with one point of the dividers coincident with the standard curve the other can be made to trace any daily rating curve desired.

In applying and modifying the indirect method for shifting channels as above outlined judgment must be used, especially for long-time intervals not covered by discharge measurements, or for radical changes in the stream bed caused by sudden floods. For illustrative examples of stations of this type see Water-Supply Papers 247, 249, and 267.

The computations have, as a rule, been carried to three significant figures. Computation machines, Crelle's tables, and the 20-inch slide rule have been generally used. All computations are carefully checked.

After the computations have been completed they are entered in tables and carefully studied and intercompared to eliminate or account for all gross errors so far as possible. Missing periods are filled in, so far as feasible, by means of comparison with adjacent streams. The attempt is made to complete years or periods of discharge, thus eliminating fragmentary and disjointed records. Full notes accompanying such estimates follow the daily and monthly discharge tables.

For most of the northern stations estimates have been made of the monthly discharge during ice periods. These are based on measurements under ice conditions wherever available, on daily records of temperature and precipitation obtained from the United States Weather Bureau, on climate and crop reports, on observers' notes of conditions, and on a careful and thorough intercomparison of results with adjacent streams. Although every care possible is used in making these estimates they are often very rough, the data for some of them being so poor that the estimates are liable to as much as 25 to 50 per cent error. It is believed, however, that estimates of this character are better than none at all, and serve the purpose of indicating in a relative way the proportionate amount of flow during the ice period. These estimates are, as a rule, included in the annual discharge. The large error of the individual months has a relatively small effect on the annual total, and it is for many purposes desirable to have the yearly discharge computed even though some error is involved in doing so.

ACCURACY AND RELIABILITY OF FIELD DATA AND COMPARATIVE RESULTS.

Practically all discharge measurements made under fair conditions are well within 5 per cent of the true discharge at the time of observation. Inasmuch as the errors of meter measurements are largely compensating, the mean rating curve, when well defined, is much more accurate than the individual measurements. Numerous tests and experiments have been made to test the accuracy of current-meter work. These show that it compares very favorably with the results from standard weirs, and, owing to simplicity of methods, usually gives results that are much more reliable than those from stations at dams, where uncertainty regarding the coefficient and complicated conditions of flow prevail.

The work is, of course, dependent on the reliability of the observers. With relatively few exceptions, the observers perform their work honestly. Care is taken, however, to watch them closely and to inquire into any discrepancies. It is, of course, obvious that one gage reading a day does not always give the mean height for that day. As an almost invariable rule, however, errors from this source are compensating and virtually negligible in a period of one month, although a single day's reading may, when taken by itself, be considerably in error.

The effort is made to visit every station at least once each year for the purpose of making a measurement to determine the constancy of conditions of flow since the last measurement made during the preceding year, and also to check the elevation of the gage. On account of lack of funds or for other causes, some stations were not visited during the current year. If conditions of flow have been reasonably permanent up to the time of the last preceding measurement, it is considered best to publish values of discharge on the basis of the latest verified rating curve rather than to omit them altogether, although it should be distinctly understood that such records are at times subject to considerable error. This is also true, although to a less degree, of the period of records since the date of the last measurement of the current year. As a rule the accuracy notes are based on the assumption that the rating curve used is strictly applicable to the current year.

In order to give engineers and others information regarding the probable accuracy of the computed results, footnotes are added to the daily discharge tables, stating the probable accuracy of the rating tables used, and an accuracy column is inserted in the monthly discharge table. For the rating tables "well defined" indicates in general that the rating is probably accurate within 5 per cent; "fairly well defined," within 10 per cent; "poorly defined" or "approximate," within 15 to 25 per cent. These notes are very general and are based on the plotting of the individual measurements with reference to the mean rating curve.

The accuracy column in the monthly discharge table does not apply to the maximum or minimum nor to any individual day, but to the monthly mean. It is based on the accuracy of the rating, the probable reliability of the observer, and knowledge of local conditions. In this column, A indicates that the mean monthly flow is probably accurate within 5 per cent; B, within 10 per cent; C, within 15 per cent; D, within 25 per cent. Special conditions are covered by footnotes.

USE OF THE DATA.

In general the policy is followed of making available for the public the base data which are collected in the field each year by the survey engineers. This is done to comply with the law, but also for the express purpose of giving to any engineer the opportunity of examining the computed results and of changing and adjusting them as may seem best to him. Although it is believed that the rating tables and computed monthly discharges are as good as the base data up to and including the current year will warrant, it should always be borne in mind that the additional data collected at each station from year to year nearly always throw new light on data already collected and published, and hence allow more or less improvement in the computed results of earlier years. It is therefore expected that the engineer who makes serious use of the data given in these papers will verify all ratings and make such adjustments in earlier years as may seem necessary. The work of compiling, studying, revising, and republishing data for different drainage basins for 5 or 10 year periods or more is carried on by the United States Geological Survey so far as the funds for such work are available.

The values in the table of monthly discharge are so arranged as to give only a general idea of the conditions of flow at the station, and it is not expected that they will be used for other than preliminary estimates. The daily discharges are published to allow a more detailed study of the variation in flow and to determine the periods of deficient flow.

COOPERATIVE DATA.

Cooperative data of various kinds and also data regarding the run-off at many stations maintained wholly by private funds are incorporated in the surface water-supply reports of the United States Geological Survey.

Many stations throughout the country are maintained for specific purposes by private parties who supply the records gratuitously to the United States Geological Survey for publication. When such records are supplied by responsible parties and appear to be reasonably accurate they are verified, so far as possible, and estimated values of accuracy are given. Records clearly known to be worthless or misleading are not published. As it is, however, impossible to completely verify all such records furnished—because of lack of funds or for other causes—they are published for what they are worth, as they are of value as a matter of record and afford at least approximate information regarding stream flow at the particular localities. The Survey does not, however, assume any responsibility for inaccuracies found in such records, although most of them are believed to be reasonably good.

COOPERATION AND ACKNOWLEDGMENTS.

Special acknowledgments are due for cooperative assistance as follows:

The Uinta irrigation survey of the United States Indian Office paid the salary and expenses of a hydrographer, amounting to about $2,000, for the stations in the Uinta Reservation, Utah, and that vicinity.

The Central Colorado Power Co. paid the salary and expenses of a hydrographer on cooperative work in the Grand River drainage basin for several months during 1909 and has otherwise materially assisted in the work.

The United States Reclamation Service has furnished all the field data for the stations in the Gunnison drainage basin in Colorado, has paid the expense of maintaining the station on the Grand at Palisades, Colo., and has paid the expense of all hydrometric work in the vicinity of the Strawberry Valley in the Duchesne River drainage basin.

The Territorial engineer of New Mexico, Mr. Vernon L. Sullivan, cooperated in the maintenance of all stations in the Colorado River basin in New Mexico from the sum of $2,500 set aside annually for this work in the Territory by the Territorial legislature, the $1,000 allotted by the Atchison, Topeka & Santa Fe Railway Co. for the work in New Mexico, and from other funds pertaining to his office.

The State engineer of Colorado, Mr. Charles W. Comstock, has assisted materially in carrying on the work in this drainage basin.

Under the provisions of a formal contract between the Director of the United States Geological Survey and the State engineer of Utah, dated July 1, 1909, the sum of $2,000 was to be expended by each party for hydrometric work in the State of Utah during the fiscal year immediately following. The expenditure of this money was so divided as to cover the salary and expenses of hydrographers, the pay of gage readers, the cost of construction, and the cost of preparing the data in the manner most suitable to expedite the work.

The State engineer of Utah, Mr. Caleb Tanner, paid the salaries of the observers at five stations in the Duchesne River drainage in cooperation with the United States Indian Office and the United States Geological Survey.

Mr. Thomas Lyons, of Gila, N. Mex., paid the expenses of a hydrographer and the salary of the gage observer at the station on Gila River at Red Rock and has otherwise contributed to the work.

Mr. R. E. Vickery, of Grand Junction, Colo., has borne the larger part of the expense of maintaining the stations on West Divide and West Mamm creeks.

Assistance was rendered or records furnished by the United States Weather Bureau, the Denver Reservoir & Irrigation Co., the Denver Union Water Co., Prof. George J. Lyon, Mr. G. H. Matthes, Mr. E. C. Jansen, Mr. Stanley Krajicek, Mr. Jay Turley, Mr. M. C. Hinderlider, Mr. G. W. Vallery, Mr. H. F. Robinson, Mr. R. M. Jones, and others.

DIVISION OF WORK.

The field data in the Grand River drainage basin were collected under the direction of W. B. Freeman, district engineer, by J. B. Stewart, G. H. Russell, C. L. Chatfield, W. H. Snelson, the last two of whom are engineers of the Central Colorado Power Co. and the United States Reclamation Service, respectively.

The field data in the Duchesne River drainage basin in the Uinta Indian Reservation were collected under the direction of W. B. Freeman, district engineer, by R. H. Fletcher, who was under the more immediate supervision of H. C. Means, superintendent of irrigation, United States Indian Office.

All field data for the State of Utah, except in the Uinta Indian Reservation, were collected under the direction of E. C. LaRue, district engineer, by E. A. Porter.

The field data for the San Juan and Gila River drainage basins in New Mexico were collected under the general direction of W. B. Freeman, district engineer, by J. B. Stewart, but under the more immediate supervision of Vernon L. Sullivan, Territorial engineer, assisted by C. D. Miller.

The field data for all stations in Arizona, together with the estimates of discharge, except those for Little Colorado River at St. Johns, Ariz., have been furnished by the United States Reclamation Service.

The field data for stations in California have been collected under the direction of W. B. Clapp, district engineer for California, by W. V. Hardy, H. A. Jones, W. F. Martin, and A. H. Hoebig, jr.

The ratings, special estimates, and studies of the completed data were made by W. B. Freeman, R. H. Bolster, and E. S. Fuller. The computations were made and the completed data prepared for publication under the direction of R. H. Bolster, assistant engineer, by R. C. Rice, J. G. Mathers, H. D. Padgett, M. E. McChristie, J. J. Phelan, H. J. Jackson, J. B. Stewart, L. T. King, and M. I. Walters. The report has been edited by Mrs. B. D. Wood.

GAGING STATIONS IN COLORADO RIVER BASIN.

The following is a list of gaging stations maintained in Colorado River basin by the United States Geological Survey and cooperative parties. The stations are arranged by river basins, in downstream order, as explained on page 13, tributaries being indicated by indention. Data for these stations have been published in the reports listed in tables on page 11.

Green River (head of Colorado) at Green River, Wyo., 1895–1906.
Green River at Jensen, Utah, 1903–1906.
Green River at Ouray, Utah, 1904–5.
Green River at Greenriver, Utah, 1894–1899 and 1905–1909.
Colorado River at Hardyville, Ariz., 1905–1907.
Colorado River at Mohave City, Ariz., 1902–3.
Colorado River at Yuma, Ariz., 1895–1909.
    Salton Sea near Salton, Cal., 1904–1909.
        Alamo River near Brawley, Cal., 1908–9.
        New River near Brawley, Cal., 1908–9.
Green River:
    Newfork River at Cora, Wyo., 1905.
        Pine Creek at Pinedale, Wyo., 1904–1906.
        Pole Creek at Fayette, Wyo., 1904–1906.
        Fall Creek at Fayette, Wyo., 1904–5.
        Boulder Creek at Boulder (Newfork), Wyo., 1904–1906.
        Eastfork River at Newfork, Wyo., 1905–6.
    Green River (Black Fork) at Granger, Wyo., 1896–1900.
    Yampa River at Steamboat Springs, Colo., 1904–1906.
    Yampa River at Craig, Colo., 1901–2, 1904–1906.
    Yampa River at Maybell, Colo., 1904–5.
        Elk River at Trull, Colo., 1904–1906.
        Elk Head Creek at Craig, Colo., 1906.
        Fortification Creek at Craig, Colo., 1905–6.
        Williams River at Hamilton, Colo., 1904–1906.
        Milk Creek at Axial, Colo., 1904–5.
        Little Snake River at Maybell, Colo., 1904.
    Ashley Creek at Vernal, Utah, 1900–1904.
        Ashley Creek (Dry Fork) at Vernal, Utah., 1904.
    White River (North Fork) at Buford, Colo., 1903–1906.
    White River at Meeker, Colo., 1901–1906.
    White River at White River City, Colo., 1895.
    White River at Rangely, Colo., 1904–5.
    White River at Dragon, Utah, 1906.
    White River at Ouray, Utah, 1904.
        Marvine Creek near Buford, Colo., 1903–1906.
        White River (South Fork) near Buford, Colo., 1903–1906.
    Duchesne River (North Fork) above Forks, Utah, 1904.
    Duchesne River at Myton, Utah, 1899–1909.
        Duchesne River (West Fork) above Forks, Utah, 1904.
        Rock Creek (East Creek) 10 miles above mouth, Utah, 1904.
        Strawberry River, above mouth of Indian Creek, in Strawberry Valley, Utah, 1909.
        Strawberry River, below mouth of Indian Creek, in Strawberry Valley, Utah, 1903–1906 and 1908–9.
        Strawberry River at Theodore, Utah, 1908–9.
            Indian Creek in Strawberry Valley, Utah, 1905–6 and 1909.
                Trail Hollow Creek in Strawberry Valley, Utah, 1909.
        Currant Creek, 13 miles above mouth, Utah, 1904.
        Currant Creek, 3 miles above mouth, Utah, 1904.
            Red Creek above Narrows, Utah, 1904.

Green River—Continued.
    Duchesne River at Myton, Utah—Continued.
        Lake Fork (West Fork), 10 miles above Forks, Utah, 1904.
        Lake Fork below Forks, Utah, 1904, 1907–1909.
        Lake Fork near Myton, Utah, 1900–1904, 1907–1909.
            Lake Fork (East Fork), 8 miles above Forks, Utah, 1904.
        Unita River near Whiterocks, Utah, 1899–1904, 1907–1909.
        Uinta River at Fork Duchesne, Utah, 1899–1904, 1906–1909.
        Uinta River at Ouray School, Utah, 1899–1904.
            Whiterocks River near Whiterocks, Utah, 1899–1904, 1907–1909.
    Price River near Helper, Utah, 1904–1909.
    San Rafael River near Greenriver, Utah, 1909.
        Cottonwood Creek near Orangeville, Utah, 1909.
            Ferron Creek near Ferron, Utah, 1909.
        Huntington Creek near Huntington, Utah, 1909.
    Grand River (North Fork) near Grand Lake, Colo., 1904–1909.
    Grand River near Granby, Colo., 1908–9.
    Grand River at Sulphur Springs, Colo., 1904–1909.
    Grand River near Kremmling, Colo., 1904–1909.
    Grand River near Wolcott, Colo., 1906–1908.
    Grand River at Shoshone, Colo., 1897.
    Grand River at Glenwood Springs, Colo., 1899–1909.
    Grand River near Palisades, Colo., 1902–1909.
    Grand River near Grand Junction, Colo., 1895–1900.
        North Inlet to Grand Lake at Grand Lake, Colo., 1905–1909.
        Grand Lake Outlet at Grand Lake, Colo., 1904–1909.
        Grand River (South Fork) near Lehman, Colo., 1907–8.
        Fraser River at Granby (Coulter), Colo., 1904–1909.
        Fraser River at upper station near Fraser, Colo., 1908–9.
        Fraser River at lower station near Fraser, Colo., 1907–1909.
            Big Jim Creek near Fraser, Colo., 1907–1909.
                Little Jim Creek near Fraser, Colo., 1907–1909.
            Vasquez Creek at upper station near Fraser, Colo., 1908–9.
            Vasquez Creek at lower station near Fraser, Colo., 1907–1909.
            Elk Creek near Fraser, Colo., 1907–1909.
            St. Louis Creek at upper station near Fraser, Colo., 1908–9.
            St. Louis Creek at lower station near Fraser, Colo., 1908–9.
            North Ranch Creek at upper station near Rollins Pass, Colo., 1908–9.
            North Ranch Creek at lower station near Rollins Pass, Colo., 1907–1909.
                Middle Ranch Creek at upper station near Arrow, Colo., 1908–9.
                Middle Ranch Creek at lower station near Arrow, Colo., 1907–1909.
                    South Ranch Creek at upper station near Arrow, Colo., 1908–9.
                    South Ranch Creek at lower station near Arrow, Colo., 1907–1909.
    Williams Fork near Sulphur Springs, Colo., 1904–1909.
    Troublesome River at Troublesome, Colo., 1904–5.
    Muddy River at Kremmling, Colo., 1904–5.
    Blue River near Kremmling, Colo., 1904–1908.
        Tenmile Creek near Kokomo, Colo., 1904.
        Tenmile Creek near Uneva Lake, Colo., 1903.
    Eagle River near Eagle, Colo., 1905–1907.
    Eagle River at Gypsum, Colo., 1907–1909.
    Roaring Fork near Emma, Colo., 1908–9.

Green River—Continued.
    Grand River near Grand Junction, Colo.—Continued.
        Roaring Fork at Glenwood Springs, Colo., 1906–1909.
            Frying Pan River at Basalt, Colo., 1908–9.
            Crystal River near Carbondale (Sewell), Colo., 1908–9.
        West Divide Creek at Hostutler's ranch, near Raven, Colo., 1909.
        West Divide Creek near Raven, Colo., 1909.
        West Mamm Creek near Rifle, Colo., 1909.
        Gunnison River near Iola, Colo., 1900–1903.
        Gunnison River near Cimarron, Colo., 1903–1905.
        Gunnison River at River Portal (east portal of Gunnison Tunnel), Colo., 1905–1909.
        Gunnison River near Cory, Colo., 1903–1905.
        Gunnison River at Roubideau, Colo., 1897.
        Gunnison River at Whitewater, Colo., 1897, 1901–1906.
        Gunnison River near Grand Junction, Colo., 1895, 1897–1899.
            Taylor River near Almont, Colo., 1905.
                East River at Almont, Colo., 1905.
            Cimarron Creek at Cimarron, Colo., 1903–1905.
            Gunnison River (North Fork) near Hotchkiss, Colo., 1903–1906.
            Uncompahgre River near Colona, Colo., 1903–1906.
            Uncompahgre River near Ouray, Colo., 1908.
            Uncompahgre River at Fort Crawford, Colo., 1895–1899, 1908.
            Uncompahgre River at Montrose, Colo., 1900, 1903–1909.
            Uncompahgre River at Delta, Colo., 1903–1909.
        Dolores River near Dolores, Colo., 1895–1903.
        San Miguel River near Fall Creek, Colo., 1895–1899.
Colorado River:
    Fremont River near Thurber, Utah, 1909.
        Muddy Creek near Emery, Utah, 1909.
    Escalante Creek near Escalante, Utah, 1909.
    San Juan River near Arboles, Colo., 1895–1899.
    San Juan River at Turley, N. Mex., 1907–8.
    San Juan River at Blanco, N. Mex., 1908–9.
    San Juan River near Bloomfield, N. Mex., 1909.
    San Juan River near Farmington, N. Mex., 1904–1906.
        Piedra River near Arboles, Colo., 1895–1899.
        Los Pinos River at Ignacio, Colo., 1899–1903.
        Animas River at Silverton, Colo., 1903.
        Animas River at Durango, Colo., 1895–1905.
        Animas River at Aztec, N. Mex., 1904, 1907–1909.
        Animas River near Farmington, N. Mex., 1904–5.
            Florida River near Durango, Colo., 1899, 1901–1903.
        La Plata River at Hesperus, Colo., 1904–1906.
        La Plata River at La Plata, N. Mex., 1905–1909.
        Mancos River at Mancos, Colo., 1898–1900.
    Little Colorado at St. Johns, Ariz., 1906–1909.
    Little Colorado at Woodruff, Ariz., 1905–1908.
    Little Colorado at Holbrook, Ariz., 1905–1909.
        Silver Creek at Snowflake, Ariz., 1906–1908.
        Silver Creek at Canyon Station, Ariz., 1906.
        Woodruff ditch at Woodruff, Ariz., 1906.

Colorado River—Continued.
  Little Colorado at Holbrook, Ariz.—Continued.
    Chevelon Fork near Winslow, Ariz., 1906–1908.
    Clear Creek near Winslow, Ariz., 1906–1909.
  Virgin River at Virgin, Utah, 1909.
    Santa Clara River near Central, Utah, 1909.
    Santa Clara River near St. George, Utah, 1909.
    Muddy River near Moapa, Nev., 1904–1906, and 1909.
  Gila River near Cliff, N. Mex., 1904–1907.
  Gila River near Redrock, N. Mex., 1908–9.
  Gila River at San Carlos, Ariz., 1899–1905.
  Gila River near Buttes, Ariz., 1889–1890, and 1895–1899.
  Gila River at Dome (Gila City), Ariz., 1903–1906.
    San Francisco River at Alma, N. Mex., 1904–1907, and 1909.
    San Pedro River at Charleston, Ariz., 1904–1906.
    San Pedro River near Dudleyville, Ariz., 1890.
    Santa Cruz River near Nogales, Ariz., 1907–1909.
    Santa Cruz River and ditches at Tucson, Ariz., 1905–1909.
      Queens Creek at Whitlow's, Ariz., 1896.
    Salt River at Roosevelt, Ariz., 1901–1907.
    Salt River below mouth of Cherry Creek, near Roosevelt, Ariz., 1906.
    Salt River 50 miles above Phoenix, Ariz., 1890.
    Salt River at Arizona dam, Ariz., 1888–1891.
    Salt River at McDowell, Ariz., 1888–1909.
      Tonto Creek at Roosevelt, Ariz., 1901–1904.
      Verde River at McDowell, Ariz., 1888–1909.
Canal stations in Colorado River Basin:
  Imperial Canal (main) near Calexico, Cal., 1904–5.
  Boundary Canal near Calexico, Cal., 1905.
  Wisteria Canal near Calexico, Cal., 1905.
  Imperial Canal 10 miles below Yuma, Ariz., Mexican boundary line, 1903–1905.
  Holt Canal at Calexico, Cal., 1904–5.
  Hemlock Canal at Calexico, Cal., 1904–5.
  Alamo Channel near Calexico, Cal., 1904.
  Alamitos Canal near Calexico, Cal., 1904–5.

GENERAL FEATURES OF COLORADO RIVER BASIN.

Colorado River is formed in the southeastern part of Utah by the junction of Grand and Green rivers. The Green is larger than the Grand and is the upward continuation of the Colorado. Including the Green, the Colorado is about 2,000 miles long. The region drained is about 800 miles long, ranges in width from 300 to 500 miles, and contains about 300,000 square miles. It comprises the southwestern part of Wyoming, the western part of Colorado, the eastern half of Utah, practically all of Arizona, and small portions of California, Nevada, New Mexico, and old Mexico. Most of this area is arid, the mean annual rainfall being about 8½ inches. The streams receive their supply from the melting snows on the high mountains of Wyoming, Utah, and Colorado.

The basin comprises two district portions. The lower third is but little above the level of the sea, though here and there ranges of mountains rise to elevations of 2,000 to 6,000 feet. This part of the valley is bounded on the north by a line of cliffs which present a bold and in many places vertical step of hundreds or thousands of feet to the table-land above. The upper two-thirds of the basin stands from 4,000 to 8,000 feet above sea level, and is bordered on the east, west, and north by ranges of snow-clad mountains, which attain altitudes ranging from 8,000 to 14,000 feet above sea level. Through this plateau the Colorado and its tributaries have cut narrow gorges or canyons in which they flow at almost inaccessible depths. At points where lateral streams enter, the canyons are broken by narrow transverse valleys, diversified by bordering willows, clumps of box elder, and small groves of cottonwood. The whole upper basin of the Colorado is traversed by a labyrinth of these canyons, most of which are dry during the greater portion of the year, and carry water only during the melting of the snow and the brief period of the autumnal and spring rains.

GREEN RIVER AND THE MAIN COLORADO RIVER.

DESCRIPTION.

Green River and its tributaries[1] drain an area rudely triangular in outline, bounded on the north and east by the Wind River Mountains and the ranges forming the Continental Divide, on the south and east by the White River Plateau and the Roan or Book Cliffs, and on the north and west by the Gros Ventre and Wyoming mountains and the great Wasatch Range. The greatest length of the basin, north and south, is about 370 miles. In an east-west direction it measures at its widest part about 240 miles. The total drainage area is approximately 41,000 square miles, and altitudes range from 14,000 feet in the high mountains to about 3,800 feet at the mouth of the Grand.

The area includes a large part of western Wyoming, northwestern Colorado, and eastern Utah. The Uinta and Uncompahgre Indian reservations are located in this basin in northeastern Utah.

The river heads on the western slope of the Wind River Mountains in western Wyoming, its ultimate source being a number of small lakes fed by the glaciers and immense snow deposits always to be found on Fremont and neighboring peaks. For perhaps 25 miles the river flows northwestward through the mountains. It then turns abruptly and runs in a general southerly direction across western Wyoming into Utah. A few miles below the Wyoming-Utah boundary another sharp turn carries the river eastward near the east end

---

[1] The geology of this basin is described in the Eleventh Ann. Rept. U. S. Geol. and Geog. Survey Terr., for 1877, pp. 509–646. Information in regard to the hydrography is contained in the first to fourth annual reports of the Reclamation Service and in United States Geological Survey reports.

of the range. It then flows southward in Colorado for about 25
miles, turns back into Utah, and continues to flow in a southwesterly
and southerly direction until it unites with the Grand to form the
Colorado. Its length, measured roughly along the course, is approximately 425 miles.

The topography of the headwater region is rugged in the extreme.
The Wind River Range on the east, and the Gros Ventre and Wyoming ranges on the southwest and west gradually close in as they
extend southward, forming a basin approximately 7,450 square miles
in extent above the discontinued gaging station at Green River,
Wyo.

The upper part of this basin is very narrow, but southward the valley opens out; near Fontanelle, Wyo., it is several miles wide, and its
benches and rolling table-lands extend westward to the foothills of
the Wyoming Range and eastward to the bluffs which hug the east
bank of the river. At Green River the valley is again narrow—only
a few hundred yards in width—and for some distance southward the
river runs between bluffs standing so close together that no flood
plain is seen. Throughout much of its course in Utah the Green
flows through a succession of long, deep, narrow canyons with walls
ranging in height from a few hundred to as many thousand feet, separated by short valleys containing small tracts of arable lands.

In its upper course the Green receives as tributaries numerous
streams heading in the Wind River, Gros Ventre, and Wyoming
ranges of mountains, some of them extending so far back into the
abrupt, ragged canyons that they dovetail with streams flowing in
the opposite direction. The most important of these tributaries
are New Fork River, Big Sandy Creek, Labarge Creek, Fontenelle
Creek, Black Fork, and Henry Fork. South of the Uinta Mountains
the first large stream flowing into the Green is the Yampa, which
comes in from the east at the point where the Green turns westward
to reenter Utah after its southward journey in Colorado. Farther
south Ashley Creek and Duchesne and White rivers discharge their
waters into the Green, Ashley Creek and the Duchesne from the
west and the White from the east. Below this point the only tributaries of importance are Price, Minnie Maud, and San Rafael rivers,
which enter from the west, the San Rafael at a point about 32 miles
above the junction of the Green and the Grand.

In the 41,000 square miles which comprise the total drainage area
of Green River there are considerably over 5,000 square miles of
timbered land in addition to a considerable woodland area. The
principal species of mountain timber are the Engelmann spruce and
lodgepole pine.

Except for the timber in the high mountains at the headwaters of
Green River in Wyoming, the upper portion of the stream is not very

extensively forested. The timbered land in that section includes probably 1,500 square miles, with an average stand of about 4,000 feet b. m. per acre. Numerous tracts of irrigated and cultivated land extend from the Wyoming line up the river to elevations of 7,000 feet.[1] In the Green River basin in Utah above the mouth of the Duchesne there are about 600 square miles of timbered land, with an average stand of nearly 3,000 feet b. m. per acre. In the drainage basins of the White and Yampa rivers in Colorado there are nearly 2,000 square miles of timbered area and woodland.

Over the plains portions of the basin, which includes considerably over half of it, the average annual precipitation is probably less than 10 inches annually; over much of the remainder the rainfall averages between 10 and 15 inches; and in only a very small area in the high mountains does the annual precipitation exceed 20 inches.

Throughout this basin the winters are severe and most of the streams have a heavy ice cover for several months. There is usually an abundance of snow in the high mountains, but the winters on the plains are frequently open.

The oldest and most extensive irrigation development in this basin is on the upper Green River in Wyoming. Recently large irrigation systems have been constructed in the Duchesne River basin. Considerable irrigation is practiced around Vernal, Utah, and also in the vicinity of Greenriver, Utah, along the line of the Denver & Rio Grande Railroad. Along White and Yampa rivers in Colorado meadow irrigation is extensively practiced, and projects are now on foot for the irrigation of 200,000 or 300,000 acres of land in that section.

Excellent reservoir sites are found on the headwaters of the Green River and its upper tributaries, and also along Yampa and White rivers on Ashley Creek, and other tributaries in the northwestern corner of Utah, and at the headwaters of Duchesne River. A very considerable portion of the flow could be equalized by storage.

The waters of this stream and its tributaries are practically unused except for irrigation. Not a water-power plant of any importance exists in the whole drainage area of Green River, though splendid opportunities are presented at the headwaters of many of the tributary streams above all irrigation diversions. Theoretically it would be possible at the present time, by utilizing known storage sites, to develop about 1,500,000 horsepower in the basin of the Green. From Wells, Wyo., to the Wyoming State line, a distance of 225 miles, the stream has an average fall of 11 feet per mile; and from the Wyoming State line to the mouth of Minnie Maud Creek, a distance of 200 miles the average fall is 7 feet to the mile.

---

[1] See also description of New Fork drainage basin in Water-Supply Paper 175, pp. 21-22.

From the junction of Green and Grand rivers the Colorado flows
southwestward, passes across the northwestern corner of Arizona,
then turns to the south and for the remainder of its course forms a
part of the southeastern boundary of Nevada and California and the
western boundary of Arizona. It empties into the Gulf of California
about 60 miles below Yuma, Ariz. The canyons through which it
flows are world famed and need not here be described.

The Colorado has been called the Nile of America, and like the Nile
it is subject to an annual summer rise which comes at the time
when the water is most needed for irrigation. It is interesting to
compare the Colorado with the Nile and the Susquehanna. The Nile
is similar in type; the Susquehanna shows the difference in flow between arid and humid regions. In the comparison a normal year,
based on a 10-year record for Colorado and Susquehanna rivers and
such data as could be found in regard to the Nile, have been used.
The Colorado has been taken as the standard of comparison.

The Nile has 5.7 times the drainage area and the Susquehanna
about one-eighth the area of the Colorado.

The rainfall in the Nile basin is 3.8 times greater; that in the Susquehanna basin is 4.5 times greater. The run-off per square mile
from the Nile basin is 1.9 times greater; that from the Susquehanna
basin is 37 times greater. The ratio of run-off to rainfall in the Nile
basin is 2 times smaller; that of the Susquehanna basin is 8.2 times
greater.

The discharge of the Nile is 10.8 times greater; that of the Susquehanna is 4.5 times greater.

The maximum flow of the Colorado is from 70,000 to 110,000 second-
feet and occurs in May, June, or July; for the Nile it is about 353,000
second-feet and occurs about the 1st of September; for the Susquehanna it is from 200,000 to 400,000 second-feet and occurs during
March, April, and May.

The minimum flow of the Colorado is from 2,500 to 3,000 second-
feet and occurs during January and February; that of the Nile is
about 14,500 second-feet and occurs about the end of May; for the
Susquehanna it is from 2,500 to 5,000 second-feet and occurs in
September and October.

The mean flow of the Colorado is about 10,700 second-feet; for the
Nile it is about 115,800 second-feet; for the Susquehanna it is about
43,000 second-feet.

The water of the Colorado carries an immense amount of sediment,
reaching as high as 2,000 parts of sediment to 100,000 parts of water.
Prof. R. H. Forbes, in Bulletin 44 of the University of Arizona
Agricultural Experiment Station, says:

On the basis of the profile, constructed from available data for the volume of flow
of the Colorado and of the year's silt determinations made in the laboratory, it is esti-

mated conservatively that the river during 1900 brought down about 61,000,000 tons of sedimentary material, which, condensed to the form of solid rock, is enough to cover 26.4 square miles 1 foot deep, or to make about 164 square miles of recently settled, submerged mud 1 foot deep, reckoning the whole amount of mud for the year to average 6.2 times the bulk of the solid sediment.

A comparatively small amount of land is irrigated by the waters of the Colorado because the stream and its tributaries are situated so far below the level of the irrigable lands as to render their diversion extremely difficult or impracticable. Two pumping plants are in operation to lift water for irrigation at Yuma, and several at other points on the river above Yuma. The Imperial canal diverts water from the river at a point about 10 miles, by river, below Yuma.

The principal tributaries of the Colorado below the Grand and Green are San Juan, Little Colorado, Williams Fork, and Gila rivers, which enters from the east, and Virgin River, which enters from the west. With the exception of Williams Fork, these streams and their tributaries are described in other parts of this report.

GREEN RIVER AT GREENRIVER, UTAH.

This station, which is located at the Rio Grande Western Railway bridge, at Greenriver railroad station, near Elgin post office (originally called Blake), in latitude 39° N., longitude 110° 9′ W., in the San Rafael quadrangle of the United States Geological Survey, was established October 21, 1894, discontinued in November, 1896, and reestablished in February, 1905.

Price River enters from the west about 16 miles above the station. Several irrigation projects are completed and being promoted in this drainage basin. The last diversion above the station is about 10 miles upstream; no water is diverted below the station.

At low and medium stages measurements are made from a ferry-boat, at a point about 450 feet above the bridge. At high stages measurements are made from the bridge.

The bed of the river at the bridge is mainly rock; but at low stages it is overlain in places with silt, which scours out and thus causes the changes in conditions of flow, necessitating new ratings at frequent intervals.

A new bridge was erected at this point between the periods of maintenance of the original station and the present one. The present datum, as near as can be learned, is 1.68 feet below the original datum; but owing to change in conditions of flow, due to the relocation of the bridge piers, it is impossible to utilize the early measurements in studies of new discharge curves. The datum of the present chain gage has remained the same since its establishment.

A careful determination, in 1909, of the angle, which the bridge makes with the main current, necessitated a correction of 15 per cent in all discharge measurements made at the bridge from 1905 to 1909. The daily and monthly estimates of discharge have been revised and supersede those previously published.

Ice usually exists at the station during December, January, and February. Conservative monthly estimates during these periods have been obtained by a consideration of the general behavior of the river at this station and by the aid of climatologic data.

Daily gage height, in feet, of Green River at Greenriver, Utah, for 1909.

[L. H. Greene, observer.]

Day.	Jan.	Feb.	Mar.	Apr.	May.	June.	July.	Aug.	Sept.	Oct.	Nov.	Dec.
1.............	4.3	5.45	4.9	7.55	7.9	11.2	12.5	7.95	8.8	6.2	5.7	6.2
2.............	4.5	5.3	4.8	7.35	8.3	11.4	12.25	7.85	8.9	6.2	5.65	5.95
3.............	4.6	5.3	4.9	7.2	8.35	11.3	12.15	7.65	8.45	6.2	5.6	5.75
4.............	4.6	5.1	5.0	7.0	8.25	11.15	12.05	7.6	7.8	6.15	5.65	5.6
5.............	4.6	5.0	5.3	6.95	8.05	11.1	11.9	7.5	7.65	6.1	5.7	5.2
6.......:	4.6	5.0	5.2	6.8	7.65	11.45	11.75	7.4	7.85	6.1	5.7	4.85
7.............	4.8	4.9	5.35	7.3	7.7	12.05	11.8	7.3	8.2	6.1	5.7	4.65
8.............	4.9	4.8	5.55	8.0	8.05	12.5	11.65	7.3	8.0	6.0	5.7	4.55
9.............	5.0	4.8	5.75	8.1	8.5	12.85	11.65	7.35	7.8	6.0	5.7	4.45
10.............	5.1	4.7	5.9	7.65	9.15	13.65	11.6	7.3	7.6	6.0	5.7	4.3
11.....\.....	5.1	4.7	6.0	7.1	9.4	14.3	11.3	7.2	7.6	5.95	5.7	4.25
12.............	5.2	4.8	6.1	7.0	9.55	14.8	10.85	7.1	7.85	5.95	5.7	4.45
13.............	5.2	4.8	6.0	6.8	10.05	15.15	10.65	7.1	8.0	5.8	5.7	4.55
14.......:.....	5.3	4.85	6.1	6.35	9.95	14.9	9.95	7.0	7.9	5.9	5.65	4.75
15.............	5.3	4.8	6.25	6.2	10.2	14.25	9.65	7.0	7.6	5.95	5.6	5.0
16.............	5.2	4.9	6.1	6.3	9.95	13.8	9.45	7.75	7.35	6.0	5.6	5.0
17.............	5.0	5.0	5.95	6.45	9.85	13.25	9.25	8.25	7.25	6.0	5.6	5.7
18.............	5.0	5.0	5.8	6.55	9.7	12.8	9.0	8.05	7.15	6.1	5.5	6.0
19.............	5.1	4.95	5.95	6.95	9.6	12.65	8.9	8.0	7.0	6.05	5.5	6.1
20.............	5.1	4.9	6.05	7.45	9.7	12.85	8.85	7.9	7.0	5.9	5.45	6.2
21.............	5.05	5.0	6.75	8.05	9.7	13.0	8.65	7.75	6.9	5.85	5.4	6.3
22.............	5.1	5.1	7.7	8.6	9.9	13.2	8.35	7.9	6.8	5.8	5.4	6.35
23.:............	5.1	5.2	9.1	8.3	10.25	13.15	8.4	7.8	6.7	5.7	5.5	6.45
24.............	5.35	5.1	11.15	8.0	10.4	13.3	8.55	7.6	6.6	5.7	5.55	6.5
25.............	5.4	5.0	9.8	7.65	10.6	13.55	8.35	7.35	6.6	5.75	5.6	6.6
26.............	5.15	4.9	9.2	7.35	10.75	13.5	8.3	7.25	6.5	5.8	5.5	6.6
27.............	5.35	4.9	8.85	7.0	11.1	13.5	8.2	7.2	6.5	5.75	5.5	6.6
28.............	5.6	4.85	8.55	7.35	11.05	13.45	8.2	7.1	6.4	5.7	5.6	6.6
29.............	5.75	......	8.35	7.25	10.65	12.9	8.1	7.0	6.35	5.7	5.6	6.6
30.............	5.8	......	7.95	7.0	10.8	12.75	8.1	7.15	6.3	5.7	5.8	6.6
31.............	5.7	......	7.7	......	10.9	......	8.0	7.4	......	5.7	......	6.6

NOTE.—Gage heights affected by ice after December 11, 1909.

Daily discharge, in second-feet, of Green River at Greenriver, Utah, for 1905–1909.

Day.	Mar.	Apr.	May.	June.	July.	Aug.	Sept.	Oct.	Nov.	Dec.
1905.										
1	1,760	2,900	6,360	22,500	13,400	3,840	1,870	6,190	1,990	1,870
2	1,870	3,180	6,870	21,400	12,900	3,720	1,870	4,180	2,060	1,760
3	1,870	3,280	8,920	22,800	12,400	3,720	1,870	3,500	1,930	1,550
4	2,050	2,900	10,100	24,500	12,100	3,720	1,870	3,390	1,820	1,370
5	2,120	2,720	11,600	27,000	11,800	3,720	2,060	2,990	2,060	1,370
6	2,120	2,720	12,400	28,400	11,300	3,720	2,060	2,900	2,190	1,290
7	2,340	2,720	12,100	28,800	10,800	3,500	3,080	2,640	2,120	1,290
8	2,720	2,720	10,800	30,200	10,300	3,280	2,640	2,560	2,120	1,220
9	2,900	2,900	9,380	32,000	9,850	3,080	2,480	2,340	2,260	1,220
10	3,500	2,720	8,920	33,900	9,380	3,080	2,640	2,410	2,120	1,220
11	3,840	2,720	8,040	33,500	8,920	2,900	2,810	2,560	1,990	1,290
12	3,720	2,810	8,040	30,200	8,260	2,810	2,810	2,410	1,930	1,330
13	3,390	3,180	8,920	31,300	7,630	2,720	2,480	2,260	1,870	1,370
14	3,180	3,720	10,600	31,300	7,440	2,720	2,340	2,340	1,820	1,370
15	3,080	4,060	9,620	29,500	6,700	2,720	2,060	2,120	1,990	1,370
16	3,080	4,700	8,700	28,800	6,520	2,720	1,990	2,120	1,990	1,290
17	3,080	4,430	8,040	24,200	6,360	2,640	1,870	2,120	1,990	1,370
18	3,080	4,430	8,040	24,500	6,030	2,480	1,990	2,120	1,990	1,370
19	3,080	5,400	8,040	23,800	5,870	2,260	2,120	1,990	1,870	1,370
20	3,280	5,400	8,920	23,100	5,710	2,260	1,990	1,870	1,870	1,370
21	3,180	4,840	10,800	21,800	5,550	2,260	2,190	1,870	1,990	......
22	3,610	4,700	13,700	20,400	5,550	2,260	2,260	1,990	2,560	......
23	3,720	4,970	16,000	19,700	5,250	2,260	2,410	2,060	2,340	......
24	3,500	5,110	18,100	18,800	5,250	1,990	2,260	1,990	2,260	......
25	3,280	5,400	20,100	18,400	4,970	1,990	2,560	1,990	2,120	......
26	3,280	5,400	21,800	17,200	4,840	2,190	2,640	1,990	1,990	......
27	3,280	4,970	23,100	16,600	4,700	2,190	2,900	1,990	2,060	......
28	3,280	5,110	24,200	15,400	4,430	1,990	2,990	1,990	2,190	......
29	3,280	5,710	23,100	14,800	4,300	1,990	4,060	1,990	2,120	......
30	3,080	6,360	23,100	14,000	4,180	1,930	6,030	1,990	1,990	......
31	3,080	......	22,800	......	4,180	1,870	......	1,990	......	......

Day.	Feb.	Mar.	Apr.	May.	June.	July.	Aug.	Sept.	Oct.	Nov.	Dec.
1906.											
1	1,740	1,870	12,100	14,200	33,900	14,200	7,840	6,700	3,720	3,080	1,700
2	1,820	1,870	11,600	13,400	35,400	14,800	8,260	7,440	3,500	5,870	1,820
3	1,700	1,870	9,380	13,100	36,800	15,400	7,840	6,870	3,500	4,700	2,120
4	1,700	2,190	8,470	12,600	39,100	15,700	8,040	6,360	3,500	4,970	2,990
5	1,650	2,260	8,040	12,600	41,400	15,700	7,240	6,360	3,500	4,970	3,280
6	1,700	2,340	7,630	12,400	41,000	15,700	7,060	6,520	3,280	4,560	3,080
7	1,990	2,410	7,060	12,900	40,200	15,700	6,870	6,520	3,280	4,060	3,080
8	1,600	2,480	5,550	13,400	41,000	15,700	6,520	6,030	3,180	3,720	3,280
9	1,650	2,640	5,550	13,700	42,100	15,100	6,190	5,550	3,080	3,720	2,810
10	1,500	2,640	5,400	14,000	42,100	15,100	6,030	5,250	3,080	3,720	2,560
11	1,600	2,640	6,030	17,800	39,900	15,700	5,710	5,110	3,080	3,720	2,410
12	1,550	3,180	7,060	20,100	38,300	15,700	5,250	4,700	3,080	3,610	2,640
13	1,500	3,950	6,870	21,700	30,200	15,400	5,250	4,430	3,080	3,500	2,560
14	1,500	5,250	6,870	25,200	27,700	15,100	5,250	4,180	2,900	3,500	2,410
15	1,000	5,550	7,240	30,200	26,300	15,100	5,400	4,060	2,900	3,280	2,560
16	1,000	5,400	8,700	32,000	25,200	14,800	4,970	4,560	2,900	3,280	2,480
17	1,550	4,970	8,470	31,300	24,500	14,500	4,970	4,300	2,900	3,280	2,260
18	1,550	4,430	8,260	30,000	25,000	14,500	4,840	4,700	2,900	3,180	2,120
19	1,700	4,180	8,470	31,300	25,600	14,000	4,430	5,250	2,720	3,080	1,990
20	1,760	3,720	8,700	30,600	25,200	14,200	4,430	4,560	2,720	2,810	1,820
21	1,820	2,990	8,260	31,000	26,300	13,700	4,700	4,560	2,720	2,480	1,760
22	1,820	3,390	9,850	31,700	25,900	12,900	5,550	5,400	2,900	2,260	1,820
23	1,820	4,060	11,600	32,400	24,900	12,100	6,700	4,700	2,900	1,870	1,820
24	1,820	4,970	12,600	33,100	25,600	10,800	7,440	4,430	2,900	1,760	1,820
25	1,820	6,030	13,100	33,100	25,900	10,300	6,360	4,430	2,900	1,870	2,190
26	1,820	9,620	14,000	33,100	26,700	10,300	5,870	4,180	2,900	2,260	2,500
27	1,820	14,800	13,700	32,400	27,400	9,620	5,710	3,950	2,810	2,410	2,560
28	1,760	21,800	14,500	33,100	24,500	9,380	6,300	3,840	2,720	2,340	2,560
29	......	20,400	16,900	34,200	24,500	8,920	6,520	3,720	2,560	2,060	2,720
30	......	18,800	15,400	34,400	25,600	8,260	7,060	3,720	2,720	1,820	2,640
31	......	16,600	......	36,500	......	8,040	6,700	......	2,900	......	2,900

Daily discharge, in second-feet, of Green River at Greenriver, Utah, for 1905–1909—Con.

Day.	Jan.	Feb.	Mar.	Apr.	May.	June.	July.	Aug.	Sept.	Oct.	Nov.	Dec.
1907.												
1	2,900	2,720	8,040	9,380	15,100	31,600	38,400	19,300	6,900	3,450	3,000	1,890
2	2,810	2,640	7,440	8,700	15,700	30,200	38,400	18,100	6,900	3,110	3,000	1,740
3	2,720	2,560	7,440	7,840	16,600	29,800	38,100	17,100	6,900	3,000	3,000	1,890
4	2,640	2,560	6,700	7,060	16,600	29,800	39,200	16,500	6,900	3,220	3,000	1,820
5	2,560	2,720	6,030	6,870	16,000	33,000	39,900	16,200	7,940	3,690	3,000	1,740
6	2,560	2,990	5,550	6,030	15,100	36,600	41,400	15,200	6,500	3,820	3,000	1,740
7	2,340	3,080	5,110	6,030	14,800	40,700	42,100	14,600	6,310	3,940	3,000	1,600
8	2,120	3,720	5,250	7,240	14,200	42,500	42,900	14,000	6,120	3,940	3,000	1,600
9	2,340	4,300	4,840	8,700	14,000	44,800	42,900	14,000	6,120	3,690	3,000	1,600
10	2,560	5,250	4,180	9,150	13,100	46,600	42,100	13,400	5,760	3,690	2,790	1,600
11	2,560	6,870	4,180	10,100	13,100	48,100	42,100	12,200	5,420	4,070	2,790	1,470
12	2,560	7,840	4,300	10,600	14,200	47,700	41,400	11,900	5,100	4,480	2,790	1,470
13	2,410	6,870	4,430	10,300	15,700	46,600	40,300	11,103	4,780	4,780	2,790	1,470
14	2,410	5,870	4,430	11,600	21,100	44,800	38,100	11,100	4,630	5,260	2,790	1,470
15	2,410	5,250	4,180	14,800	23,800	42,900	36,600	10,800	4,480	5,260	2,790	1,470
16	2,560	4,970	4,180	16,900	24,900	40,700	33,700	10,600	4,480	4,630	2,590	1,350
17	2,560	5,110	4,180	19,400	24,900	39,600	31,900	9,790	4,340	4,070	2,590	1,350
18	2,480	4,700	3,950	21,800	23,800	38,800	30,200	9,060	4,200	3,690	2,500	1,350
19	2,410	4,700	3,720	22,800	22,800	37,700	27,700	8,600	3,820	3,690	2,400	1,350
20	2,340	4,700	3,720	24,900	22,800	37,700	25,300	8,380	3,690	3,450	2,400	1,350
21	2,480	4,700	3,720	24,500	26,700	38,100	23,900	7,940	3,690	3,340	2,400	1,240
22	2,640	5,710	6,520	24,500	30,900	37,700	22,600	7,720	3,450	3,220	2,220	1,240
23	2,260	6,360	4,180	20,100	34,800	37,700	21,300	7,300	3,450	3,220	2,220	1,240
24	2,120	6,520	7,240	17,500	38,400	37,000	20,300	7,100	3,450	3,220	2,220	1,240
25	1,990	5,400	12,900	16,300	41,000	37,000	20,300	7,300	3,220	3,220	2,050	1,240
26	1,820	5,870	13,400	16,000	42,500	36,300	20,300	7,510	3,220	3,220	2,050	1,240
27	2,060	6,030	13,400	15,100	41,800	37,700	20,300	9,300	3,220	3,220	1,890	1,350
28	2,260	7,440	12,900	15,100	42,900	37,300	20,300	8,600	3,220	3,220	1,890	1,350
29	2,480	....	12,100	15,100	40,700	38,400	19,700	8,160	3,220	3,000	1,890	1,350
30	2,560	....	11,100	15,100	37,000	38,400	19,700	7,720	3,220	3,000	1,890	1,350
31	2,720	....	10,100	....	29,800	....	19,000	7,300	....	3,000	....	1,350

Day.	Feb.	Mar.	Apr.	May.	June.	July.	Aug.	Sept.	Oct.	Nov.	Dec.
1908.											
1		1,740	4,200	9,060	14,000	14,400	5,730	5,000	6,120	3,220	800
2		1,740	3,940	8,600	13,400	13,900	6,510	4,820	5,000	3,220	800
3		1,890	3,690	8,160	13,400	12,800	7,540	4,820	4,480	2,950	800
4		2,400	3,450	8,160	11,900	12,800	8,220	4,480	4,150	2,950	800
5		2,400	3,450	8,380	11,400	12,800	8,890	4,320	3,820	2,950	750
6		2,590	3,450	8,830	13,400	12,800	7,780	4,150	3,660	2,700	800
7		2,590	3,450	9,060	13,400	13,300	5,540	3,660	3,510	2,700	895
8		2,590	3,450	9,300	14,000	13,300	5,000	3,080	3,220	2,470	....
9		2,590	3,450	9,540	13,400	12,800	4,820	2,950	3,220	2,470	....
10		2,790	3,450	10,300	14,000	12,800	8,000	2,700	2,950	2,260	....
11		2,790	3,450	10,800	13,400	12,800	6,710	2,470	2,950	2,260	....
12		2,790	3,450	11,700	14,300	12,500	4,820	2,470	2,950	2,260	....
13		2,790	3,450	13,700	16,200	12,500	5,360	2,260	2,950	2,260	....
14		3,000	3,820	14,300	19,700	12,000	8,000	2,260	2,950	2,260	....
15		3,000	4,340	14,600	20,300	11,000	7,340	2,260	2,700	2,070	....
16		3,450	5,100	14,000	22,900	10,800	7,340	2,070	2,700	1,980	....
17		4,630	5,120	12,200	23,600	10,800	7,560	2,070	2,700	1,900	....
18		4,780	5,940	11,700	24,300	10,300	7,560	2,070	3,080	1,750	....
19		4,200	6,700	11,100	24,300	9,820	7,130	2,070	3,820	1,750	....
20		3,940	7,510	11,100	25,000	9,350	7,130	1,900	4,650	1,600	....
21		4,340	8,830	11,700	25,000	8,890	7,560	2,070	4,650	1,460	....
22		5,100	10,300	12,500	22,400	8,440	7,560	5,170	4,150	1,460	....
23	1,740	5,940	11,400	13,100	21,800	8,000	7,560	3,820	3,820	1,600	....
24	1,740	5,420	12,500	13,400	22,800	8,000	7,130	3,510	3,510	1,600	....
25	1,740	5,100	12,800	14,300	21,800	7,560	6,920	4,820	3,510	1,900	....
26	1,890	5,100	12,800	12,800	20,800	6,510	7,340	4,150	3,510	1,900	....
27	1,890	4,780	11,900	12,500	19,200	6,120	6,710	3,220	3,510	2,070	....
28	1,740	4,340	11,400	13,700	18,600	5,540	6,510	3,680	3,220	2,070	....
29	1,740	4,070	10,800	14,300	17,100	4,820	6,310	3,220	3,220	1,900	....
30		3,940	9,800	14,000	16,200	5,730	5,920	5,300	3,220	830	....
31		3,940	....	14,000	....	4,820	5,360	....	3,220	....	....

Daily discharge, in second-feet, of Green River at Greenriver, Utah, for 1905–1909—Con.

Day.	Jan.	Feb.	Mar.	Apr.	May.	June.	July.	Aug.	Sept.	Oct.	Nov.	Dec.
1909.												
1......	930	2,580	1,600	10,500	12,300	33,400	42,600	12,500	17,400	4,820	3,220	4,820
2......	1,100	2,260	1,460	9,580	14,400	34,800	40,800	12,000	18,000	4,820	3,080	3,980
3......	1,210	2,260	1,600	8,890	14,700	34,100	40,100	11,000	15,300	4,820	2,950	3,360
4......	1,210	1,900	1,750	8,000	14,200	33,000	39,400	10,800	11,800	4,650	3,080	2,950
5......	1,210	1,750	2,260	7,780	13,100	32,700	38,300	10,300	11,000	4,480	3,220	2,070
6......	1,210	1,750	2,070	7,130	11,000	35,200	37,200	9,820	12,000	4,480	3,220	1,530
7......	1,460	1,600	2,360	9,350	11,300	39,400	37,600	9,350	13,900	4,480	3,220	1,270
8......	1,600	1,460	2,820	12,800	13,100	42,600	36,600	9,350	12,800	4,150	3,220	1,160
9......	1,750	1,460	3,360	13,300	15,600	45,200	36,600	9,580	11,800	4,150	3,220	1,060
10......	1,900	1,330	3,820	11,000	19,500	51,100	36,200	9,350	10,800	4,150	3,220	930
11......	1,900	1,330	4,150	8,440	21,200	55,900	34,100	8,890	10,800	3,980	3,220	895
12......	2,070	1,460	4,480	8,000	22,100	59,600	31,000	8,440	12,000	3,980	3,220	......
13......	2,070	1,460	4,150	7,130	25,400	62,200	29,600	8,440	12,800	3,510	3,220	......
14......	2,260	1,530	4,480	5,360	24,800	60,400	24,800	8,000	12,300	3,820	3,080	......
15......	2,260	1,460	5,000	4,820	26,400	55,600	22,800	8,000	10,800	3,980	2,950	......
16......	2,070	1,600	4,480	5,170	24,800	52,200	21,500	11,500	9,580	4,150	2,950	......
17......	1,750	1,750	3,980	5,730	24,100	48,200	20,200	14,100	9,120	4,150	2,950	......
18......	1,750	1,750	3,510	6,120	23,100	44,800	18,600	13,100	8,660	4,480	2,700	......
19......	1,900	1,680	3,980	7,780	22,400	43,700	18,000	12,800	8,000	4,320	2,700	......
20......	1,900	1,600	4,320	10,100	23,100	45,200	17,700	12,300	8,000	3,820	2,580	......
21......	1,820	1,750	6,920	13,100	23,100	46,300	16,500	11,500	7,560	3,660	2,470	......
22......	1,900	1,900	11,300	16,200	24,400	47,800	14,700	12,300	7,130	3,510	2,470	......
23......	1,900	2,070	19,200	14,400	26,800	47,400	15,000	11,800	6,710	3,220	2,700	......
24......	2,360	1,900	33,000	12,800	27,800	48,500	15,900	10,800	6,310	3,220	2,820	......
25......	2,470	1,750	23,800	11,000	29,200	50,400	14,700	9,580	6,310	3,360	2,950	......
26......	1,980	1,600	19,900	9,580	30,200	50,000	14,400	9,120	5,920	3,510	2,700	......
27......	2,360	1,600	17,700	8,000	32,700	50,000	13,900	8,890	5,920	3,360	2,700	......
28......	2,950	1,530	15,900	9,580	32,400	49,600	13,900	8,440	5,540	3,220	2,950	......
29......	3,360	....	14,700	9,120	29,600	45,600	13,300	8,000	5,360	3,220	2,950	......
30......	3,510	....	12,500	8,000	30,600	44,400	13,300	8,660	5,170	3,220	3,510	......
31......	3,220	....	11,300		31,300	....	12,800	9,820	....	3,220		......

NOTE.—These daily discharges are based on rating curves applicable as follows (except as noted): Feb. 16, 1905, to May 21, 1907, well defined between 2,000 and 37,000 second-feet; May 22, 1907, to June 21, 1908, fairly well defined between 5,000 and 39,000 second-feet; June 22, 1908, to Dec. 31, 1909, well defined between 2,500 and 39,000 second-feet.

Discharge estimated for ice periods as follows: Dec. 21 to 30, 1905, 1,200 second-feet per day; Jan. 1 to 31, 1906, 1,400 second-feet per day; Feb. 1 to 12, 1906, 1,500 second-feet per day; Jan. 1 to 31, 1908, 1,300 second-feet per day; Feb. 1 to 10, 1908, 1,400 second-feet per day; Feb. 11 to 22, 1908, 1,500 second-feet per day; Dec. 8 to 31, 1908, 800 second-feet per day; Dec. 12 to 31, 1909, 800 second-feet per day.

These estimates for ice periods are based on the observers' notes and the climatological data of the United States Weather Bureau.

Monthly discharge of Green River at Greenriver, Utah, for 1905–1909.

[Drainage area, 38,200 square miles.]

Month.	Discharge in second-feet.				Run-off.		Accuracy.
	Maximum.	Minimum.	Mean.	Per square mile.	Depth in inches on drainage area.	Total in acre-feet.	
1905.							
March....................	3,840	1,760	2,990	0.078	0.09	184,000	A.
April....................	6,360	2,720	4,070	.106	.12	242,000	A.
May.....................	24,200	6,360	12,900	.338	.39	793,000	A.
June....................	33,900	14,000	24,300	.636	.71	1,450,000	B.
July....................	13,400	4,180	7,640	.200	.23	470,000	A.
August..................	3,840	1,870	2,730	.071	.08	168,000	A.
September...............	6,030	1,870	2,510	.066	.07	149,000	A.
October.................	6,190	1,870	2,480	.065	.07	152,000	A.
November................	2,560	1,820	2,050	.054	.06	122,000	A.
December................	1,870	....	1,320	.035	.04	81,200	C.
The period.............	....	....	....	....	....	3,810,000	

Monthly discharge of Green River at Greenriver, Utah, for 1905–1909—Continued.

Month.	Discharge in second-feet.				Run-off.		Accuracy.
	Maximum.	Minimum.	Mean.	Per square mile.	Depth in inches on drainage area.	Total in acre-feet.	
1906.							
January			a 1,400	0.037	0.04	86,100	D.
February			b 1,620	.042	.04	90,000	C.
March	21,800	1,870	6,110	.160	.18	376,000	B.
April	16,900	5,400	9,580	.251	.28	570,000	A.
May	36,500	12,400	24,800	.649	.75	1,520,000	A.
June	42,100	24,500	31,300	.819	.91	1,860,000	A.
July	15,700	8,040	13,400	.351	.40	824,000	A.
August	8,260	4,430	6,170	.162	.19	379,000	A.
September	7,440	3,720	5,080	.133	.15	302,000	A.
October	3,720	2,560	3,020	.079	.09	186,000	A.
November	5,870	1,760	3,260	.085	.09	194,000	A.
December	3,280	1,700	2,430	.064	.07	149,000	B.
The year	42,100		9,050	.237	3.19	6,540,000	
1907.							
January	2,900	1,820	2,440	.064	.07	150,000	B.
February	7,840	2,560	4,910	.129	.13	273,000	A.
March	13,400	3,720	6,760	.177	.20	416,000	A.
April	24,900	6,030	14,000	.367	.41	833,000	A.
May	42,900	13,100	24,700	.647	.75	1,520,000	B.
June	48,100	29,800	38,800	1.02	1.17	2,310,000	B.
July	42,900	19,000	31,600	.828	.95	1,940,000	B.
August	19,300	7,100	11,200	.293	.34	689,000	A.
September	7,940	3,220	4,820	.126	.14	287,000	A.
October	5,260	3,000	3,670	.096	.11	226,000	A.
November	3,000	1,890	2,560	.067	.07	152,000	B.
December	1,890	1,240	1,470	.038	.04	90,400	C.
The year	48,100	1,240	12,300	.322	4.38	8,890,400	
1908.							
January	1,820	1,350	1,300	.034	.04	79,900	C.
February	1,890	1,600	1,530	.040	.05	88,000	C.
March	5,940	1,740	3,570	.093	.11	220,000	A.
April	12,800	3,450	6,590	.172	.19	392,000	A.
May	14,600	8,160	11,600	.304	.35	713,000	A.
June	25,000	11,400	18,100	.474	.53	1,080,000	A.
July	14,400	4,820	10,300	.270	.31	633,000	A.
August	8,890	4,820	6,810	.179	.21	419,000	A.
September	5,300	1,900	3,380	.088	.10	201,000	B.
October	6,120	2,700	3,580	.094	.11	220,000	B.
November	3,220	830	2,160	.057	.06	129,000	C.
December	1,460	750	801	.021	.02	49,300	
The year	25,000	750	5,810	.152	2.08	4,220,000	
1909.							
January	3,510	930	1,980	.052	.06	122,000	B.
February	2,580	1,330	1,720	.045	.05	95,500	B.
March	33,000	1,460	8,120	.213	.25	499,000	B.
April	16,200	4,820	9,290	.243	.27	553,000	A.
May	32,700	11,000	22,400	.586	.68	1,380,000	A.
June	62,200	32,700	46,300	1.21	1.35	2,760,000	C.
July	42,600	12,800	25,200	.660	.76	1,550,000	A.
August	14,100	8,000	10,300	.270	.31	633,000	A.
September	18,000	5,170	9,960	.261	.29	593,000	A.
October	4,820	3,220	3,930	.103	.12	242,000	A.
November	3,510	2,470	2,980	.078	.09	177,000	A.
December	4,820	a 800	1,290	.034	.04	79,300	C.
The year	62,400	a 800	12,000	.314	4.27	8,680,000	

a Estimated. b One-half month estimated.

NOTE.—See notes under 1909 daily discharge table for estimates during ice periods, 1905 to 1909.

COLORADO RIVER AT YUMA, ARIZ.

This station, which is located in the town of Yuma, Ariz., 1½ miles below the mouth of Gila River and 10 miles by river above the Mexican border, furnishes information concerning the amount of water available for irrigation along the lower Colorado River. Records of river height have been kept by the Southern Pacific Co. since April 1, 1878.

The records given herewith are furnished by the United States Reclamation Service, through F. L. Sellew, project engineer, Yuma,. Ariz.

As the bed of the stream is composed of silt and sand and is very unstable frequent measurements are necessary to properly determine the daily discharge. Neither bank is subject to overflow. Previous to May 31, 1903, discharge measurements were made from the railroad bridge. On that date a cable station was established at a point 600 feet below the bridge, and all measurements are now made from a car, except during highest floods, when a boat is used. At flood stages a large part of the water flows through an old channel and does not pass under the cable. At such times this overflow water is measured at the point where it passes under the railway trestle, one-third mile north of the main channel.

The staff gage is in two sections, the upper section reading above 24 feet being the original gage established in 1876. It is located at the railroad bridge, 600 feet above the cable station. The elevation of the zero of the gage is 137.4 feet above sea level.

Discharge measurements of Colorado River at Yuma, Ariz., in 1909.

[By R. L. North and N. B. Conway.]

Date.	Gage height.	Discharge.	Date.	Gage height.	Discharge.
	Feet.	*Sec.-ft.*		*Feet.*	*Sec.-ft.*
Jan. 1	18.70	7,800	Apr. 1	24.90	43,800
4	18.40	6,000	3	23.40	33,500
6	18.60	6,800	5	22.60	29,100
8	18.30	6,100	8	21.40	24,100
12	18.30	5,800	10	21.20	21,600
15	18.80	6,200	12	21.80	25,600
18	19.40	7,400	15	21.95	26,400
20	19.60	11,900	17	21.50	22,500
22	19.20	9,900	20	21.30	22,000
25	20.40	15,300	22	21.20	21,300
27	20.90	21,300	24	22.65	32,900
30	20.50	17,600	26	24.20	43,700
Feb. 1	21.80	25,100	28	24.65	46,100
3	20.70	18,100	30	23.90	41,900
5	20.10	15,500	May 3	22.70	32,400
9	19.70	12,300	5	23.20	38,600
11	19.70	12,000	7	23.40	40,400
15	19.80	11,400	10	23.00	37,300
17	20.00	11,800	12	23.50	42,500
19	20.30	12,800	14	24.20	49,000
23	20.20	12,200	17	25.20	60,400
25	20.90	16,800	19	25.70	65,400
27	20.00	12,000	21	26.20	68,500
Mar. 1	19.90	11,400	24	25.95	64,700
3	20.20	12,000	26	25.50	65,400
5	20.30	12,900	29	25.80	71,400
8	20.20	11,100	June 1	26.50	74,600
10	20.30	11,600	3	26.90	76,700
13	20.75	15,700	7	27.10	78,700
15	20.90	17,500	9	27.05	79,000
17	21.10	18,700	12	26.45	77,100
19	20.80	16,100	14	26.80	80,400
23	20.80	14,600	16	27.60	90,700
25	21.50	18,200	18	28.80	114,900
27	21.50	17,700	20	29.60	126,000
29	21.50	16,900	22	30.30	139,500

Discharge measurements of Colorado River at Yuma, Ariz., in 1909—Continued.

Date.	Gage height.	Discharge.	Date.	Gage height.	Discharge.
	Feet.	*Sec.-ft.*		*Feet.*	*Sec.-ft.*
June 24	30.75	149,500	Oct. 12	16.30	13,500
26	30.65	145,000	12	16.30	13,300
July 1	29.40	130,200	14	16.30	12,700
3	29.40	132,400	14	16.30	12,900
6	29.00	132,000	16	16.50	13,700
8	28.20	116,900	16	16.50	13,700
10	27.00	102,900	19	16.40	13,100
13a	25.50	86,300	19	16.40	13,300
15a	24.60	75,300	21	16.40	12,900
17a	23.80	68,400	21	16.40	12,900
19a	22.20	62,600	23	16.40	12,700
22a	20.00	42,900	23	16.40	12,600
24a	19.40	35,600	26	16.30	12,000
26a	18.90	34,400	26	16.30	12,000
29a	19.25	46,500	28	16.30	11,400
31a	19.80	51,900	30	16.00	11,100
Aug. 3a	19.50	44,300	Nov. 1	16.00	11,000
5a	18.90	38,700	1	16.00	10,900
7a	18.60	36,100	4	15.90	10,000
10a	18.00	28,400	6	15.80	9,600
12a	18.40	30,400	9	15.80	9,300
14a	18.30	28,500	11	15.80	8,700
17a	19.30	42,500	13	15.80	9,100
19a	19.80	42,300	16	15.90	9,000
21a	19.70	39,400	18	15.90	9,100
24a	20.95	53,800	20	15.90	9,400
26a	21.20	50,600	23	15.90	9,300
28a	20.20	47,200	27	16.00	9,500
31a	19.80	42,400	30	15.85	9,000
Sept. 2a	19.45	39,000	Dec. 2	15.90	9,600
4a	21.10	65,200	4	16.00	9,800
7a	20.80	50,400	7	16.10	10,800
11a	22.30	76,000	9	16.10	11,600
15a	22.50	79,900	11	16.00	10,900
18a	19.90	40,000	14	16.00	10,200
21	18.90	33,600	16	15.70	9,300
23	18.40	29,400	18	15.35	7,600
28	17.60	22,600	21	15.10	6,300
30	17.40	21,300	23	15.20	6,600
Oct. 2	17.10	19,200	25	15.30	6,600
5	16.80	17,200	28	15.00	5,700
7	16.70	16,100	30	14.80	4,800
9	16.50	14,500			

a Acoustic meter used and the coefficient of reduction was taken as 90 per cent.

Daily gage height, in feet, of Colorado River, at Yuma, Ariz., for 1909.

[R. L. North and N. B. Conway, observers.]

Day.	Jan.	Feb.	Mar.	Apr.	May.	June.	July.	Aug.	Sept.	Oct.	Nov.	Dec.
1	18.65	21.7	19.9	24.8	23.45	26.6	29.4	20.0	19.5	17.35	16.0	15.8
2	18.6	21.1	20.2	24.1	23.0	26.7	29.45	19.8	19.5	17.1	16.0	15.95
3	18.45	20.6	20.25	23.5	22.7	26.9	29.4	19.6	21.25	16.9	15.85	16.0
4	18.35	20.2	20.55	23.05	22.85	27.0	29.35	19.05	21.75	16.9	15.9	16.05
5	18.55	20.1	20.3	22.65	23.25	27.0	29.15	18.9		16.8	15.8	16.0
6	18.6	20.05	20.3	22.15	23.45	27.1	29.05	18.85	21.05	16.8	15.8	16.0
7	18.4	19.85	20.35	22.0	23.35	27.15	28.6	18.5	20.9	16.7	16.1	16.1
8	18.25	19.75	20.15	21.55	23.05	27.15	28.1	18.45	20.4	16.6	15.8	16.2
9	18.3	19.7	20.25	21.65	23.0	26.95	27.6	18.3	21.65	16.4	15.8	16.1
10	18.3	19.65	20.35	21.3	23.0	26.65	26.95	18.05		16.4	15.8	16.0
11	18.35	19.7	20.35	21.75	23.15	26.4	26.55	18.05	22.45	16.3	15.75	16.0
12	18.3	19.7	20.5	21.8	23.55	26.45	26.1	18.35	23.3	16.3	15.75	16.0
13	18.45	19.9	20.85	21.8	23.95	26.6	25.6	18.35	23.7	16.3	15.8	16.0
14	18.7	19.8	20.85	21.95	24.2	26.9	25.05	18.35	23.5	16.35	15.8	16.0
15	18.8	19.9	20.9	22.0	24.5	27.25	24.65	18.85	22.3	16.35	15.8	16.0
16	18.9	20.1	20.9	21.85	24.9	27.6	24.2	20.4	21.2	16.5	15.9	15.7
17	19.05	20.05	21.05	21.45	25.3	28.25	23.65	19.35	20.35	16.4	15.9	15.5
18	19.6	20.5	20.85	21.25	25.6	28.8	22.8	19.1	19.85	16.4	15.9	15.3
19	19.75	20.35	20.8	21.35	25.8	29.3	22.05	19.85	19.55	16.45	15.9	15.25
20	19.6	20.3	20.85	21.35	26.2	29.7	21.35	19.95	19.2	16.4	15.9	15.2

Daily gage height, in feet, of Colorado River, at Yuma, Ariz., for 1909—Continued.

Day.	Jan.	Feb.	Mar.	Apr.	May.	June.	July.	Aug.	Sept.	Oct.	Nov.	Dec.
21	19.35	20.15	20.65	21.25	26.2	30.05	20.7	19.8	18.85	16.4	15.95	15.15
22	19.2	20.2	20.65	21.25	26.3	30.25	20.05	20.65	18.6	16.4	15.9	15.2
23	19.1	20.3	21.0	21.55	26.2	30.5	19.8	20.45	18.35	16.4	15.9	15.25
24	19.25	20.45	21.3	22.8	25.95	30.75	19.45	21.0	18.2	16.3	15.95	15.3
25	20.4	20.75	21.55	23.7	25.75	30.75	19.1	21.1	18.05	16.3	16.05	15.3
26	20.35	20.25	21.6	24.3	25.5	30.65	18.9	21.15	17.8	16.3	16.0	15.3
27	20.85	20.0	21.5	24.65	25.55	30.45	19.1	20.5	17.7	16.3	16.0	15.15
28	20.55	19.9	21.5	24.65	25.75	30.1	18.95	20.1	17.6	16.25	15.95	15.0
29	20.55	....	21.55	24.15	25.85	29.85	19.35	19.35	17.55	16.1	15.95	14.9
30	20.5	....	22.15	23.85	26.0	29.65	19.75	19.45	17.4	16.0	15.85	14.75
31	22.0	....	24.3	....	26.35	....	19.85	19.9	....	16.0	....	14.6

Daily discharge, in second-feet, of Colorado River at Yuma, Ariz., for 1909.

Day.	Jan.	Feb.	Mar.	Apr.	May.	June.	July.	Aug.	Sept.	Oct.	Nov.	Dec.
1	7,800	25,100	11,400	43,800	39,400	75,100	130,200	51,600	39,800	20,700	10,900	8,800
2	7,100	21,100	13,200	37,600	36,900	75,600	133,000	48,500	39,300	19,200	10,900	9,800
3	6,100	18,100	12,000	33,500	32,400	76,700	132,400	43,300	47,800	18,800	10,100	9,900
4	6,000	14,900	12,700	31,100	33,600	77,200	133,700	39,200	66,200	17,800	9,800	10,100
5	6,900	15,500	12,900	29,400	39,200	77,200	132,300	38,700	65,700	17,200	9,600	9,900
6	6,800	15,300	12,900	26,600	41,700	78,700	133,000	39,500	56,200	17,000	9,600	10,000
7	6,000	14,400	13,400	25,800	40,000	79,200	124,700	33,000	51,000	16,100	9,500	10,800
8	6,100	14,000	11,100	24,700	37,200	79,200	115,400	34,200	52,000	15,300	9,400	11,900
9	6,100	12,300	12,000	25,100	36,800	78,700	110,000	32,300	55,200	14,500	9,300	11,600
10	6,100	11,900	11,600	22,800	37,400	77,700	102,000	29,300	67,300	13,900	9,000	10,900
11	6,200	12,000	11,800	28,500	38,500	76,900	98,500	25,000	77,500	13,600	8,300	10,900
12	5,800	12,000	12,600	25,600	43,000	77,100	93,500	29,500	87,700	13,500	8,300	10,700
13	5,900	13,800	15,700	25,600	47,200	78,500	88,100	29,600	93,200	13,100	9,100	10,400
14	6,100	11,400	16,600	26,600	49,000	81,300	80,800	29,600	91,500	13,100	8,800	10,200
15	6,200	11,400	17,500	26,700	51,800	86,200	76,200	34,900	77,800	12,800	8,400	9,900
16	6,300	12,000	17,500	25,900	55,500	90,700	71,700	47,400	63,700	13,700	9,000	9,300
17	6,400	11,800	18,700	22,100	61,500	103,700	66,300	42,800	50,700	13,300	9,700	8,400
18	7,400	13,500	17,200	20,300	65,000	114,900	62,700	39,500	39,600	13,000	9,100	7,300
19	8,100	12,800	16,100	21,200	66,400	124,900	60,200	42,700	37,800	13,600	9,200	7,000
20	11,900	12,800	16,500	22,100	70,400	127,500	54,500	42,300	35,500	13,200	9,400	6,800
21	10,600	12,300	14,800	21,900	68,500	134,500	48,700	40,200	33,300	12,900	9,900	6,500
22	9,900	12,500	14,800	21,600	69,100	138,700	43,500	47,700	31,100	12,800	9,400	6,700
23	9,400	12,200	14,200	23,800	68,500	144,200	40,300	48,500	28,000	12,700	9,300	6,800
24	10,200	13,700	17,200	34,100	64,700	149,500	36,000	54,100	27,700	12,200	9,600	6,800
25	15,300	16,800	18,200	41,300	61,700	148,000	34,500	52,400	26,500	12,100	10,500	6,600
26	15,100	12,500	18,700	44,400	65,400	145,000	34,400	50,200	24,400	12,000	9,800	6,800
27	21,300	12,000	17,700	46,800	66,400	142,700	39,200	47,600	23,600	11,700	9,500	6,200
28	17,100	11,500	17,700	46,100	70,400	138,300	40,700	46,500	22,600	11,200	9,400	5,700
29	17,100	....	16,900	43,400	72,400	135,200	47,700	40,800	22,300	11,000	9,500	5,300
30	17,600	....	21,700	41,600	72,300	133,100	51,800	40,800	21,300	11,000	9,000	4,600
31	31,500	....	35,900	....	73,900	....	52,800	43,100	....	11,000	....	4,100

NOTE.—These discharges were obtained by the indirect method for shifting channels.

Monthly discharge of Colorado River at Yuma, Ariz., for 1909.

[Drainage area, 225,000 square miles.]

Month.	Discharge in second-feet.				Run-off.	
	Maximum.	Minimum.	Mean.	Per square mile.	Depth in inches on drainage area.	Total in acre-feet.
January	31,500	5,800	10,000	0.044	0.05	615,000
February	25,100	11,400	13,900	.062	.06	772,000
March	35,900	11,100	15,900	.071	.08	978,000
April	46,800	20,300	30,300	.135	.15	1,800,000
May	73,900	32,400	54,100	.240	.28	3,330,000
June	149,500	75,100	105,000	.467	.52	6,250,000
July	133,700	34,400	79,600	.354	.41	4,890,000
August	54,100	25,000	40,800	.181	.21	2,510,000
September	93,200	21,300	48,500	.216	.24	2,890,000
October	20,700	11,000	14,000	.062	.07	861,000
November	10,900	8,300	9,440	.042	.05	562,000
December	11,900	4,100	8,410	.037	.04	517,000
The year	149,500	4,100	35,800	.159	2.16	26,000,000

SALTON SINK.

SALTON SEA NEAR SALTON, CAL.

Salton Sink originally formed a part of the Colorado Desert, which extends northwestward almost 100 miles from the California-Mexico boundary line and comprises an area of nearly 2,000 square miles. This desert comprises two fertile valleys, one to the northwest of the sink, in Riverside County, known as the Coachella Valley, and the other to the southeast of the sink, in Imperial County, called the Imperial Valley. Salton Sea, which now partly fills the sink, lies between the two valleys, being partly in Riverside County and partly in Imperial County. It is about 160 miles southeast of Los Angeles, 90 miles northwest of Yuma, and 50 miles north of Calexico. The longer diameter of the sea trends northwest and southeast. On December 31, 1908, its surface was 206 feet below mean sea level, its length was nearly 45 miles, its maximum width about 15 miles, its minimum width 9.5 miles, its maximum depth 67.5 feet, and its superficial area about 443 square miles.

During the high water of the summer of 1891 the Colorado overflowed into Salton Sink to such an extent as to endanger the Southern Pacific Railroad at its lowest point. In the summer of 1905, after a succession of winter and spring floods in Gila River followed by an exceptionally heavy summer flow in the Colorado, the flood into the sink was repeated on a much larger scale. The old river channel occupied by Alamo River was transformed into a deep wide gorge, and another channel, now called New River, was formed. The flood did great damage to the tracks of the Southern Pacific Railroad, to the plant of the New Liverpool Salt Co. below Mecca, and to the ranches in the vicinity.

Gage-height records kept by the New Liverpool Salt Co. from November, 1904, to February 26, 1906, show the actual depth of the water above the lowest portion of the sink. February 23, 1906, the Government installed a gage at the same datum, about half a mile west of Salton railroad station and 3 miles southeast of the old Salton station. This gage was destroyed by waves. The Southern Pacific Co. had graduated a trestle bent across Salt Creek about 2½ miles east of Salton, using the company's datum; the zero of this gage is 273.5 feet below mean sea level as determined from United States Geological Survey bench marks, or at an elevation of 280.3 feet below sea level according to the Southern Pacific Co.

Practically all the water received by Salton Sea enters through Alamo and New rivers, but chiefly through the former. These rivers run through Imperial Valley and are the drainage channels for the excess and waste water from the irrigation system and from the power plants.

The following tables show the fluctuation of the surface of Salton Sea:

Daily gage height, in feet, of Salton Sea near Salton, Cal., for 1909.

[J. A. Jeffreys and Benj. C. Kedel, observers.]

Day.	Jan.	Feb.	Mar.	Apr.	May.	June.	July.	Aug.	Sept.	Oct.	Nov.	Dec.
1	67.45	67.4	67.25	67.0	66.7	66.25	65.9	65.3	65.35	64.7	64.1	63.65
2	67.45	67.4	67.25	67.0	66.7	66.25	65.95	65.25	65.3	64.65	64.1	63.65
3	67.45	67.4	67.25	67.0	66.7	66.25	65.85	65.2	65.25	64.6	64.1	63.65
4	67.45	67.4	67.25	67.0	66.7	66.2	65.8	65.2	65.3	64.6	64.05	63.65
5	67.45	67.4	67.2	66.95	66.7	66.2	65.8	65.2	65.25	64.55	64.05	63.6
6	67.45	67.4	67.2	66.95	66.7	66.15	65.75	65.15	65.25	64.55	64.05	63.6
7	67.4	67.35	67.2	66.95	66.7	66.1	65.75	65.35	65.3	64.5	64.05	63.55
8	67.4	67.35	67.2	66.95	66.7	66.1	65.75	65.3	65.25	64.5	64.0	63.55
9	67.4	67.35	67.2	66.95	66.65	66.15	65.7	65.3	65.25	64.5	64.0	63.6
10	67.4	67.3	67.15	66.95	66.65	66.15	65.7	65.25	65.25	64.45	63.95	63.55
11	67.45	67.3	67.15	66.9	66.65	66.15	65.7	65.25	65.2	64.45	63.95	63.55
12	67.45	67.3	67.15	66.9	66.65	66.15	65.7	65.25	65.15	64.45	63.9	63.55
13	67.45	67.3	67.1	66.9	66.65	66.15	65.65	65.2	65.1	64.4	63.9	63.55
14	67.45	67.3	67.1	66.9	66.65	66.1	65.65	65.2	65.05	64.4	63.9	63.55
15	67.45	67.3	67.1	66.9	66.65	66.1	65.65	65.2	65.05	64.4	63.85	63.55
16	67.45	67.3	67.1	66.9	66.6	66.1	65.65	65.25	65.05	64.35	63.8	63.5
17	67.45	67.3	67.1	66.9	66.6	66.05	65.6	65.15	65.0	64.35	63.8	63.5
18	67.45	67.3	67.1	66.85	66.6	66.05	65.55	65.2	65.0	64.3	63.75	63.5
19	67.45	67.3	67.05	66.85	66.6	66.0	65.55	65.2	65.0	64.3	63.75	63.5
20	67.45	67.3	67.05	66.85	66.6	66.0	65.55	65.15	64.95	64.3	63.75	63.5
21	67.45	67.3	67.05	66.8	66.6	66.0	65.55	65.2	64.9	64.3	63.75	63.45
22	67.4	67.3	67.05	66.8	66.6	66.0	65.55	65.15	64.9	64.25	63.7	63.45
23	67.4	67.3	67.05	66.8	66.55	66.0	65.55	65.15	64.9	64.25	63.7	63.45
24	67.4	67.3	67.05	66.8	66.55	65.95	65.5	65.15	64.9	64.25	63.75	63.45
25	67.4	67.3	67.05	66.8	66.5	65.95	65.45	65.15	64.85	64.25	63.75	63.4
26	67.4	67.25	67.05	66.75	66.45	65.95	65.45	65.1	64.85	64.2	63.75	63.4
27	67.4	67.25	67.05	66.75	66.45	65.95	65.4	65.1	64.8	64.2	63.7	63.4
28	67.4	67.25	67.05	66.75	66.4	65.95	65.35	65.1	64.8	64.2	63.7	63.4
29	67.4	......	67.05	66.75	66.35	65.9	65.35	65.1	64.75	64.2	63.7	63.4
30	67.4	......	67.0	66.7	66.3	65.9	65.35	65.15	64.75	64.2	63.65	63.4
31	67.4	......	67.0	......	66.3	......	65.3	65.25	......	64.2	......	63.4

Monthly rise of Salton Sea near Salton, Cal., for 1904–1909.

Month.	Monthly rise.	Total rise.	Month.	Monthly rise.	Total rise.	Month.	Monthly rise.	Total rise.
1904.	*Feet.*	*Feet.*	**1906.**	*Feet.*	*Feet.*	**1908.**	*Feet.*	*Feet.*
November	0.6	......	July	8.6	66.5	March	−0.2	72.0
December	.2	0.8	August	2.9	69.4	April	− .4	71.6
1905.			September	.9	70.3	May	.6	71.0
January	1.4	2.2	October	1.2	71.5	June	− .5	70.5
February	1.6	3.8	November	− .2	71.3	July	− .5	70.0
March	.8	4.6	December	1.2	72.5	August	− .6	69.4
April	1.2	5.8				September	− .8	68.6
May	1.0	6.8	**1907.**			October	− .7	67.9
June	2.2	9.0	January	2.8	75.3	November	− .3	67.6
July	4.4	13.4	February	.7	76.0	December	− .2	67.4
August	2.2	15.6	March	− .1	75.9			
September	1.2	16.8	April	− .3	75.6	**1909.**		
October	1.4	18.2	May	− .5	75.1	January	.0	67.4
November	1.6	19.8	June	− .4	74.7	February	− .15	67.25
December	2.9	22.7	July	− .2	74.5	March	− .25	67.0
			August	− .3	74.2	April	− .3	66.7
1906.			September	− .7	73.5	May	− .4	66.3
January	1.1	23.8	October	− .4	73.1	June	− .4	65.9
February	1.8	25.6	November	− .5	72.6	July	− .6	65.3
March	2.7	28.3	December	− .3	72.3	August	− .05	65.25
April	5.6	33.9				September	− .5	64.75
May	8.7	42.5	**1908.**			October	− .55	64.2
June	15.4	57.9	January	.0	72.3	November	− .55	63.65
			February	− .1	72.2	December	− .25	63.4

ALAMO RIVER NEAR BRAWLEY, CAL.

During 1908 discharge measurements were made on Alamo River at a highway bridge 3½ miles east of Brawley, Cal., by H. R. Edwards, engineer for the New Liverpool Salt Co.[1]   During 1909 measurements were made by engineers of the United States Geological Survey.   On June 24, 1909, a continuous record of gage heights was commenced at this point.   The staff gage is spiked vertically to a pile in the left abutment of the bridge.   The datum of the gage has remained the same during the maintenance of the station.   All discharge measurements are made from the bridge.

The data obtained at this station, together with those obtained on New River, show the amount of waste water reaching Salton Sea and are of value in connection with experiments being made by the United States Weather Bureau for determining the evaporation from Salton Sea.

Conditions for obtaining accurate discharge data are poor.   The channel is constantly scouring or filling as the stage fluctuates. Both banks are high and well above overflow.

Discharge measurements of Alamo River near Brawley, Cal., in 1909.

Date.	Hydrographer.	Width.	Area of section.	Gage height.	Discharge.
		Feet.	*Sq. ft.*	*Feet.*	*Sec -ft.*
Jan. 14	Hardy and Jones	55	91	5.01	213
June 24	W. F. Martin	55	128	5.40	401
July 30	....do	56	120	5.26	290
Sept. 12	A. H. Koebig, jr.	60	175	6.00	675
Dec. 31	....do	55	82	4.70	164

Daily gage height, in feet, of Alamo River near Brawley, Cal., for 1909.

[Mrs. Flora Helmar, observer.]

Day.	June.	July.	Aug.	Sept.	Oct.	Nov.	Dec.	Day.	June.	July.	Aug.	Sept.	Oct.	Nov.	Dec.
1		5.95	5.5	8.0	5.8	5.4	5.65	16		4.7	5.95	5.4	5.0	5.4	5.85
2		5.8	5.45	7.6	5.75	5.35	5.45	17		4.8	7.65	5.2	5.05	5.35	5.55
3		5.9	5.3	7.1	5.65	5.45	5.4	18		4.95	7.65	5.15	5.2	5.5	5.45
4		5.85	5.35	6.8	5.45	5.45	5.4	19		5.05	8.3	5.0	5.2	5.5	5.35
5		5.9	5.2	6.3	5.75	5.3	5.45	20		4.85	8.1	5.25	5.2	5.65	5.2
6		5.7	5.3	4.95	5.55	5.3	5.45	21		4.8	7.9	5.7	5.25	5.45	5.15
7		5.5	5.3	5.1	5.6	5.25	5.4	22		4.95	8.4	5.6	5.25	5.6	5.3
8		5.4	5.35	5.55	5.65	5.35	5.4	23		4.75	8.35	5.35	5.2	5.35	5.2
9		5.35	5.4	6.3	5.65	5.15	5.5	24	5.5	4.65	8.4	5.4	5.2	5.3	5.2L
10		5.4	5.35	6.2	5.35	5.2	5.8	25	5.65	4.65	8.4	5.45	5.15	5.4	5.55
11		5.65	5.15	6.0	5.25	5.25	5.8	26	5.8	5.45	8.5	5.55	5.15	5.55	5.55
12		5.6	5.0	6.0	5.25	5.2	6.1	27	5.65	4.9	8.4	5.65	5.2	5.6	5.65
13		5.45	5.05	6.4	5.2	5.25	6.1	28	5.9	4.85	7.8	5.8	5.1	5.75	5.35
14		5.15	5.3	5.85	5.1	5.35	5.8	29	5.95	4.95	5.95	5.85	5.2	5.8	5.0
15		4.8	5.45	5.8	5.05	5.35	5.8	30	5.95	5.2	5.95	6.2	5.3	5.8	4.85
								31		5.3	7.15		5.3		4.55

[1] These measurements were published in Water-Supply Paper 249, p. 52.

Daily discharge, in second-feet, of Alamo River near Brawley, Cal., for 1909.

Day.	June.	July.	Aug.	Sept.	Oct.	Nov.	Dec.	Day.	June.	July.	Aug.	Sept.	Oct.	Nov.	Dec.
1....	......	604	410	1,480	539	368	474	16....	......	176	604	368	237	368	560
2....	......	539	389	1,310	518	348	389	17....	......	194	1,330	293	250	348	432
3....	......	582	328	1,100	474	389	368	18....	......	226	1,330	278	278	348	389
4....	......	560	348	969	389	389	368	19....	......	250	1,610	237	293	410	348
5....	......	582	293	754	518	328	389	20....	......	204	1,530	310	293	474	293
6....	......	496	328	226	432	328	389	21....	......	194	1,440	496	310	389	278
7....	......	410	328	263	453	310	368	22....	......	226	1,660	453	310	453	328
8....	......	368	348	432	474	348	368	23....	......	185	1,640	348	293	348	293
9....	......	348	368	754	474	278	410	24....	410	168	1,660	368	293	328	310
10....	......	368	348	711	348	293	539	25....	474	389	1,660	389	278	368	432
11....	......	474	278	625	310	310	539	26....	539	389	1,700	432	278	432	432
12....	......	453	237	625	310	293	668	27....	474	214	1,660	474	293	453	474
13....	......	389	250	797	293	310	668	28....	582	204	1,400	539	263	518	348
14....	......	278	328	560	263	348	539	29....	604	226	604	560	293	539	237
15....	......	194	389	539	250	348	539	30....	604	293	604	711	328	539	204
								31....	......	328	1,120	......	328	......	153

NOTE.—These discharges are based on a rating curve that is fairly well defined between discharges of 100 and 1,050 second-feet.

Monthly discharge of Alamo River near Brawley, Cal., for 1909.

Month.	Discharge in second-feet.			Run-off (total in acre-feet).	Accu-racy.
	Maximum.	Minimum.	Mean.		
June 24–30..	604	410	529	7,340	B.
July..	604	168	339	20,800	B.
August...	1,700	237	856	52,600	B.
September..	1,480	226	580	34,500	B.
October..	539	237	344	21,200	B.
November..	539	278	377	22,400	B.
December..	668	153	404	24,800	B.
The period...........................				184,000	

NEW RIVER NEAR BRAWLEY, CAL.

During 1908 discharge measurements were made at a wagon bridge over New River, 1½ miles west of Brawley, Cal., by H. R. Edwards, engineer for the New Liverpool Salt Co.[1] During 1909 measurements were made by engineers of the United States Geological Survey. On June 24, 1909, a continuous record of gage heights was begun at this point. The staff gage is spiked vertically to the third bridge pile from the right bank. The datum of the gage has remained the same during the maintenance of the station. At high stages discharge measurements are made from the bridge, but at medium and low stages measurements are made by wading near the bridge.

The data obtained at this station, together with those obtained on Alamo River, show the amount of waste water reaching Salton Sea and are of value in connection with experiments being made by the United States Weather Bureau to determine the evaporation from Salton Sea.

---

[1] These measurements were published in Water-Supply Paper 249, p. 52.

Conditions for obtaining accurate discharge data are exceedingly poor. The great amount of fine silt carried by this stream causes continual changes in the channel. The current is light at low stages. Floods occur at long intervals and are extremely torrential.

Conditions at this station during 1909 were fairly good up to the middle of August, when heavy rains fell in the Imperial Valley and surrounding country. A considerable flood occurred on New River, washing out the earth approaches to the bridge, and changing the channel so completely that measurements made prior to August are not comparable with those that will be made later. Probably the channel was fairly stable after October 1, 1909, but sufficient discharge measurements have not been made to define the new rating curve. Estimates of flow are, therefore, withheld.

Discharge measurements of New River near Brawley, Cal., in 1909.

Date.	Hydrographer.	Width.	Area of section.	Gage height.	Discharge.
		Feet.	*Sq. ft.*	*Feet.*	*Sec.-ft.*
Jan. 8	Hardy and Jones..	23	33	....	46
Jan. 14	H. A. Jones..	25.1	32	6.00	47
June 24	W. F. Martin..	58	66	6.31	89
July 30	.....do..	48.9	53	6.18	63

NOTE.—All measurements were made from downstream side of wagon bridge.

Mean daily gage height, in feet, of New River near Brawley, Cal., for 1909.

[Herschell Darnell, observer.]

Day.	June.	July.	Aug.	Sept.	Oct.	Nov.	Dec.	Day.	June.	July.	Aug.	Sept.	Oct.	Nov.	Dec.
1....	......	6.15	6.2	6.8	5.0	4.9	4.8	16...	......	6.2	6.85	5.0	4.7	4.9	4.8
2....	......	6.1	6.2	6.15	4.9	4.85	4.9	17...	......	6.25	6.8	5.0	4.7	4.9	4.8
3....	......	6.1	6.25	5.5	4.95	4.85	4.8	18...	......	6.25	6.8	5.0	4.7	4.9	4.95
4....	......	6.15	6.25	5.5	4.8	4.9	4.85	19...	......	6.0	7.1	5.0	4.7	4.85	4.85
5....	......	6.1	6.25	5.5	4.8	4.9	4.9	20...	......	5.9	7.05	5.0	4.7	4.9	4.9
6....	......	6.1	5.9	5.6	4.8	4.9	4.85	21...	......	5.9	6.75	5.0	4.7	4.9	5.15
7....	......	6.1	5.9	5.6	4.8	4.9	4.8	22...	......	5.85	6.6	5.0	4.7	4.9	5.0
8....	......	6.15	5.9	5.3	4.8	4.85	4.9	23...	......	5.8	6.6	5.0	4.7	4.9	5.0
9....	......	6.2	5.9	5.3	4.8	4.8	4.9	24...	6.3	5.8	6.3	5.0	4.7	4.85	4.9
10....	......	6.2	5.9	5.3	4.7	4.8	4.9	25...	6.3	6.0	5.95	4.95	4.7	4.9	5.0
11....	......	6.2	5.9	5.2	4.7	4.8	4.9	26...	6.3	6.05	......	5.0	4.7	4.9	4.9
12....	......	6.2	6.2	5.1	4.7	4.8	4.8	27 ..	6.25	6.1		4.95	4.7	4.8	4.9
13....	......	6.2	7.15	5.0	4.7	4.8	4.8	28...	6.3	6.15		5.05	4.7	4.9	5.0
14....	......	6.2	7.0	5.1	4.7	4.8	4.8	29...	6.25	6.2		5.0	4.65	4.9	5.0
15....	......	6.2	6.9	5.0	4.7	4.9	4.8	30...	6.2	6.2	a9.2	4.9	4.5	4.8	5.1
								31...	......	6.2	8.75	......	4.5	......	5.1

a Estimated.   Maximum gage height was 12.5 feet.

TRIBUTARY BASINS.

DUCHESNE RIVER BASIN.

DESCRIPTION.

Duchesne River rises in the high peaks of the Uinta and Wasatch mountains in northeastern Utah, flows for about 100 miles in a general southeasterly direction, and enters Green River at Ouray,

Utah, about 3 miles above the mouth of White River. The stream has a total drainage area of 4,000 square miles. Altitudes range from 4,700 feet at the mouth of the river to more than 13,000 feet at the summits of the highest peaks.

The principal tributaries of the Duchesne are Rock (East) Creek, Strawberry River, Lake Fork, and Uinta River.

The drainage basin of the upper Duchesne proper is mountainous. The stream emerges from the mountains at the mouth of Rock Creek, at an elevation of about 6,000 feet. From the mouth of Strawberry River down to Lake Fork the valley of the Duchesne is about 2 miles in average width, and is bordered by sandstone bluffs approximately 200 feet high. The bluffs on the northern side of the river are capped by heavy deposits of coarse gravel and cobblestones. The general course of the stream throughout this stretch and on down to the mouth of the Uinta is easterly. Along the lower course of the stream the plateaus on each side of the stream valley are comparatively low and can be easily reached by irrigation canals from the main stream.

Strawberry River, the main upper tributary of the Duchesne, drains an area of about 1,200 square miles. The stream rises in the Uinta Mountains, and enters the Duchesne at Theodore. Its flow is derived chiefly from melting snow, except during the late summer, when the flow comes from small springs well distributed over the entire drainage basin. The upper stream basin has numerous tributaries, particularly from the north and west. Among the most important may be mentioned Indian Creek, Bryant's Fork, Mud Creek, Horse Creek, Sugar Springs, and Co-op Creek. They are all short and fall rapidly until they reach Strawberry Valley, through which they flow sluggishly in well-defined channels. The main stream traverses the valley from north to south and is very sluggish. Indian Creek drains a small portion of the southern slopes of the Uinta Mountains. Its basin comprises smooth, rolling hills, fairly well timbered with pine and aspen. The normal flow is derived chiefly from springs. The greater part of the precipitation is in the form of snow, which covers the ground for six or eight months of each year. As its average elevation is 7,500 feet above sea level, Strawberry Valley is not well suited for agricultural development, but is excellently adapted to grazing. At the mouth of the river, about 35 miles below Strawberry Valley, the elevation is about 5,500 feet, and the fall in that distance is, therefore, nearly 2,000 feet.

Rock Creek, Lake Fork, the Uinta, and its most important tributary, the Whiterocks, head in a series of small lakes in the Uinta Mountains. These lakes are fed by snow that exists the year round in the canyons and on the high slopes. All these streams drain areas mountainous and difficult of access in their upper portions, and all of them emerge from their canyons at an elevation of about 7,000 feet. Rock Creek continues its course in a narrow valley, but the others spread out so that their valleys are comparatively wide, and the adjoining benches relatively low.

The drainage area of the Duchesne includes about 1,400 square miles of forest reserve, of which about 1,000 square miles may be classed as timbered land with an average stand of over 3,000 feet b. m. to the acre. The principal species of timber are Engelmann spruce and lodgepole pine. The timbered land is distributed through the areas of the various tributaries about as follows: Upper Duchesne, 120 square miles; Rock Creek, 130 square miles; Strawberry River, 380 square miles; Lake Fork, 190 square miles; Uinta (above Whiterocks), 120 square miles; Whiterocks, 70 square miles.

Little information is available as to the precipitation in this basin. In the plains portion of the area the average rainfall is probably less than 10 inches; in the middle portion, comprising considerably over half the area, it probably averages between 10 and 15 inches; in only a small part in the high mountains is there an annual precipitation in excess of 20 inches, and at Fort Duchesne, at an elevation of 5,000 feet, a record extending over several years shows a mean annual rainfall of only 7 inches.

The winters are very severe throughout this basin. In the high mountains the snowfall is heavy, and in many places the snow lies through the whole year. In the hills above an elevation of 7,000 feet there is very considerable snowfall, which usually forms in drifts in canyons, and not infrequently the snow lies for extended periods in the valleys and plateaus of the more open country. All the streams in this region are usually covered with thick ice from about December 1 to April 1 of each year.

At the present time the water in these streams is unused except for irrigation. The United States Indian Service has constructed a series of canals diverting water for irrigation from Lake Fork, the Uinta, and Whiterocks, and the Duchesne proper. The private canal systems now in operation are small, but eventually several hundred thousand acres of land below an elevation of 6,500 feet will be brought under irrigation. Practically no storage is used in connection with any of the irrigation systems now in operation or under construction. As the mountain drainage areas of all the main tributaries are studded with lakes, reservoir sites can easily be found where water can be stored for the irrigation of the valley lands (Pl. IV, B). It is believed that the entire flow from the drainage basin can be equalized by storage.

The United States Reclamation Service is constructing a tunnel, with a capacity of 500 second-feet, which will divert water from a

A. DAM SITE FOR RESERVOIR ON WILLIAMS FORK, GRAND RIVER, COLO.

B. RESERVOIR SITE ON WEST FORK OF LAKE FORK ABOVE FORKS, UTAH.

100,000-acre reservoir on the upper Strawberry across the divide to the headwaters of the Spanish Fork, there to be used for irrigation.

At the present time there are no water-power plants in this drainage basin, though with proper storage 200,000 horsepower could be developed. Very little water will be diverted for irrigation above an elevation of 6,500 feet, and as most of the reservoir sites are at an elevation of 8,000 feet or more, good opportunities for power development exist above irrigation diversions. Some of the streams have falls of 100 to 150 feet or more to the mile along these stretches.

None of the records of stream flow in this basin extend back of 1899, and they are not continuous since that time. The driest year for which records are available was 1900, although 1902 was almost as low. The year of greatest average run-off was 1907.

DUCHESNE RIVER AT MYTON,[1] UTAH.

This station, which is located at the highway bridge at Myton, Utah, about 3 miles below the mouth of Lake Fork and about 15 miles above the mouth of the Uinta River, was established October 26, 1899, to determine the amount of water available for storage and irrigation. The records show practically the entire run-off of the Duchesne Basin above the mouth of Uinta River.

Ditches built by the United States Indian Office divert water from this stream and its tributaries for irrigation on the Uinta Reservation. Water is also diverted by private enterprise for irrigation outside the limits of the reservation.

Results at this station are affected by ice for about four months each year, and during this period it is usually impossible to apply open-channel ratings. The discharge has also been more or less affected by eddies about the wooden crib piers and by drift lodged against them.

The datum of the gage remained practically constant from the establishment of the station until June 6, 1909, when the river cut a new channel around the bridge and the bridge station was abandoned. A new chain gage was established July 9, 1909, about one-fourth of a mile upstream from the bridge and at a different datum. This gage was replaced on August 9, 1909, by another chain gage, 100 feet upstream on right bank, at the same datum.

There was no bridge or cable from which discharge measurements could be made during the latter part of 1909.

---

[1] Described in the earlier reports as the Price Road Bridge station.

Discharge measurements of Duchesne River at Myton, Utah, in 1909.

Date.	Hydrographer.	Width.	Area of section.	Gage height.	Discharge.
		Feet.	*Sq. ft.*	*Feet.*	*Sec.-ft.*
Jan. 5 a	R. H. Fletcher...................................	100	274	6.44	460
19 a	....do..	114	313	6.62	574
Feb.15 a	....do..	105	265	6.72	483
25 a	....do..	105	267	6.61	396
Mar.15 a	....do..	100	276	6.62	462
29	....do..	100	598	5.78	664
Apr.14	....do..	100	583	5.69	558
27	....do..	102	639	6.68	1,240
May 12	....do..	108	847	8.44	3,340
26	....do..	108	872	8.50	3,380
Dec.22 b	....do..	110	362	6.00	630

a Made through ice.
b Made through ice. Gage height refers to different datum.

NOTE.—On Jan. 5 ice was 1.6 feet thick under gage; Jan. 19 it was 1.8 feet thick; Feb. 25, 2 feet thick; Mar. 15, 2.8 feet thick; Mar. 29, ice on edges and floating. Dec. 22, about 1 foot thick, with slush and anchor ice.

Daily gage height, in feet, of Duchesne River at Myton, Utah, for 1909.

[Alice Todd, observer.]

Day.	Apr.	May.	June.	July.	Aug.	Sept.	Oct.	Nov.	Dec.
1...................	5.75	6.65	8.55	....	6.3	7.15	5.4	5.3	5.15
2...................	5.85	6.50	8.8	....	6.2	7.0	5.4	5.3	5.15
3...................	6.05	6.65	9.3	....	6.1	6.4	5.4	5.3	5.15
4...................	6.25	6.8	9.8	....	6.0	6.2	5.5	5.3	5.15
5...................	6.0	7.1	10.1	....	5.9	6.2	5.5	5.3	5.15
6...................	5.8	7.45	10.35	....	5.9	6.3	5.5	5.3	5.15
7...................	5.8	7.55	....	....	5.9	6.4	5.4	5.3	5.15
8...................	5.7	7.7	....	....	5.9	6.4	5.4	5.3	....
9...................	5.65	7.95	....	....	5.9	6.15	5.4	5.3	....
10..................	5.7	8.15	....	7.15	5.85	6.1	5.4	5.3	....
11..................	5.7	8.3	....	7.0	5.85	6.1	5.45	5.3	....
12..................	5.7	8.4	....	6.9	5.8	6.1	5.4	5.3	....
13..................	5.65	8.35	....	6.8	5.8	6.0	5.4	5.3	....
14..................	5.75	8.1	....	6.7	5.8	5.9	5.4	5.3	....
15..................	5.8	8.0	....	6.6	5.8	5.9	5.4	5.2	....
16..................	5.9	8.1	....	6.6	5.8	5.8	5.4	5.2	....
17..................	6.05	8.1	....	6.5	5.8	5.7	5.4	5.15	....
18..................	6.25	8.0	....	6.5	5.85	5.65	5.4	5.1	....
19..................	6.3	8.0	....	6.5	5.9	5.65	5.4	5.1	....
20..................	6.45	8.2	....	6.6	6.05	5.6	5.4	5.25	....
21..................	6.4	8.4	....	6.55	6.1	5.6	5.35	5.3	....
22..................	6.3	8.7	....	6.4	6.2	5.6	5.35	5.3	....
23..................	6.3	8.8	....	6.4	6.25	5.55	5.35	5.3	....
24..................	6.25	8.8	....	6.45	6.1	5.55	5.35	5.3	....
25..................	6.35	8.7	....	6.4	6.0	5.55	5.35	5.25	....
26..................	6.4	8.6	....	6.45	5.9	5.5	5.35	5.2	....
27..................	6.7	8.6	....	6.6	5.85	5.5	5.35	5.2	....
28..................	6.9	8.8	....	6.7	5.8	5.5	5.35	5.15	....
29..................	7.0	9.0	....	6.6	5.7	5.5	5.35	5.15	....
30..................	6.8	8.85	....	6.5	5.7	5.45	5.3	5.15	....
31..................	....	8.6	....	6.45	6.05	....	5.3	....	....

NOTE.—Ice conditions during January, February, March, and from about Dec. 8 to 31; gage washed out June 6. New gage referred to new datum, installed on July 10.

Daily discharge, in second-feet, of Duchesne River at Myton, Utah, for 1909.

[Alice Todd, observer.]

Day.	Apr.	May.	June.	July.	Aug.	Sept.	Oct.	Nov.	Dec.
1	610	1,190	3,500	....	1,960	3,270	875	775	645
2	662	1,080	4,000	....	1,820	3,020	875	775	645
3	780	1,190	5,130	....	1,700	2,100	875	775	645
4	908	1,310	6,430	....	1,570	1,820	980	775	645
5	750	1,560	7,310	....	1,440	1,820	980	775	645
6	635	1,920	8,080	....	1,440	1,960	980	775	645
7	635	2,020	....	....	1,440	2,100	875	775	645
8	585	2,200	....	....	1,440	2,100	875	775	....
9	562	2,530	....	....	1,440	1,760	875	775	....
10	585	2,820	....	3,270	1,380	1,700	875	775	....
11	585	3,050	....	3,020	1,380	1,700	928	775	....
12	585	3,220	....	2,860	1,320	1,700	875	775	....
13	562	3,140	....	2,700	1,320	1,570	875	775	....
14	610	2,740	....	2,550	1,320	1,440	875	775	....
15	635	2,600	....	2,400	1,320	1,440	875	685	....
16	690	2,740	....	2,400	1,320	1,320	875	685	....
17	780	2,740	....	2,240	1,320	1,200	875	645	....
18	908	2,600	....	2,240	1,380	1,150	875	605	....
19	940	2,600	....	2,240	1,440	1,150	875	605	....
20	1,040	2,890	....	2,400	1,630	1,090	875	730	....
21	1,010	3,220	....	2,320	1,700	1,090	825	775	....
22	940	3,790	....	2,100	1,820	1,090	825	775	....
23	940	4,000	....	2,100	1,890	1,040	825	775	....
24	908	4,000	....	2,170	1,700	1,040	825	775	....
25	975	3,790	....	2,100	1,570	1,040	825	730	....
26	1,010	3,590	....	2,170	1,440	980	825	685	....
27	1,230	3,590	....	2,400	1,380	980	825	685	....
28	1,390	4,000	....	2,550	1,320	980	825	645	....
29	1,480	4,430	....	2,400	1,200	980	825	645	....
30	1,310	4,100	....	2,240	1,200	928	775	645	....
31	....	3,590	....	2,170	1,630	....	775	....	....

NOTE.—These discharges are based on rating tables applicable as follows: Apr. 1 to June 6, fairly well defined between 500 and 4,400 second-feet; July 10 to Dec. 7, defined by 1910 measurements between 300 and 1,200 second-feet.

Monthly discharge of Duchesne River at Myton, Utah, for 1909.

[Drainage area, 2,750 square miles.]

Month.	Discharge in second-feet.				Run-off.		Accuracy.
	Maximum.	Minimum.	Mean.	Per square mile.	Depth in inches on drainage area.	Total in acre-feet.	
April	1,480	562	841	0.306	0.34	50,000	A.
May	4,430	1,080	2,850	1.04	1.20	175,000	A.
June 1-6	....	....	5,740	2.09	.47	68,300	B.
July 10-31	3,270	2,100	2,410	.876	.72	105,000	D.
August	1,960	1,200	1,490	.542	.62	91,600	C.
September	3,270	928	1,520	.553	.62	90,400	C.
October	980	775	866	.315	.36	53,200	C.
November	775	605	731	.266	.30	43,500	C.
December	....	....	a 637	.232	.27	39,200	D.
The period	....	....	....	....	....	716,000	

a Discharge estimated from Dec. 8 to 31 on basis of ice measurement made Dec. 22.

STRAWBERRY RIVER ABOVE MOUTH OF INDIAN CREEK IN STRAWBERRY VALLEY, UTAH.

A station was established on Strawberry River September 15, 1909, in the narrows below Strawberry Valley, about 3 miles above the mouth of Indian Creek and about half a mile below the dam site of the Strawberry Valley project, to determine the amount of water available for storage from Strawberry River for the Strawberry Valley project of the United States Reclamation Service. The station takes the place of the one previously maintained below Indian Creek. The new station is about 35 miles northeast of Thistle, Utah, the nearest railroad point.

The vertical staff gage is on the right bank directly underneath the cable from which discharge measurements are made.

Neither bank is liable to overflow except in extreme high water. The stream bed consists of coarse gravel and although rough is believed to be fairly permanent. The measuring conditions during the open-water season are good.

As to the lower station the river is frozen and deeply covered with snow about five months of the year. The flow during the winter is, however, fairly constant and a fair estimate of it may be made.

The following discharge measurement was made by E. S. Fuller:

September 15, 1909: Width, 44.5 feet; area, 54.4 square feet; gage height, 2.70 feet; discharge, 41.0 second-feet. Made by wading.

Daily gage height, in feet, and discharge, in second-feet, of Strawberry River above mouth of Indian Creek in Strawberry Valley, Utah, for 1909.

[J. C. Warfield, observer.]

Day.	October.		November.		Day.	October.		November.	
	Gage height.	Discharge.	Gage height.	Discharge.		Gage height.	Discharge.	Gage height.	Discharge.
1		37	2.75	43	16		37		41
2		38		43	17	2.6	37		41
3	2.65	39		42	18		37		41
4		40		42	19		38		41
5	2.7	41		41	20		38		41
6		41	2.7	41	21		39		41
7	2.7	41		41	22	2.65	39	3.1	41
8		41		41	23		39		41
9	2.7	41		41	24		39		41
10		41		41	25	2.65	39		41
11	2.7	41		41	26		39		41
12		40	2.7	41	27		40		41
13	2.65	39		41	28		40		41
14		39		41	29		41		41
15		38		41	30	2.7	41		41
					31		42		

NOTE.—The daily discharges are based on a rating curve that is not well defined. Ice after Nov. 15.

Monthly discharge of Strawberry River in Strawberry Valley, Utah, for 1909.

Month.	Discharge in second-feet.			Run-off (total in acre-feet).
	Maximum.	Minimum.	Mean.	
October...	42	37	39. 4	2,420
November..	43	41	41. 2	2,450
December..			40. 0	2,460

NOTE.—These estimates are approximate.

STRAWBERRY RIVER BELOW MOUTH OF INDIAN CREEK IN STRAWBERRY VALLEY, UTAH.[1]

This station was originally located above the junction of Indian Creek and Strawberry River, where it was maintained from May 12, 1903, to July 12, 1906. On October 14, 1908, the station was reestablished at a point about 200 feet below the mouth of Indian Creek, where it was maintained until September 30, 1909, when it was discontinued and separate·records were started on Indian Creek and Strawberry River. All of these stations are located at the lower end of Strawberry Valley, about 25 miles northeast of Thistle, the nearest railway point.

The records show the amount of water available for storage in the Strawberry Valley reservoir. Winters in this region are very severe, the river being frozen and covered deeply with snow about five months. The flow during this period, however, is fairly constant and a fair estimate of the winter flow may be made. Conditions during the open-water season are favorable for accurate measurements.

The gage is a vertical staff located 20 feet downstream from the cable and on the right bank. Discharge measurements are made from a car and cable.

Discharge measurements of Strawberry River below mouth of Indian Creek in Strawberry Valley, Utah, in 1909.

Date.	Hydrographer.	Width.	Area of section.	Gage height.	Discharge.
		Feet.	*Sq. ft.*	*Feet.*	*Sec.-ft.*
June 6	E. S. Fuller...	64	273	11. 2	1,380
July 7	.....do...	39. 5	58. 8	6. 61	196
Sept. 12	.....do...	39	37. 6	6. 1	80. 6

[1] This station was called "Strawberry River in Strawberry Valley, Utah," in previous reports.

Daily gage height, in feet, of Strawberry River below mouth of Indian Creek in Strawberry Valley, Utah, for 1909.

[J. C. Warfield, observer.]

Day.	May.	June.	July.	Aug.	Sept.	Day.	May.	June.	July.	Aug.	Sept.
1......		9.9			6.55	16......		8.5			
2......			6.8	6.2		17......			6.35	6.2	6.05
3......		10.7			6.2	18......		8.3			
4......			6.9	6.2		19......			6.4	6.3	6.05
5......		11.0			6.15	20......		10.6	8.0		
6......		11.2	6.65	6.2		21......			6.45	6.15	6.0
7......			6.6		6.2	22......		7.65			
8......		11.0				23......			6.4	6.2	6.0
9......			6.55	6.2	6.15	24......		7.4			
10......	11.6	10.1				25......			6.3	6.15	6.0
11......			6.5		6.1	26......		7.2			
12......		9.2		6.2		27......			6.3	6.1	6.0
13......	12.5		6.45		6.1	28......		7.0			
14......		8.9				29......			6.25		6.0
15......	10.8		6.4	6.15	6.1	30......		6.9		6.05	
						31......	10.0		6.2		

NOTE.—Ice in January, February, March, and April.

Daily discharge, in second-feet, of Strawberry River below mouth of Indian Creek in Strawberry Valley, Utah, for 1909.

Day.	May.	June.	July.	Aug.	Sept.	Day.	May.	June.	July.	Aug.	Sept.
1......	20	1,020	251	101	182	16......	1,260	660	142	96	72
2......	20	1,140	239	101	142	17......	1,250	635	136	101	67
3......	205	1,250	251	101	101	18......	1,240	610	142	112	67
4......	390	1,290	263	101	96	19......	1,230	572	147	124	67
5......	575	1,330	234	101	90	20......	1,220	535	152	107	62
6......	760	1,390	204	101	96	21......	1,200	492	158	90	56
7......	945	1,360	193	101	101	22......	1,190	448	152	96	56
8......	1,130	1,330	188	101	96	23......	1,170	416	147	101	56
9......	1,320	1,200	182	101	90	24......	1,160	385	136	96	56
10......	1,500	1,080	176	101	84	25......	1,140	360	124	90	56
11......	1,580	960	170	101	78	26......	1,120	335	124	84	56
12......	1,660	842	164	101	78	27......	1,110	311	124	78	56
13......	1,750	803	158	98	78	28......	1,100	287	118	75	56
14......	1,510	764	152	94	78	29......	1,080	275	112	71	56
15......	1,270	712	147	90	78	30......	1,070	263	106	67	58
						31......	1,050		101	124	

NOTE.—These discharges are based on a rating curve that is fairly well defined. Discharge interpolated for days when gage was not read.

Monthly discharge of Strawberry River below mouth of Indian Creek in Strawberry Valley, Utah, for 1909.

Month.	Discharge in second-feet.			Run-off (total in acre-feet).	Accuracy.
	Maximum.	Minimum.	Mean.		
May..........................	1,750	20	1,070	65,800	D.
June.........................	1,390	263	768	45,700	C.
July..........................	263	101	164	10,100	C.
August.......................	124	67	97.0	5,960	C.
September....................	182	56	78.8	4.690	C.

STRAWBERRY RIVER AT THEODORE, UTAH.

This station, which is located at the west boundary of the Theodore town site, along the wagon road to Heber, about 1¼ miles above the junction of the Strawberry with Duchesne River, about half a mile upstream from the mouth of Indian Canyon and about 18 miles below the mouth of Currant Creek, was established June 10, 1908, to determine the run-off of the lower Strawberry and the amount of water available for irrigation in that section. The drainage area above the station is nearly 1,200 square miles.

The chain gage is located about 50 feet downstream from cable from which discharge measurements are made. The datum of the gage has remained constant since the station was established.

Very good results should be obtained at this station except at extremely high stages, when the stream overflows the left bank, rendering it impossible to make gagings.

Discharge measurements of Strawberry River at Theodore, Utah, in 1909.

[By R. H. Fletcher.]

Date.	Width.	Area of section.	Gage height.	Dis-charge.	Date.	Width.	Area of section.	Gage height.	Dis-charge.
	Feet.	*Sq. ft.*	*Feet.*	*Sec.-ft.*		*Feet.*	*Sq. ft.*	*Feet.*	*Sec.-ft.*
Jan. 6.......	55	49	3.40	116	Aug. 10.......	62	149	3.50	341
21.......	52	60	4.00	147	26.......	61	136	3.40	267
Mar. 30.......	60	94	2.65	217	Sept. 9.......	62	140	3.38	325
Apr. 13.......	58	90	2.65	208	26.......	61	119	3.06	242
28.......	66	225	4.96	989	Oct. 8.......	62	121	2.97	228
May 25.......	128	475	7.80	a 3,920	25.......	61	113	2.91	209
July 11.......	63	189	4.15	540	Nov. 13.......	61	104	2.75	168
22.......	63	169	3.76	440	23.......	61	131	3.22	267

a Discharge estimated from extension of area and velocity curves.

Daily gage height, in feet, of Strawberry River at Theodore, Utah, for 1909.

[E. S. Winslow and M. M. Smith, observers.]

Day.	Apr.	May.	June.	July.	Aug.	Sept.	Oct.	Nov.	Dec.
1................	2.7	4.35	7.4	4.8	3.5	6.1	2.95	2.85	3.0
2................	2.9	4.3	7.5	4.65	3.5	4.05	2.95	2.85	2.95
3................	3.1	4.35	7.8	4.6	3.5	3.65	2.95	2.85	3.05
4................	3.3	4.8	8.0	4.7	3.4	3.55	3.05	2.85	3.2
5................	2.9	5.5	8.15	4.6	3.4	3.6	3.0	2.85	3.6
6................	2.8	6.2	8.25	4.5	3.4	3.55	3.0	2.85	3.3
7................	2.7	6.55	8.3	4.4	3.5	4.0	3.0	2.85	3.3
8................	2.7	6.9	8.3	4.3	3.5	3.6	2.95	2.8	3.7
9................	2.65	7.2	8.1	4.2	3.5	3.35	2.95	2.8	4.0
10................	2.7	7.5	7.8	4.15	3.5	3.3	2.95	2.85	3.85
11................	2.7	7.7	7.55	4.25	3.7	3.2	2.95	2.85	3.8
12................	2.65	7.9	7.4	4.0	3.55	3.25	2.95	2.8	3.35
13................	2.65	7.65	7.25	3.95	3.55	3.2	2.95	2.9	3.6
14................	2.8	7.4	7.15	3.9	3.55	3.2	2.95	2.85	3.75
15................	2.85	7.3	7.05	3.9	3.5	3.2	2.95	2.85	3.7
16................	2.9	7.45	6.85	3.8	3.55	3.15	2.95	2.85	3.5
17................	3.15	7.4	6.7	3.75	3.85	3.15	2.95	3.15	3.85
18................	3.4	7.25	6.6	3.7	3.9	3.1	2.95	3.1	3.4
19................	3.55	7.3	6.45	3.8	3.95	3.1	2.95	3.0	3.7
20................	3.7	7.45	6.35	4.1	3.65	3.2	2.9	3.0	3.8
21................	3.6	7.6	6.1	3.85	3.6	3.05	2.9	3.05	3.75
22................	3.55	7.8	5.9	3.75	3.8	3.05	2.9	3.15	3.65
23................	3.55	7.95	5.75	3.8	3.6	3.05	2.9	3.15	3.75
24................	3.65	7.95	5.6	3.85	3.55	3.05	2.9	3.1	3.6
25................	3.75	7.75	5.5	3.75	3.5	3.05	2.9	3.1	3.7
26................	3.9	7.5	5.4	3.7	3.4	3.0	2.9	3.1	3.75
27................	4.4	7.55	5.2	3.7	3.4	3.0	2.85	3.1	3.6
28................	4.9	7.75	5.05	3.6	3.35	3.0	2.85	2.95	3.6
29................	4.95	7.9	4.95	3.6	3.3	3.0	2.85	2.65	3.65
30................	4.55	7.8	4.9	3.55	3.3	3.0	2.85	2.95	3.85
31................	....	7.5	....	3.5	3.55	....	2.85	....	4.0

NOTE.—Ice in January, February, March, and from December 4 to 31.

Daily discharge, in second-feet, of Strawberry River at Theodore, Utah, for 1909.

Day.	Apr.	May.	June.	July.	Aug.	Sept.	Oct.	Nov.	Dec.
1	220	685	2,260	870	352	1,520	216	194	228
2	290	660	2,330	800	352	542	216	194	216
3	368	670	2,550	776	352	398	216	194	240
4	453	880	2,700	823	316	367	240	194	....
5	290	1,220	2,820	776	316	382	228	194	....
6	253	1,570	2,910	730	316	367	228	194	....
7	220	1,760	2,950	686	352	522	228	194	....
8	220	1,950	2,950	644	352	382	216	183	....
9	205	2,120	2,780	602	352	303	216	183	....
10	220	2,330	2,550	582	352	290	216	194	....
11	220	2,480	2,360	623	414	205	216	194	....
12	205	2,620	2,260	522	367	278	216	183	....
13	205	2,440	2,160	503	367	265	216	205	....
14	253	2,260	2,090	484	367	265	216	194	....
15	272	2,190	2,030	484	352	265	216	194	....
16	290	2,300	1,920	448	367	263	216	194	....
17	389	2,260	1,840	431	466	263	216	263	....
18	497	2,160	1,780	414	484	251	216	251	....
19	564	2,190	1,700	448	503	251	216	228	....
20	634	2,300	1,650	562	398	265	205	228	....
21	587	2,400	1,520	466	382	240	205	240	....
22	564	2,550	1,410	431	448	240	205	263	....
23	564	2,660	1,340	448	382	240	205	263	....
24	610	2,660	1,260	466	367	240	205	251	....
25	658	2,510	1,210	431	352	240	205	251	....
26	732	2,330	1,160	414	316	240	205	251	....
27	840	2,360	1,060	414	316	228	194	251	....
28	960	2,510	989	382	303	228	194	216	....
29	970	2,620	941	382	290	228	194	154	....
30	780	2,550	917	367	290	228	194	216	....
31	....	2,330	....	352	367	....	194	....	....

NOTE.—These discharges are based on rating curves applicable as follows: April 1 to April 26, well defined between 100 and 410 second-feet; April 27 to May 5, indirect method for shifting channels used; May 6 to Dec. 3, 1909; well defined between 160 and 960 second-feet.

Monthly discharge of Strawberry River at Theodore, Utah, for 1909.

Month.	Discharge in second-feet.			Run-off (total in acre-feet).	Accuracy.
	Maximum.	Minimum.	Mean.		
April	970	205	451	26,800	C.
May	2,660	660	2,080	128,000	D.
June	2,950	917	1,950	116,000	D.
July	870	352	541	33,300	C.
August	503	290	365	22,400	B.
September	1,520	228	335	19,900	B.
October	240	194	212	13,000	B.
November	263	154	214	12,700	B.
The period	.....	.....	.....	372,000	

INDIAN CREEK IN STRAWBERRY VALLEY, UTAH.

This station, which was established April 5, 1905, discontinued July 12, 1906, and reestablished October 1, 1909, is located just above the mouth of the creek. The station was originally located about 250 feet above the mouth of the creek, but was reestablished in 1909, about half a mile farther upstream, in T. 4 S., R. 11 W. It is about 25 miles northeast of Thistle, Utah.

This point is below all tributaries, Trail Hollow Creek entering a few hundred feet above the new station. No water is diverted above the station. The records are of value to the United States Reclamation Service in connection with the Strawberry Valley project, which proposes to divert the waters of Indian Creek across a low pass into the Strawberry Valley reservoir.

The staff gage is driven vertically into the bed of the creek and braced to the right bank about 10 feet above a new footbridge from which measurements are made.

The river is frozen over and covered with a deep layer of snow during about five months of the year. The winter flow, however, is fairly constant and a fair estimate may be made. The open-water measuring conditions are excellent.

Discharge measurements of Indian Creek in Strawberry Valley, Utah, in 1909.

Date.	Hydrographer.	Width.	Area of section.	Gage height.	Discharge.
		Feet.	*Sq. ft.*	*Feet.*	*Sec.-ft.*
Sept. 12	E. S. Fuller...	22	20.7	1.60	31.3
15	.....do...	21.5	18.5	1.55	29.7

Daily gage height, in feet, and discharge, in second-feet, of Indian Creek in Strawberry Valley, Utah, for 1909.

[J. C. Warfield, observer.]

Day.	October.		November.		Day.	October.		November.	
	Gage height.	Discharge.	Gage height.	Discharge.		Gage height.	Discharge.	Gage height.	Discharge.
1...............	....	23	1.4	23	16...............	....	23	....	23
2...............	....	24	....	23	17...............	1.4	23	....	23
3...............	1.45	25	....	23	18...............	....	23	....	23
4...............	....	25	....	23	19...............	....	23	....	23
5...............	1.45	25	....	23	20...............	....	23	....	23
6...............	....	26	1.4	23	21...............	....	23	....	23
7...............	1.5	27	....	23	22...............	1.4	23	1.48	23
8...............	....	27	....	23	23...............	....	23	....	23
9...............	1.5	27	....	23	24...............	....	23	....	23
10...............	....	26	....	23	25...............	....	23	....	23
11...............	1.45	25	....	23	26...............	1.4	23	....	23
12...............	....	25	1.4	23	27...............	....	23	....	23
13...............	1.45	25	....	23	28...............	....	23	....	23
14...............	....	25	....	23	29...............	....	23	....	23
15...............	....	24	....	23	30...............	1.4	23	....	23
					31...............	....	23	....	

NOTE.—The daily discharges are based on a rating curve that is not well defined. Discharge interpolated for days when gage was not read. Ice conditions after Nov. 15.

Monthly discharge of Indian Creek in Strawberry Valley, Utah, for 1909.

Month.	Discharge in second-feet.			Run-off (total in acre-feet).
	Maximum.	Minimum.	Mean.	
October...	27	23	24.1	1,480
November...	23	23	23.0	1,370
December...	....	....	a 20.0	1,230

a Estimated.

NOTE.—These estimates are approximate.

TRAIL HOLLOW CREEK IN STRAWBERRY VALLEY, UTAH.

This station, which was established October 1, 1909, to determine the amount of water entering Indian Creek below the proposed point of diversion, is located just above the mouth of the stream. No water is at present diverted above the station. The records are of value to the United States Reclamation Service in that they show the portion of the flow of Indian Creek which can not be diverted into Strawberry Valley in connection with the Strawberry Valley project.

The staff gage is driven vertically into the bed of the stream and braced to the left bank. Discharge measurements can best be made by wading at ordinary stages, but during high water they can be made from a bridge made of two logs that is about 15 feet above the gauge.

The stream is frozen over and deeply covered with snow during about five months of the year.

The following discharge measurement was made by E. S. Fuller:

September 12, 1909: Width, 5.5 feet; area, 5.75 square feet; gage height, 3.40 feet; discharge, 4.84 second-feet. Made by wading.

Daily gage height, in feet and discharge, in second-feet, of Trail Hollow Creek in Strawberry Valley, Utah, for 1909.

[J. C. Warfield, observer.]

Day.	October.		November.		Day.	October.		November.	
	Gage height.	Discharge.	Gage height.	Discharge.		Gage height.	Discharge.	Gage height.	Discharge.
1	....	3.6	3.15	3.4	16	....	3.6	....	....
2	....	3.6	....	3.4	17	3.2	3.6	....	....
3	3.2	3.6	....	3.4	18	....	3.6	....	....
4	....	3.6	....	3.4	19	....	3.5	....	....
5	3.2	3.6	....	3.4	20	....	3.5	....	....
6	....	3.6	3.15	3.4	21	....	3.4	....	....
7	3.2	3.6	....	3.4	22	3.15	3.4	3.7	....
8	....	3.6	....	3.4	23	....	3.4	....	....
9	3.2	3.6	....	3.4	24	....	3.4	....	....
10	....	3.6	....	3.4	25	....	3.4	....	....
11	3.2	3.6	....	3.4	26	3.15	3.4	....	....
12	....	3.6	3.15	3.4	27	....	3.4	....	....
13	3.2	3.6	....	3.4	28	....	3.4	....	....
14	....	3.6	....	3.4	29	....	3.4	....	....
15	....	3.6	....	3.4	30	3.15	3.4	....	....
					31	....	3.4	....	....

NOTE.—These discharges are based on a curve that is not well defined. Discharges interpolated for days when gage was not read. Ice conditions after Nov. 15.

Monthly discharge of Trail Hollow Creek in Strawberry Valley, Utah, for 1909.

Month.	Discharge in second-feet.			Run-off (total in acre-feet).
	Maximum.	Minimum.	Mean.	
October	3.6	3.4	3.52	216
November	....	....	a 3.40	202
December	....	....	a 3.40	209

a Estimated.

NOTE.—These estimates are approximate.

LAKE FORK BELOW FORKS NEAR WHITEROCKS, UTAH.

This station, which is located about 500 feet downstream from the junction of the East and West forks, on the old Indian trail from Spanish Fork to Whiterocks, Utah, about 30 miles west of White-rocks, was established on May 10, 1907, but a fragmentary record was maintained at the same place during 1904.

The station is above all present diversions and furnishes valuable data for determining the run-off and showing the amount of water available for irrigation below and storage above.

No important tributaries enter between this station and the mouth of the stream, and none on either branch for some distance above. The drainage area above the station is about 300 square miles.

The flow of this stream could doubtless be equalized at compara-tively small expense by utilizing the storage facilities afforded by a number of the small lakes and reservoir sites found on both branches of the stream above the station. As both of the main tributaries have rapid fall, opportunities for power development are presented above all irrigation diversions.

The stream is icebound for several months each year.

The chain gage established May 10, 1907, has no relation whatever to the 1904 gage. Still another chain gage and datum have been used since September 1, 1907. This gage is located about 100 feet upstream from the cable from which discharge measurements are made.

As the stream bed is rough and the current is swift at high and moderate stages, the results obtained at this station can be considered only fair or approximate except at low stages, when they are fairly good.

Discharge measurements of Lake Fork below forks near Whiterocks, Utah, in 1909.

[By R. H. Fletcher.]

Date.	Width.	Area of section.	Gage height.	Dis-charge.	Date.	Width.	Area of section.	Gage height.	Dis-charge.
	Feet.	*Sq. ft.*	*Feet.*	*Sec.-ft.*		*Feet.*	*Sq. ft.*	*Feet.*	*Sec.-ft.*
Jan. 13 a..	48	75	2.56	185	Aug. 27 ...	67	122	2.54	612
May 3...	54	79	1.75	243	Sept. 9...	74	138	2.88	777
24...	70	118	2.42	547	25...	62	96	2.18	333
June 16...	107	308	4.80	2,850	Oct. 7...	61	92	2.06	326
July 6...	100	216	3.80	1,530	26...	58	75	1.78	257
21...	84	144	2.75	768	Nov. 15...	57	76	1.64	224
Aug. 11...	68	121	2.65	632	24...	57	74	1.66	226

a Ice conditions. Measurement made partly from cable and partly from ice surface.

Daily gage height, in feet, of Lake Fork below forks near Whiterocks, Utah, for 1909.

[Charles and Paul J. Elliott, observers.]

Day.	Apr.	May.	June.	July.	Aug.	Sept.	Oct.	Nov.	Dec.
1	1.5	1.7	2.5	4.25	2.3	3.9	2.0	1.65	1.6
2	1.5	1.7	2.85	4.2	2.55	3.8	2.0	1.65	1.65
3	1.5	1.75	3.55	4.2	2.45	3.	2.05	1.65	1.55
4	1.45	1.9	4.35	4.1	2.3	3.	2.1	1.65	1.5
5	1.5	2.0	4.75	4.0	2.3	3.	2.1	1.55	1.3
6	1.5	2.0	4.85	3.95	2.35	3.2	2.1	1.6	1.3
7	1.4	2.05	4.9	3.65	2.4	3.0	2.3	....	1.5
8	1.4	2.1	4.5	3.4	2.4	3.	2.0	....	1.5
9	1.4	2.2	4.4	3.4	2.45	2.	2.0	....	....
10	1.35	2.2	4.15	3.2	2.65	2.	1.95	1.65	....
11	1.4	2.2	4.2	3.2	2.45	2.8	1.95	1.6	....
12	1.4	2.1	4.35	3.1	2.7	2.6	1.9	1.6	....
13	1.4	2.05	4.55	3.0	2.7	2.	1.9	ᵃ1.6	....
14	1.5	2.0	4.5	2.9	2.45	2.	1.9	1.65	2.55
15	1.5	2.0	4.6	2.8	2.45	2.	1.9	1.6	2.6
16	1.6	2.0	4.7	2.8	2.6	2.5	1.85	1.6	2.8
17	1.7	2.0	4.9	2.65	3.05	2.4	1.85	1.6	2.8
18	1.7	2.0	4.95	2.7	3.1	2.4	1.85	1.6	2.7
19	1.7	2.1	4.9	2.9	2.95	2.35	1.85	1.65	....
20	1.7	2.2	4.3	2.9	3.0	2.3	1.8	1.7	....
21	1.65	2.3	4.2	2.7	3.25	2.25	1.8	1.7	....
22	1.6	2.5	4.35	2.65	3.0	2.2	1.8	1.7	....
23	1.6	2.5	4.6	2.7	3.0	2.	1.8	1.65	....
24	1.6	2.4	4.85	2.65	2.75	2.	1.8	1.65	....
25	1.6	2.3	4.6	2.6	2.65	2.5	1.8	1.6	....
26	1.75	2.3	4.4	2.9	2.6	2.1	1.8	1.6	....
27	1.9	2.45	4.5	3.0	2.55	2.0	1.7	1.55	....
28	1.8	2.6	4.5	2.65	2.4	2.5	1.7	1.6	....
29	1.85	2.5	4.55	2.5	2.4	2.5	1.7	1.6	....
30	1.75	2.45	4.3	2.4	2.3	2.	1.7	1.6	....
31	....	2.4	....	2.3	2.65	....	1.7	....	....

NOTE.—Ice during January, February, and March, and from Dec. 9 to 31.

Daily discharge, in second-feet, of Lake Fork below forks near Whiterocks, Utah, for 1909.

Day.	Jan.	Feb.	Mar.	Apr.	May.	June.	July.	Aug.	Sept.	Oct.	Nov.	Dec.
1	160	180	170	196	236	550	2,120	445	1,710	317	226	215
2	160	180	170	196	236	775	2,060	580	1,400	317	226	226
3	160	180	170	196	248	1,350	2,060	522	1,120	335	226	206
4	170	180	170	187	287	2,240	1,940	445	955	352	226	196
5	170	180	170	196	317	2,790	1,820	445	955	352	206	160
6	175	170	170	196	317	2,930	1,760	470	1,040	352	215	160
7	175	170	170	178	335	3,000	1,450	495	955	445	ᵃ218	196
8	175	170	170	178	352	2,440	1,210	495	880	317	ᵃ221	196
9	180	170	170	178	395	2,310	1,210	522	810	317	ᵃ224	195
10	180	170	170	169	395	2,000	1,040	642	810	302	226	200
11	180	160	180	178	395	2,060	1,040	522	740	302	215	190
12	185	160	180	178	352	2,240	955	675	675	287	215	180
13	185	160	180	178	335	2,510	880	675	675	287	215	195
14	180	160	180	196	317	2,440	810	522	610	287	226	195
15	180	160	180	196	317	2,580	740	522	550	287	215	195
16	185	150	180	215	317	2,720	740	610	550	274	215	195
17	185	150	175	236	317	3,000	642	918	495	274	215	190
18	185	150	170	236	317	3,070	675	955	495	274	215	180
19	190	150	170	236	352	3,000	810	845	470	274	226	175
20	190	150	170	236	395	2,180	810	880	445	260	236	165
21	190	160	160	226	445	2,060	675	1,080	420	260	236	180
22	190	160	160	215	550	2,240	642	880	395	260	236	185
23	190	160	160	215	550	2,580	675	880	395	260	226	185
24	190	160	160	215	495	2,930	642	708	395	260	226	185
25	190	160	160	215	445	2,580	610	642	374	260	215	185
26	190	170	150	248	445	2,310	810	610	352	260	215	185
27	190	170	150	287	522	2,440	880	580	352	236	206	185
28	190	170	150	260	610	2,440	642	495	335	236	215	185
29	190	......	150	274	550	2,510	550	495	335	236	215	185
30	190	......	170	248	522	2,180	495	445	335	236	215	185
31	190	......	180	......	495	......	445	642	......	236	......	185

ᵃ Interpolated.

NOTE.—Discharges Apr. 1 to Dec. 8 are based on a rating curve that is fairly well defined between 215 and 3,140 second-feet.

Discharges Jan. 1 to Mar. 31 and Dec. 9 to 31 estimated by hydrograph comparison of various streams in the Uinta River drainage.

Monthly discharge of Lake Fork below forks near Whiterocks, Utah, for 1909.

Month.	Discharge in second-feet.			Run-off (total in acre-feet).	Accuracy.
	Maximum.	Minimum.	Mean.		
January			182	11,200	D.
February			165	9,160	D.
March			168	10,300	D.
April	287	169	212	12,600	B.
May	610	236	393	24,200	B.
June	3,070	550	2,350	140,000	B.
July	2,120	445	1,030	63,300	B.
August	1,080	445	634	39,000	B.
September	1,710	335	668	39,700	B.
October	445	236	289	17,800	B.
November	236	206	220	13,100	B.
December			188	11,600	D.
The year			542	392,000	

LAKE FORK NEAR MYTON,[1] UTAH.

This station, which is located about 3 miles above Myton, Utah, was originally established July 3, 1900; was discontinued at the end of the season of 1903, and was reestablished in June, 1907. Several discharge measurements were, however, made in 1904.

As the station is only about half a mile above the junction of the stream with Duchesne River, the records show the amount of water which Lake Fork contributes to the Duchesne, and in connection with the records obtained at the station on Lake Fork below the forks, which is about 20 miles upstream and above all present diversions, they indicate also the amount of water diverted for irrigation along the stream. No important tributaries enter between the two stations.

Several canal systems built by the United States Indian Office take water from this stream above the station for irrigation. Some private canal systems are proposed or in operation. As at all other stations in this basin, the stream is icebound for several months during the winter season.

The gage was in the same position and the same datum was used from 1900 to 1904, inclusive. During 1907 and 1908 three distinct gages and datums were used—from June 13 to 30, 1907, from August 18 to December 31, 1907, and during 1908.

The results obtained during 1908 and 1909 have been very satisfactory, except for a few difficulties in the gage readings. Previous records are not so good.

Gage heights from June 15 to 21 are too uncertain, owing to the settlement of the gage, to warrant publication. On June 22 the gage was removed to the opposite bank and reinstalled at a different datum. Gage heights, beginning June 22, are therefore not comparable with those of previous dates.

---

[1] Described in early reports as "Lake Fork at mouth."

Discharge measurements of Lake Fork near Myton, Utah, in 1909.

[By R. H. Fletcher.]

Date.	Width.	Area of section.	Gage height.	Dis- charge.	Date.	Width.	Area of section.	Gage height.	Dis- charge.
	Feet.	*Sq. ft.*	*Feet.*	*Sec.-ft.*		*Feet.*	*Sq. ft.*	*Feet.*	*Sec.-ft.*
Jan. 7 a ..	66	107	4.55	178	June 23....	70	429	6.90	2,470
20 a ..	66	128	4.68	207	July 10....	66	234	3.90	724
Feb. 16 a ..	66	113	4.60	149	23....	66	185	3.18	398
26 a ..	66	116	4.65	180	Aug. 9....	66	162	2.80	286
Mar. 16 a ..	66	116	4.00	183	26....	66	183	3.14	385
29 b ..	65	118	3.59	144	Sept. 8....	67	230	3.89	726
Apr. 12....	65	107	3.46	113	26....	66	149	2.76	258
28....	65	134	3.90	191	Oct. 8....	66	142	2.61	226
May 12 c ..	65	152	....	265	23....	66	125	2.35	165
25....	65	167	4.45	327	Nov. 12....	66	122	2.25	143
June 9....	70	375	7.50	2,070	22....	66	135	2.48	178
22....	70	360	6.04	1,860	Dec. 22....	64	124	3.15	183

a Measurement made through ice.
b Measurement from cable; some ice in section.
c Gage out of order.

NOTE.—Ice on Jan. 7, 1.2 feet thick; Jan. 20, 1.4 feet thick; Feb. 16, 1.8 feet thick; Feb. 26, 2.0 feet thick; Mar. 16, 2.0 feet thick; Dec. 22, 1.0 foot thick.

Daily gage height, in feet, of Lake Fork near Myton, Utah, for 1909.

[James E. Pitts, observer.]

Day.	Apr.	May.	June.	July.	Aug.	Sept.	Oct.	Nov.	Dec.
1	3.6	3.8	4.55	6.35	2.6	5.65	2.5	2.2	2.35
2	3.6	3.7	4.8	5.8	2.85	5.05	2.5	2.2	2.35
3	3.65	3.7	6.8	5.7	2.8	4.4	2.45	2.2	2.35
4	3.6	3.8	7.9	5.45	2.65	4.0	2.45	2.2	2.1
5	3.5	3.9	7.85	5.2	2.6	3.85	2.65	2.2	2.1
6	3.45	4.05	....	5.0	2.7	4.0	2.65	2.2	2.3
7	3.5	4.1	8.5	4.7	2.9	4.1	2.6	2.2	2.3
8	3.5	4.1	7.75	4.3	3.0	3.9	2.6	2.2	2.45
9	3.5	4.15	7.3	4.05	2.8	3.65	2.55	2.2	2.45
10	3.5	....	6.9	3.95	2.85	3.7	2.55	2.2	2.5
11	....	....	7.0	3.8	3.0	3.55	2.55	2.2	2.3
12	3.5	....	7.2	3.6	3.0	3.5	2.5	2.2	2.5
13	3.5	....	7.4	3.5	3.0	3.4	2.45	2.2	2.5
14	3.55	....	7.5	3.4	3.0	3.3	2.45	2.25	2.5
15	3.6	....	....	3.3	2.75	3.2	2.4	2.3	2.6
16	3.6	....	....	3.2	3.0	3.2	2.35	2.3	2.5
17	3.7	4.0	....	3.15	4.55	3.05	2.4	2.3	2.4
18	3.7	4.0	....	3.15	3.85	3.05	2.4	2.35	2.4
19	3.8	4.05	....	3.3	3.6	3.0	2.35	2.5	2.25
20	3.8	4.2	....	3.4	3.55	3.0	2.35	2.5	2.55
21	3.8	4.35	....	3.3	3.5	2.9	2.35	2.5	2.8
22	3.7	4.45	6.0	3.15	3.6	2.85	2.35	2.4	2.8
23	3.7	4.6	6.75	3.05	3.55	2.85	2.4	2.45	....
24	3.6	4.6	7.1	3.3	3.5	2.8	2.4	2.45	2.5
25	....	4.5	6.9	3.15	3.15	2.7	2.3	2.45	3.0
26	3.7	4.45	6.5	3.2	3.1	2.75	2.3	2.4	3.1
27	....	4.5	6.4	3.7	3.0	2.7	2.3	2.4	3.1
28	3.9	4.65	6.35	3.3	3.0	2.65	2.3	2.35	3.1
29	3.9	4.7	6.45	3.05	2.85	2.55	2.3	2.35	3.1
30	3.8	4.7	6.25	2.9	2.8	2.45	2.3	2.3	3.1
31	....	4.5	....	2.75	3.2	....	2.3	....	3.1

NOTE.—Ice conditions during Jan., Feb., Mar., and on Dec. 15 and from Dec. 19 to 31. Owing to settlement of the gage, gage heights June 15 to 21 are too uncertain to warrant publishing. New gage established on June 22. Gage heights after this date are not comparable with those before this date.

Daily discharge, in second-feet, of Lake Fork near Myton, Utah, for 1909.

Day.	Jan.	Feb.	Mar.	Apr.	May.	June.	July.	Aug.	Sept.	Oct.	Nov.	Dec.
1	175	185	180	142	172	358	2,080	215	1,610	190	130	160
2	175	185	180	142	156	455	1,710	285	1,250	190	130	160
3	175	185	180	149	156	1,580	1,640	270	915	180	130	160
4	175	180	180	142	172	2,360	1,490	228	735	180	130	115
5	175	180	180	130	190	2,330	1,340	215	668	228	130	115
6	175	175	180	124	221	2,580	1,220	240	735	228	130	150
7	178	175	180	130	232	2,840	1,060	300	780	215	130	150
8	180	170	180	130	232	2,250	870	335	690	215	130	180
9	180	170	180	130	244	1,930	758	270	580	202	130	180
10	180	170	180	130	240	1,650	712	285	600	202	130	190
11	180	165	180	130	235	1,720	645	335	540	202	130	150
12	185	165	180	130	230	1,860	560	335	520	190	130	190
13	185	160	180	130	226	2,000	520	335	480	180	130	190
14	185	160	180	136	222	2,070	480	335	440	180	140	190
15	185	155	180	142	218	2,140	440	255	405	170	150	190
16	190	149	183	142	214	2,140	405	335	405	160	150	190
17	190	150	180	156	210	2,680	388	990	352	170	150	170
18	195	150	175	156	210	2,920	388	668	352	170	160	170
19	200	155	170	172	221	2,680	440	560	335	160	190	140
20	207	155	160	172	256	2,140	480	540	335	160	190	170
21	205	160	160	172	296	1,720	440	520	300	160	190	180
22	205	165	160	156	325	1,840	388	560	285	160	170	183
23	205	165	155	156	375	2,360	352	540	285	170	180	180
24	200	170	155	142	375	2,620	440	520	270	170	180	180
25	200	175	155	149	340	2,470	388	388	240	150	180	180
26	195	180	150	156	325	2,180	405	370	255	150	170	180
27	195	180	150	173	340	2,110	600	335	240	150	170	180
28	195	180	150	190	395	2,080	440	335	228	150	160	180
29	190	....	144	190	415	2,140	352	285	202	150	160	180
30	190	....	140	172	415	2,000	300	270	180	150	150	180
31	190	....	140	....	340	....	255	405	....	150	....	180

NOTE.—These discharges are based on rating curves applicable as follows: Apr. 1 to June 21, fairly well defined between discharges of 120 and 2,440 second-feet. June 22 to Dec. 19, well defined between discharges of 115 and 2,920 second-feet. Discharges during Jan., Feb., Mar., June 15 to 22, and Dec. 20 to 31 estimated directly from measurements and hydrographs.

Monthly discharge of Lake Fork near Myton, Utah, for 1909.

[Drainage area, 475 square miles.]

Month.	Discharge in second-feet.				Run-off.		Accuracy.
	Maximum.	Minimum.	Mean.	Per square mile.	Depth in inches on drainage area.	Total in acre-feet.	
January	....	....	189	0.398	0.46	11,600	C.
February	....	....	168	.354	.37	9,330	C.
March	....	....	169	.356	.41	10,400	C.
April	190	124	149	.314	.35	8,870	A.
May	415	156	264	.556	.64	16,200	B.
June	2,920	358	2,070	4.36	4.86	123,000	B.
July	2,080	255	709	1.49	1.72	43,600	A.
August	990	215	383	.806	.93	23,600	A.
September	1,610	180	507	1.07	1.19	30,200	A.
October	228	150	177	.373	.43	10,900	A.
November	190	130	151	.318	.35	8,980	A.
December	....	....	171	.360	.42	10,500	A.
The year	....	....	426	.896	12.13	307,000	C.

UINTA RIVER NEAR WHITEROCKS, UTAH.

This station was originally established in connection with the investigation for the water supply of the Uinta Reservation, on September 16, 1899, and the records were continued until the latter part of 1904. It was reestablished in the same locality on August 13, 1907.

The present station is located at the highway bridge on the Government road up Uinta Canyon, usually known as the sawmill road. The bridge is about 8 miles northwest of the Indian agency at Whiterocks. Previous records were taken at points a short distance upstream from this bridge.

The station is situated about a mile below the mouth of Pole Creek. The Whiterocks comes in several miles below, but there are no other tributaries except some dry gulches which occasionally carry considerable flood water.

No water is diverted from the stream above the station, but the United States Indian Office has constructed a series of irrigation canals, which divert water at various points below. The upper reaches of this stream present excellent opportunities for storage and power development.

The results at the station are affected by ice during the winter season. Winter measurements are usually taken at riffles or open places in the channel.

The same gage was used from 1899 to 1904, inclusive. The gage established in August, 1907, was located a short distance upstream from the old gage, and at a different datum. The present chain gage has no determined relation to this last gage. It was established on October 22, 1907, and is located on the bridge about a mile downstream from the other gage.

The gage is read only when the hydrographer visits the station to make discharge measurements, and the discharges for intermediate days are estimated by comparison with the hydrographs of other streams in that locality.

As the stream bed is rough and the current swift at high and moderate stages, the discharge measurements, except at low stages, are apt to be considerably in error.

Discharge measurements of Uinta River near Whiterocks, Utah, in 1909.

[By R. H. Fletcher.]

Date.	Width.	Area of section.	Gage height.	Discharge.	Date.	Width.	Area of section.	Gage height.	Discharge.
	Feet.	*Sq. ft.*	*Feet.*	*Sec.-ft.*		*Feet.*	*Sq. ft.*	*Feet.*	*Sec.-ft.*
Jan. 15a...	66	120	2.55	194	July 15....	78	123	1.85	451
26a...	66	92	1.75	182	27....	81	148	2.35	795
Feb. 6a...	66	81	1.47	83	Aug. 3....	74	113	1.80	464
18a...	60	53	.90	103	16....	78	130	2.10	602
Mar. 9....	61	57	.95	133	30....	75	113	1.80	447
20....	66	62	1.00	136	Sept. 3....	84	160	2.30	790
Apr. 3....	66	67	1.00	156	14....	74	128	2.00	524
16....	69	71	1.10	177	28....	72	108	1.72	369
May 6....	70	92	1.34	268	Oct. 4....	73	108	1.70	372
19....	72	107	1.56	362	21....	73	86	1.45	226
29....	75	118	1.75	477	30....	71	82	1.45	196
June 3....	83	166	2.42	1,060	Nov. 8....	71	85	1.40	193
14....	84	204	2.63	1,270	18a...	67	85	1.42	193
25....	83	186	3.00	1,250	30a...	67	84	1.40	178
July 2....	85	160	2.60	809	Dec. 17a...	60	151	3.15	164

a Ice conditions.

Daily discharge, in second-feet, of Uinta River near Whiterocks, Utah, for 1909.

Day.	Jan.	Feb.	Mar.	Apr.	May.	June.	July.	Aug.	Sept.	Oct.	Nov.	Dec.
1	195	120	120	140	220	470	1,040	500	600	360	200	175
2	195	100	120	140	215	485	1,030	470	680	360	200	175
3	195	100	120	140	220	1,060	1,000	415	772	360	200	175
4	195	100	120	160	235	1,720	960	410	800	360	200	175
5	195	90	120	160	245	2,080	940	400	780	340	200	175
6	195	83	125	160	268	2,400	920	395	770	320	195	170
7	195	85	125	160	285	2,380	900	395	770	315	195	170
8	195	85	125	160	320	2,280	880	400	760	300	193	170
9	195	85	125	160	325	1,880	760	420	720	290	195	170
10	195	85	135	160	330	1,440	640	440	680	280	195	170
11	195	90	135	160	330	1,320	560	465	660	280	190	165
12	195	90	135	160	320	1,220	520	485	620	280	190	165
13	195	90	135	160	310	1,220	500	480	560	280	190	165
14	195	90	135	165	300	1,270	475	480	540	270	190	165
15	194	90	135	170	290	1,340	445	500	510	260	190	165
16	195	90	135	175	280	1,400	440	615	490	255	190	165
17	195	95	135	200	280	1,440	460	640	470	250	190	164
18	195	110	135	215	290	1,520	485	650	460	240	193	160
19	195	105	135	215	362	1,540	515	680	450	235	190	160
20	195	105	140	215	325	1,480	515	700	440	230	190	160
21	195	105	135	210	380	1,400	500	700	430	230	190	160
22	195	110	135	200	430	1,340	480	660	420	220	190	160
23	195	110	135	190	430	1,340	480	560	400	225	190	160
24	195	110	135	190	420	1,380	520	480	390	225	190	160
25	180	110	135	200	415	1,390	580	430	380	230	190	160
26	182	110	135	205	415	1,320	660	400	370	230	180	160
27	175	110	135	235	420	1,240	815	400	370	230	180	160
28	170	110	135	235	435	1,200	760	400	369	220	180	160
29	160	......	135	230	477	1,120	720	420	360	210	180	160
30	120	......	135	225	455	1,080	630	415	360	196	178	160
31	120	......	135	......	445	......	560	500	......	200	......	160

NOTE.—Daily discharges were obtained by comparison with hydrographs of Lake Fork below forks, and Uinta River at Fort Duchesne. Feb. 15 to Oct. 25 discharges, on dates of actual measurements, obtained by applying gage heights of these measurements to curves instead of using actual discharge.

Monthly discharge of Uinta River near Whiterocks, Utah, for 1909.

[Drainage area, 218 square miles.]

Month.	Discharge in second-feet.		Run-off.	
	Mean.	Per square mile.	Depth in inches on drainage area.	Total in acre-feet.
January..	187	0.858	0.99	11,500
February.......................................	98.7	.453	.47	5,480
March..	131	.601	.69	8,060
April..	183	.839	.94	10,900
May..	338	1.55	1.79	20,800
June...	1,430	6.56	7.32	85,100
July...	667	3.06	3.53	41,000
August...	494	2.27	2.62	30,400
September......................................	546	2.50	2.79	32,500
October..	267	1.22	1.41	16,400
November.......................................	191	.876	.98	11,400
December.......................................	165	.757	.87	10,100
The year...................................	391	1.80	24.40	284,000

NOTE.—The accuracy of these estimates may be classed as C.

UINTA RIVER AT FORT DUCHESNE, UTAH.

This station, which is located at the wooden highway bridge on the road to Vernal, one-fourth of a mile from Fort Duchesne, Utah, was originally established on September 4, 1899, and continued until the end of 1904. It was also maintained for a brief period during 1906, and on April 9, 1907, the station was reestablished.

The data obtained at this point show the amount of water contributed by this stream to Duchesne River, except the comparatively small amount diverted for irrigation below, and in connection with the records of the stations above on the Whiterocks and the Uinta, they show the amount of water taken for irrigation by the numerous diversions both on the Uinta and Whiterocks above the station. The upper tributaries, above irrigation diversions, afford excellent opportunities for storage and power development.

The flow of the stream is affected by ice for about four months during the winter season, and the accuracy of the results is somewhat affected by eddies around the crib piers and by deposits of sediment brought down by Deep Creek during floods.

Practically the same datum was used for the gage up to and including 1906. The present chain gage, established April 9, 1907, has an entirely different datum. This gage is fastened to the bridge from which discharge measurements are made.

Discharge measurements of Uinta River at Fort Duchesne, Utah, in 1909.

[By R. H. Fletcher.]

Date.	Width.	Area of section.	Gage height.	Discharge.	Date.	Width.	Area of section.	Gage height.	Discharge.
	Feet.	*Sq. ft.*	*Feet.*	*Sec.-ft.*		*Feet.*	*Sq. ft.*	*Feet.*	*Sec.-ft.*
Jan. 4 a..	40	64	7.50	170	July 7...	80	399	7.10	572
28 a..	40	74	6.84	127	23...	80	372	6.65	308
Feb. 10 a..	35	57	7.60	128	Aug. 7...	80	353	6.50	188
Mar. 13 a..	35	61	6.90	116	25...	80	308	6.90	386
27...	80	316	6.60	222	Sept. 6...	80	438	7.85	1,470
Apr. 6...	78	288	6.37	161	22...	80	388	6.92	4.6
29...	79	273	6.49	197	Oct. 9...	80	353	6.66	287
May 11...	80	225	6.80	378	22...	80	350	6.55	200
26...	80	230	6.82	409	Nov 11 b..	79	89	6.54	229
June 8...	80	490	8.27	2,780	20 b..	85	111	6.38	188
21...	80	444	7.82	1,460	Dec. 21 b..	105	135	7.10	171

a Ice; measurements made through holes cut in ice.   b Made by wading.

NOTE.—Ice at gage on Jan. 4, 1.6 feet thick; Jan. 28, 2.5 feet thick; Feb. 10, 3 feet thick; Dec. 21, 1.4 feet thick.

Daily gage height, in feet, of Uinta River at Fort Duchesne, Utah, for 1909.

[Bertha L. Wouldhave, observer.]

Day.	Apr.	May.	June.	July.	Aug.	Sept.	Oct.	Nov.	Dec.
1	6.45	6.4	6.85	7.55	6.5	7.6	6.7	6.5	6.5
2	6.5	6.45	7.15	7.3	6.4	7.9	6.65	6.5	6.5
3	6.5	6.45	7.8	7.3	6.5	7.55	6.65	6.5	6.4
4	6.5	6.45	8.15	7.3	6.4	7.35	6.7	6.5	6.35
5	6.4	6.5	8.75	7.3	6.4	7.6	6.7	6.5	6.3
6	6.35	6.55	8.8	7.4	6.45	7.85	6.7	6.5	6.25
7	6.4	6.65	8.65	7.25	6.45	7.65	6.7	6.5	6.5
8	6.35	6.65	8.15	7.05	6.45	7.6	6.7	6.4	6.55
9	6.4	6.65	8.25	7.0	6.45	7.5	6.65	6.4	6.6
10	6.4	6.75	7.9	6.9	6.4	7.45	6.6	6.6	6.6
11	6.4	6.75	7.75	6.85	6.4	7.4	6.6	6.5	6.7
12	6.35	6.7	7.8	6.7	6.4	7.5	6.6	6.5	6.6
13	6.35	6.65	7.95	6.6	6.6	7.3	6.6	6.5	6.55
14	6.4	6.65	7.85	6.6	6.55	7.2	6.6	6.5	6.8
15	6.4	6.55	8.05	6.6	6.45	7.15	6.55	6.5	6.8
16	6.45	6.55	7.85	6.6	6.6	7.1	6.55	6.5	6.7
17	6.5	6.55	8.15	6.5	7.1	7.1	6.5	6.4	6.7
18	6.5	6.55	8.25	6.45	7.15	7.0	6.5	6.45	6.75
19	6.5	6.55	8.1	6.6	7.0	7.0	6.5	6.4	6.6
20	6.5	6.65	7.95	6.9	6.9	6.95	6.5	6.4	6.75
21	6.5	6.75	7.85	6.75	6.9	6.9	6.5	6.4	6.7
22	6.4	6.85	7.85	6.6	6.9	6.9	6.5	6.4	7.45
23	6.4	6.85	7.8	6.6	6.8	6.9	6.5	6.45	7.6
24	6.4	6.85	7.95	6.75	6.75	6.9	6.5	6.4	7.4
25	6.4	6.85	7.95	6.6	6.8	6.8	6.5	6.45	7.3
26	6.4	6.75	7.75	6.6	6.7	6.8	6.55	6.45	7.3
27	6.4	6.85	7.7	7.1	6.7	6.7	6.5	6.6	7.3
28	6.4	7.05	7.7	6.9	6.7	6.65	6.5	6.5	7.3
29	6.6	7.05	7.65	6.65	6.6	6.65	6.5	6.45	7.25
30	6.5	7.00	7.6	6.6	6.6	6.7	6.5	6.5	7.35
31	....	6.85	....	6.5	6.8	....	6.5	....	7.25

NOTE.—Probable ice conditions Jan. 1 to Mar. 31 and Dec. 7 to 31.

Daily discharge, in second-feet, of Uinta River at Fort Duchesne, Utah, for 1909.

Day.	Jan.	Feb.	Mar.	Apr.	May.	June.	July.	Aug.	Sept.	Oct.	Nov.	Dec.
1	160	120	120	192	175	412	1,090	210	1,160	310	210	210
2	160	120	120	210	192	685	775	175	1,620	282	210	210
3	165	125	115	210	192	1,660	775	210	1,090	282	210	175
4	170	125	110	210	192	2,400	775	175	835	310	210	160
5	165	125	100	175	210	4,300	775	175	1,160	310	210	145
6	165	125	100	160	232	4,470	895	192	1,540	310	210	132
7	160	125	100	175	282	3,980	720	192	1,230	310	210	140
8	160	125	100	160	282	2,460	518	192	1,160	310	175	150
9	160	125	100	175	282	2,740	472	192	1,020	282	175	160
10	160	128	100	175	342	1,860	392	175	858	255	255	170
11	160	125	110	175	342	1,560	358	175	895	255	210	170
12	160	125	115	160	310	1,660	264	175	1,020	255	210	170
13	160	125	116	160	282	1,970	213	255	775	255	210	170
14	160	125	118	175	282	1,760	213	232	665	255	210	170
15	160	125	120	175	232	2,210	213	192	614	232	210	170
16	160	125	140	192	232	1,540	255	213	630	232	210	170
17	155	125	160	210	232	2,070	210	563	630	210	175	170
18	155	125	160	210	232	2,260	192	614	535	210	192	170
19	155	125	170	210	232	1,980	255	472	535	210	175	170
20	150	125	180	210	282	1,710	450	392	492	210	175	170
21	150	125	190	210	342	1,540	342	392	450	210	175	170
22	150	125	195	175	412	1,540	255	392	450	210	175	170
23	150	125	200	175	412	1,460	255	323	450	210	192	170
24	140	125	200	175	412	1,710	342	294	450	210	175	170
25	140	125	200	175	412	1,710	255	323	375	210	192	170
26	130	125	210	175	342	1,380	255	264	375	232	192	170
27	130	125	222	175	412	1,300	630	264	310	210	255	170
28	127	125	210	175	582	1,300	450	264	282	210	210	170
29	120	....	200	255	582	1,230	282	213	282	210	192	170
30	120	....	180	210	535	1,160	255	213	310	210	210	170
31	120	....	160	....	412	....	210	323	....	210	....	170

Monthly discharge of Uinta River at Fort Duchesne, Utah, for 1909.

[Drainage area, 672 square miles.]

Month.	Discharge in second-feet.				Run-off.		Accuracy.
	Maximum.	Minimum.	Mean.	Per square mile.	Depth in inches on drainage area.	Total in acre-feet.	
January			151	0.225	0.26	9,280	C.
February			125	.186	.19	6,940	C.
March			149	.222	.26	9,160	C.
April	255	160	187	.278	.31	11,100	B.
May	582	175	319	.475	.55	19,600	B.
June	4,470	412	1,940	2.89	3.22	115,000	C.
July	1,090	192	430	.640	.74	26,400	C.
August	614	175	272	.405	.47	16,700	B.
September	1,540	282	740	1.10	1.23	44,000	B.
October	310	210	246	.366	.42	15,100	B.
November	255	175	201	.299	.33	12,000	B.
December			168	.250	.29	10,300	D.
The year			411	.611	8.27	296,000	

NOTE.—These discharges were obtained from rating curves applicable as follows: Apr. 1 to June 15, July 16 to Aug. 15, and Sept. 16 to Dec. 6, fairly well defined between 120 and 3,500 second-feet; June 16 to July 15 and Aug. 16 to Sept. 15, not well defined. Discharges Jan., Feb., Mar., and Dec. 7 to 31 were estimated by hydrograph comparison with other streams in the vicinity of the gaging station.

WHITEROCKS RIVER NEAR WHITEROCKS, UTAH.

This station, which is located at the mouth of the canyon at the foot of "Dugway" on the road from the plateau to the river bottom, about 10 miles above the Indian agency at Whiterocks, was established April 18, 1899, and continued until the end of 1904. On April 11, 1907, it was reestablished at practically the same place.

The information contained here is valuable in connection with general studies of run-off problems and to show the amount of water available for storage and power above and for irrigation below.

The station is below all important tributaries of the Whiterocks. The first diversion for irrigation is about 3 miles below the station. Excellent storage and power sites exist above all irrigation diversions.

Like other streams in this region, the river is icebound for several months in the winter.

The same gage and datum were used from the establishment of the station until the end of 1904. A new chain gage and datum were used from April 11, 1907, to May 8, 1908, and the present chain gage, at a still different datum, has been used since May 9, 1908. Measurements are made from a cable about 100 feet downstream from the gage.

Owing to the remoteness of this gage from any dwelling, daily gage observations have not been made, and daily and monthly discharges have been obtained by comparing the relatively frequent discharge measurements with the hydrographs of other streams in that section.

As the stream bed is rather rough and the current is swift, measurements at high or medium stages are not very accurate. The daily and monthly discharge estimates, computed by the method outlined above, are necessarily only approximate.

Discharge measurements of Whiterocks River near Whiterocks, Utah, in 1909.

[By R. H. Fletcher.]

Date.	Width.	Area of section.	Gage height.	Discharge.	Date.	Width.	Area of section.	Gage height.	Discharge.
	Feet.	*Sq. ft.*	*Feet.*	*Sec.-ft.*		*Feet.*	*Sq. ft.*	*Feet.*	*Sec.-ft.*
Jan. 11a...	30	51	2.32	50	July 28....	44	60	2.12	277
27a...	31	65	2.58	73	Aug. 5....	34	47	1.80	183
Feb. 4b...	30	43	1.62	61	23....	45	73	2.20	330
Mar. 10b...	30	34	1.45	51	Sept. 2. ..	55	111	2.90	603
22	30	30	1.02	57	13....	50	82	2.38	373
Apr. 5	30	32	1.10	60	29....	35	47	1.85	177
17	31	39	1.30	102	Oct. 5....	34	46	1.84	161
May 7	32	46	1.58	145	20....	33	40	1.50	105
21	45	64	2.10	280	29....	33	33	1.45	96
31	50	76	2.30	372	Nov. 6b...	31	34	1.38	66
June 12	58	130	3.30	774	17b...	33	36	1.50	92
26	55	104	2.80	533	26b...	33	34	1.40	73
July 3	52	88	2.55	435	Dec. 16c...	33	46	1.30	67
17	37	53	1.95	211					

a Backwater effect at gage due to ice.
b Slight ice conditions.
c Considerable ice in river.

Daily discharge, in second-feet, of Whiterocks River near Whiterocks, Utah, for 1909.

Day.	Jan.	Feb.	Mar.	Apr.	May.	June.	July.	Aug.	Sept.	Oct.	Nov.	Dec.
1...............	50	65	50	80	80	370	440	180	400	180	80	60
2...............	50	65	50	80	80	400	435	180	603	180	80	60
3...............	50	65	50	65	80	440	435	180	760	170	75	60
4...............	50	61	50	60	100	600	440	180	720	170	75	60
5...............	50	55	50	62	120	1,600	480	185	680	178	70	60
6...............	50	55	50	50	125	1,960	500	180	680	160	66	60
7...............	50	50	50	50	147	2,020	400	190	700	160	60	60
8...............	50	50	50	50	160	1,900	340	190	690	160	60	60
9...............	50	50	50	60	180	1,000	260	180	660	160	60	60
10...............	50	50	51	60	185	800	200	170	600	160	70	60
11...............	50	50	50	60	180	750	·180	170	520	100	75	65
12...............	50	50	50	60	175	774	180	170	460	160	80	65
13...............	50	50	50	60	170	840	180	170	373	150	80	65
14...............	50	50	50	70	170	960	190	170	360	150	85	65
15...............	50	50	50	80	160	1,080	200	170	360	140	90	65
16...............	55	50	55	85	170	1,120	200	170	370	120	90	67
17...............	55	50	55	92	180	1,120	227	240	370	120	92	65
18...............	55	50	55	105	190	1,080	210	400	370	120	80	65
19...............	55	50	55	105	200	1,380	200	380	350	115	80	60
20...............	60	50	60	90	220	800	200	360	350	105	75	60
21...............	60	50	60	100	290	780	220	330	340	100	70	60
22...............	60	50	57	80	320	800	220	330	340	100	70	60
23...............	65	50	60	80	340	840	220	325	330	100	60	60
24...............	65	50	60	80	320	800	220	350	320	100	60	60
25...............	70	50	60	80	320	640	220	350	280	100	60	60
26...............	70	50	60	80	300	538	235	330	240	100	73	60
27...............	73	50	65	80	320	480	275	320	220	100	60	60
28...............	70	50	65	85	330	440	280	300	200	90	60	60
29...............	70	......	65	85	370	440	190	280	180	95	60	60
30...............	65	......	70	80	370	440	180	280	180	90	60	60
31...............	65	......	70	......	372	......	180	320	......	80	......	60

NOTE.—The daily discharges were obtained by comparison of hydrographs of Lake Fork below forks and Uinta River at Fort Duchesne.

Discharges on dates of actual measurements obtained by applying gage heights of these measurements to rating tables from Apr. 1 to Oct. 31.

Monthly discharge of Whiterocks River near Whiterocks, Utah, for 1909.

[Drainage area, 114 square miles.]

Month.	Discharge in second-feet.		Run-off.	
	Mean.	Per square mile.	Depth in inches on drainage area.	Total in acre-feet.
January...	56.9	0.499	0.58	3,500
February..	52.4	.460	.48	2,910
March...	55.6	..488	.56	3,420
April...	75.1	.659	.74	4,470
May...	217	1.90	2.19	13,300
June..	908	7.96	8.88	54,000
July..	269	2.36	2.72	16,500
August..	249	2.18	2.51	15,300
September...	434	3.81	4.25	25,800
October...	131	1.15	1.33	8,060
November..	71.9	.631	.70	4,280
December..	61.4	.538	.62	3,780
The year..	215	1.89	25.56	155,000

NOTE.—The accuracy of these estimates may be classed as C.

PRICE RIVER BASIN.

DESCRIPTION.

Price River rises in the Wasatch Mountains, in the southeastern part of Utah County, flows in a generally southeasterly direction and unites with Green River at a point about 14 miles above Greenriver, Utah. The main source of supply is the snow in the upper reaches of the basin, where elevations range from 8,000 to 9,000 feet. The region is extremely rough and rugged. The predominant rock is a loose and badly disintegrated sandstone. The soil is scanty and supports practically no vegetation except small groves of scrubby cedar and a few scattered pines. The original sparse underbrush and grass have been almost entirely tramped out by sheep and cattle.

The river is subject to floods in the spring and early summer, during which time it carries immense quantities of sediment. Gordon and Pleasant creeks, the principal tributaries, are both short, steep streams and enter from the west almost at right angles.

PRICE RIVER NEAR HELPER, UTAH.

This station, which was established February 21, 1904, is located at an old ford crossing in the settlement of Spring Glen, about 3 miles south of Helper, Utah, and about 350 feet west of the tracks of the Denver & Rio Grande Railroad.

This station is below Pleasant Creek and White River, the two principal tributaries above, and is above Gordon Creek, which enters about 5 miles below, and Grassy Trail Creek, which enters about 35 miles below. There are no important diversions above. Records indicate the amount of water available for the Price River Irrigation Co. and for the canals for the town of Price.

Discharge measurements are made from a car and cable.

The datum of the original chain gage remained unchanged until the gage was washed out by high water April 11, 1907. It was replaced by a temporary gage June 23, 1907, and by a permanent gage July 16, 1907. All gage heights after June 22, 1907, are referred to a new datum 0.7 foot above the original datum.

A fair estimate may be made of winter flow, though ice is usually rather heavy. The bed of the stream is somewhat shifting, but the records may, on the whole, be considered good.

Discharge measurements of Price River near Helper, Utah, in 1909.

Date.	Hydrographer.	Width.	Area of section.	Gage height.	Discharge.
		Feet.	*Sq. ft.*	*Feet.*	*Sec.-ft.*
June 19	E. A. Porter...................................	61	209	4.30	696
Sept. 23	.....do.......................................	51	67.2	3.00	101
Nov. 4	.....do.......................................	50	53.9	2.80	53.3

Daily gage height, in feet, of Price River, near Helper, Utah, for 1909.

[Andy Woolsey, observer.]

Day.	Jan.	Feb.	Mar.	Apr.	May.	June.	July.	Aug.	Sept.	Oct.	Nov.	Dec.
1	2.5	2.4	3.0	3.0	3.9	5.0	3.6	3.0	5.0	2.9	2.8	2.75
2	2.5	2.4*	2.9	3.2	3.9*	5.1	3.6	3.0	3.4	2.8	2.8	2.8
3	2.5	2.4	2.8	3.3	4.0	5.5	3.5	3.0	3.3	2.9	2.8	2.8
4	2.5	2.4	2.8	3.6	4.3	5.7	3.6	3.0	3.1	2.9	2.8	2.8
5	2.5	2.5	2.8	3.5	4.5	5.8	3.5	3.0	3.1	2.9	2.8	2.8
6	2.6	2.5	2.7	3.4	4.6	5.9	3.4	3.0	3.2	2.9	2.8	2.8
7	2.6	2.4	2.8	3.3	4.8	5.9	3.3	3.1	3.3	2.9	2.8	2.8
8	2.6	2.4	2.7	3.2	4.9	5.8	3.2	3.1	3.2	3.0	2.8	2.8
9	2.6	2.4	2.9	3.2	5.0	5.6	3.2	3.1	3.0	2.9	2.8	2.6
10	2.6	2.4	2.8	3.3	5.1	5.2	3.2	3.2	3.0	2.9	2.8	2.6
11	2.5	2.4	2.7	3.4	5.2	5.0	3.2	3.2	3.0	2.9	2.8	.....
12	2.6	2.4	2.7	3.3	5.2	4.9	3.1	3.1	3.0	2.9	2.8	2.6
13	2.6	2.4	2.7	3.3	4.9	4.8	3.1	3.1	3.0	3.0	2.8	2.7
14	2.6	2.5	2.7	3.6	4.8	4.8	3.1	3.1	3.0	2.9	2.8	2.8
15	2.6	2.5	2.6	3.7	4.7	4.6	3.1	3.1	3.0	2.9	2.8	2.8
16	2.6	2.5	2.8	3.8	4.7	4.5	3.0	3.1	3.0	2.9	2.8	2.8
17	2.5	2.5	2.8	4.0	4.7	4.4	3.0	3.3	3.0	2.8	2.8	2.8
18	2.5	2.5	2.8	4.3	4.7	4.4	3.2	3.4	3.0	2.9	2.8	2.8
19	2.5	2.5	2.9	4.2	4.8	4.3	3.4	3.5	3.0	2.9	2.8	2.8
20	2.5	2.5	2.8	4.0	4.9	4.3	3.3	3.5	3.0	2.9	2.8	2.8
21	2.5	2.5	2.8	3.9	5.4	4.2	3.2	3.4	3.0	2.9	2.8	2.8
22	2.5	2.5	2.8	3.8	5.4	4.2	3.2	3.5	3.0	2.9	2.8	2.8
23	2.5	2.5	2.9	3.7	5.5	4.1	3.0	3.4	3.0	2.9	2.8	2.8
24	2.5	2.5	2.8	3.8	5.6	4.1	3.4	3.2	3.0	2.9	2.8	.....
25	2.5	2.6	2.7	3.8	5.2	4.0	3.1	3.0	3.0	2.9	2.8	.
26	2.5	3.1	2.8	3.9	5.0	3.9	3.2	3.0	3.0	2.9	2.8	.....
27	2.5	3.0	3.0	4.2	5.1	3.9	3.1	3.0	3.0	2.9	2.8	.....
28	2.5	2.8	3.0	4.4	5.4	3.8	3.1	3.0	3.0	2.9	2.8	.....
29	2.7	......	3.1	4.4	5.4	3.8	3.1	2.9	2.9	2.9	2.8	.....
30	2.9	......	2.9	4.0	5.3	3.7	3.1	2.9	2.9	2.9	2.8	.....
31	2.5	......	3.0	......	5.0	......	3.1	3.7	......	2.8	......	.....

NOTE.—Ice conditions during January and February. Ice jams during these months affected the gage heights.

Gage heights Dec. 24 to 31 have been suppressed, as observer read the gage wrong.

Daily discharge, in second-feet, of Price River, near Helper, Utah, for 1909.

Day.	Jan.	Feb.	Mar.	Apr.	May.	June.	July.	Aug.	Sept.	Oct.	Nov.	Dec.
1	14	8	95	95	455	1,090	311	95	1,090	72	53	45
2	14	8	72	150	455	1,150	311	95	223	53	53	53
3	14	8	53	185	507	1,400	267	95	185	72	53	53
4	14	8	53	311	671	1,530	311	95	120	72	53	53
5	14	14	53	267	787	1,590	267	95	120	72	53	53
6	14	14	37	223	846	1,660	223	95	150	72	53	53
7	14	8	53	185	966	1,660	185	120	185	72	53	53
8	14	8	37	150	1,030	1,590	150	120	150	95	52	53
9	14	8	72	150	1,090	1,460	150	120	95	72	53	24
10	14	8	53	185	1,150	1,210	150	150	95	72	53	24
11	14	8	37	223	1,210	1,090	150	150	95	72	53	24
12	14	8	37	185	1,210	1,030	120	120	95	72	53	24
13	14	8	37	185	1,030	966	120	120	95	95	53	37
14	14	14	37	311	966	966	120	120	95	72	53	53
15	14	14	24	357	906	846	120	120	95	72	53	53
16	14	14	53	405	906	787	95	120	95	72	53	53
17	14	14	53	507	906	729	95	185	95	53	53	53
18	14	14	53	615	906	729	95	223	95	72	53	53
19	14	14	72	615	966	671	223	267	95	72	53	53
20	14	14	53	507	1,030	671	185	267	95	72	53	53
21	14	14	53	455	1,340	615	150	223	95	72	53	53
22	14	14	53	405	1,340	615	150	267	95	72	53	53
23	14	14	72	357	1,400	561	95	223	95	72	53	53
24	14	14	53	405	1,460	561	223	150	95	72	53	50
25	14	14	37	405	1,210	507	120	95	95	72	53	50
26	14	14	53	455	1,090	455	150	95	95	72	53	50
27	14	25	95	615	1,150	455	120	95	95	72	53	50
28	14	40	95	729	1,340	405	120	95	95	72	53	50
29	14	......	120	729	1,340	405	120	72	72	72	53	50
30	14	......	72	507	1,280	357	120	72	72	72	53	50
31	14	......	95	......	1,090	......	120	357	......	53	......	50

NOTE.—These discharges are based on a rating curve that is well defined between 25 and 200 second-feet; fairly well defined between 200 and 800 second-feet.

Discharges estimated because of ice conditions on Jan. 6 to 14, 29, 30, Feb. 25 to 28, and Dec. 24 to 31.

Monthly discharge of Price River near Helper, Utah, for 1909.

Month.	Discharge in second-feet.			Run-off (total in acre-feet).	Accuracy.
	Maximum.	Minimum.	Mean.		
January...	....	....	14.0	861	D.
February..	....	8	13.0	722	D.
March..	120	24	59.1	3,630	A.
April..	729	95	364	21,700	A.
May...	1,460	455	1,030	63,300	B.
June..	1,660	357	925	55,000	B.
July..	311	95	167	10,300	A.
August..	357	72	146	8,980	A.
September..	1,090	72	142	8,450	A.
October...	95	53	71.6	4,400	B.
November..	53	53	53.0	3,150	B.
December..	53	24	47.7	2,930	C.
The year.......................................	1,660	....	254	183,000	

SAN RAFAEL RIVER BASIN.

DESCRIPTION.

San Rafael River is formed in the western part of Emery County, crosses the central part of the county in a general southeasterly direction, and enters Green River about 16 miles below the mouth of the Price. The river has three principal branches—Ferron, Cottonwood, and Huntington creeks—which rise in the Wasatch Plateau at an altitude of about 10,000 feet above sea level. These streams fall rapidly in their upper courses and leave the plateau through almost impassable canyons cut in its eastern wall overlooking Castle Valley. They unite below Castledale, and the stream formed by their combined waters flows southeastward through the San Rafael Swell in a deep, narrow canyon, from which it emerges to flow across a low, broken country to its junction with the Green. The water of this river is derived chiefly from the melting snow on the high plateau.

SAN RAFAEL RIVER NEAR GREENRIVER, UTAH.

This station, which was established May 5, 1909, to determine the unappropriated run-off of San Rafael River, is located at the county bridge on the road from Green River to Hanksville, about 16 miles southwest of Greenriver, Utah, and about three-fourths of a mile below the Morris ranch dam. It is below all important tributaries and diversions.

The winter flow is affected by ice, and as the bed of the stream shifts somewhat frequent measurements must be made in order to get satisfactory records.

The staff gage is nailed securely to the southwest pier of the bridge, from which discharge measurements are made. The gage datum has remained unchanged since the station was established.

Discharge measurements of San Rafael River near Greenriver, Utah, in 1909.

Date.	Hydrographer.	Width.	Area of section.	Gage height.	Discharge.
		Feet.	*Sq. ft.*	*Feet.*	*Sec.-ft.*
May 5	E. A. Porter...	47	88	2. 20	223
June 9	.....do...	100	692	7. 20	3,040
20	.....do...	102	572	6. 00	2,380
Sept. 28	.....do...	57	108	1. 60	164
Nov. 23	.....do...	60	124	1. 20	178

Daily gage height, in feet, of San Rafael River near Greenriver, Utah, for 1909,

[E. F. Marshall, observer.]

Day.	May.	June.	July.	Aug.	Sept.	Oct.	Nov.	Dec.
1.....................................		4.5	4.5	1.85	7.95	1.5	1.1	1.15
2.....................................		5.15	4.35	1.75	8.95	1.45	1.1	1.35
3.....................................		5.85	4.05	1.75	3.75	1.4	1.1	1.4
4.....................................		6.85	3.9	1.65	3.7	1.35	1.1	1.1
5.....................................	2.2	7.25	4.0	1.6	4.4	1.4	1.1	1.0
6.....................................	2.5	7.55	3.75	1.75	2.75	1.3	1.1	.9
7.....................................	2.8	7.85	3.65	2.15	5.45	1.3	1.1	.9
8.....................................	2.75	7.9	3.55	3.0	3.55	1.3	1.0	....
9.....................................	2.95	7.35	3.35	3.8	2.65	1.3	1.05	....
10..:..................................	3.1	7.05	3.05	3.5	2.4	1.3	1.25	....
11.....................................	3.2	6.95	2.95	3.65	2.45	1.3	1.2	....
12.....................................	3.5	7.05	2.85	3.8	2.9	1.3	1.35	....
13.....................................	2.95	6.75	2.7	3.3	2.45	1.3	1.15	....
14.....................................	2.8	6.75	2.55	3.0	2.2	1.3	1.1	....
15.....................................	2.75	6.65	2.1	2.65	2.15	1.3	1.1	....
16.....................................	2.95	6.4	1.85	2.5	2.4	1.2	1.05	....
17.....................................	2.8	6.5	1.65	7.75	2.1	1.2	1.0	....
18.....................................	2.9	6.65	1.65	5.6	1.8	1.2	1.0	....
19.....................................	3.1	6.35	1.9	5.15	1.7	1.2	.9	....
20.....................................	3.55	5.9	2.4	4.75	1.65	1.2	.9	....
21.....................................	3.95	5.8	2.0	4.35	1.7	1.2	1.0	....
22.....................................	4.4	5.75	2.05	3.9	1.65	1.2	1.1	....
23.....................................	4.6	5.7	2.45	3.55	1.65	1.2	1.1	....
24.....................................	4.65	5.75	2.7	3.05	1.6	1.2	1.55	....
25.....................................	4.2	5.65	2.25	2.7	1.55	1.15	1.15	....
26.....................................	3.9	5.35	2.05	2.65	1.55	1.1	1.25	....
27.....................................	4.45	5.15	1.9	2.55	1.55	1.1	1.15	....
28.....................................	5.25	5.15	2.5	2.35	1.55	1.1	2.0	....
29.....................................	5.4	4.9	2.15	2.45	1.55	1.1	1.4	....
30.....................................	5.0	4.6	2.05	2.3	1.55	1.1	1.25	....
31.....................................	4.6	....	1.85	4.45	....	1.1	....	....

NOTE.—Ice in river in December.

Daily discharge, in second-feet, of San Rafael River near Greenriver, Utah, for 1909.

Day.	May.	June.	July.	Aug.	Sept.	Oct.	Nov.
1	....	1,090	1,310	230	3,850	140	140
2	....	1,490	1,220	220	4,720	135	140
3	....	1,990	1,030	220	820	130	140
4	....	2,760	950	190	800	125	140
5	221	3,080	1,000	180	1,150	130	140
6	290	3,320	880	220	450	120	140
7	374	3,560	830	300	1,830	130	140
8	359	3,610	780	560	730	130	130
9	423	3,160	700	880	420	130	135
10	474	2,920	580	750	350	130	170
11	510	2,930	540	810	360	130	160
12	620	3,020	510	880	500	130	185
13	423	2,780	460	670	360	130	150
14	374	2,780	420	550	300	130	140
15	359	2,800	300	430	290	130	140
16	423	2,600	230	390	330	120	150
17	374	2,680	200	3,730	260	120	140
18	406	2,800	200	2,000	190	120	140
19	474	2,650	250	1,750	180	120	130
20	640	2,300	370	1,400	170	120	130
21	816	2,230	270	1,160	180	120	140
22	1,040	2,180	280	890	170	120	160
23	1,150	2,150	380	730	170	120	160
24	1,180	2,180	460	540	160	120	240
25	938	2,110	330	430	150	115	170
26	792	1,890	280	420	150	110	185
27	1,060	1,740	250	380	150	140	170
28	1,560	1,740	390	330	150	140	300
29	1,660	1,570	300	360	150	140	210
30	1,390	1,320	280	310	150	140	185
31	1,150	....	230	1,180	....	140	....

NOTE.—Discharges May 5 to June 9 are based on a rating curve that is partly defined. For the remainder of the year the indirect method for shifting channels was used, with the above curve as standard.

Monthly discharge of San Rafael River near Greenriver, Utah, for 1909.

Month.	Discharge in second-feet.			Run-off (total in acre-feet).	Accuracy.
	Maximum.	Minimum.	Mean.		
May (5–31)	1,660	221	721	38,600	B.
June	3,610	1,090	2,450	146,000	B.
July	1,310	200	523	32,200	C.
August	3,730	180	745	45,800	C.
September	4,720	150	655	39,000	C.
October	140	110	128	7,870	C.
November	360	130	162	9,640	C.
The period	....	....	....	319,000	

COTTONWOOD CREEK NEAR ORANGEVILLE, UTAH.

This station, which was established May 1, 1909, is located at Johnson's ranch in the canyon about 5 miles northwest of Orangeville, Utah, and about 35 miles southwest of Price, the nearest railway point.

The station is below all important tributaries and above all diversions except Johnson's ditch, which takes out a small amount of water a short distance above the station.

Previous to August 22, 1909, the stage was recorded by measuring to the water surface from a reference point on a cottonwood tree 60 feet above the cable. A staff gage was installed August 22, 1909, at the same point, and all previous observations were connected to the datum of this gage. During the flood of August 31 this gage washed out. From September 1 to 17 the record was kept of the water depth at the gage site. From September 20 observations was made of the distance to water surface from a mark on a rock at the site. All gage heights for 1909 have been reduced to the datum of the staff gage.

Discharge measurements are made from a cable 60 feet below the gage site.

As the stream bed shifts, accurate determination of discharge is difficult. Heavy ice forms at this station during the winter months.

Discharge measurements of Cottonwood Creek near Orangeville, Utah, in 1909.

Date.	Hydrographer.	Width.	Area of section.	Gage height.	Dis-charge.
		Feet.	*Sq. ft.*	*Feet.*	*Sec.-ft.*
May 1...	E. A. Porter...................................	43	48	2.16	105
June 13 .	.....do...................................	54	127	4.66	784
July 27..	.....do...................................	47	48.4	2.61	125
Aug. 22 .	.....do...................................	46	41.6	2.53	89.8
Sept. 20.	.....do...................................	25	28	3.68	62.3
Nov. 5..	.....do...................................	14	15.1	3.31	31.8
Dec. 1...	.....do...................................	16	16.6	3.48	40.3

Daily gage height, in feet, of Cottonwood Creek near Orangeville, Utah, for 1909.

[Robert Johnson, observer.]

Day.	May.	June.	July.	Aug.	Sept.	Oct.	Nov.	Dec.
1.................................	2.15	4.75	3.95	....	5.3	3.6	3.4	3.5
2.................................	....	5.05	3.75	2.45	3.8	3.6	3.4	3.5
3.................................	2.45	5.55	3.65	2.45	5.2	....	3.3	3.5
4.................................	....	5.75	....	2.45	4.1	3.6	3.3	3.4
5.................................	....	5.75	....	2.45	....	3.5	....	....
6.................................	2.65	....	3.35	2.75	4.2	3.5	3.3	3.8
7.................................	....	5.25	3.35	3.75	....	3.5	....	....
8.................................	3.05	4.95	3.25	....	3.8	3.5	3.3	3.8
9.................................	....	....	3.15	....	3.8	3.5	3.3	3.8
10.................................	3.15	4.75	3.15	5.45	3.6	....	3.3	...ι
11.................................	2.95	5.15	....	3.95	3.6	3.5	3.3	3.9
12.................................	2.95	4.75	2.95	....	....	3.5	3.3	....
13.................................	2.85	....	2.95	2.45	3.8	3.5	3.3	3.8
14.................................	2.95	4.75	2.85	....	3.8	3.5	....	....
15.................................	3.15	4.65	2.85	....	3.7	3.5	3.4	3.5
16.................................	....	4.85	....	5.65	3.6	3.5	3.4	....
17.................................	....	5.15	2.75	5.75	3.6	....	3.4	4.1
18.................................	3.15	4.55	....	5.75	....	3.5	3.6	4.0
19.................................	3.35	4.35	2.95	5.65	....	3.5	3.4	....
20.................................	3.55	....	2.75	3.95	3.7	3.5	3.6	4.6
21.................................	3.75	4.55	2.75	5.75	3.7	3.5	....	....
22.................................	4.05	4.45	2.75	2.55	3.6	3.5	....	4.2
23.................................	....	4.65	2.75	2.5	3.6	3.5	....	....
24.................................	3.65	4.85	....	2.4	3.6	....	....	4.0
25.................................	....	4.55	....	2.4	3.6	3.5	....	....
26.................................	3.95	4.15	2.65	2.4	....	3.5	....	....
27.................................	4.45	....	2.60	2.4	3.6	3.5	....	4.2
28.................................	4.55	3.95	2.55	....	3.6	3.4	....	....
29.................................	4.65	3.65	2.55	....	3.6	3.4	....	4.0
30.................................	....	4.25	2.45	3.0	3.6	3.4	....	....
31.................................	....	....	2.45	4.0	....	....	....	....

NOTE.—Ice Dec. 4 to 31. Ice 4 inches thick Dec. 11; 9 inches on Dec. 18; 14 inches on Dec. 24, and 16 inches on Dec. 29.

Daily discharge, in second-feet, of Cottonwood Creek near Orangeville, Utah, for 1909.

Day.	May.	June.	July.	Aug.	Sept.	Oct.	Nov.	Dec.
1	105	872	513	92	500	56	36	45
2	135	994	434	90	84	56	36	45
3	1f5	1,200	410	89	460	56	27	45
4	180	1,280	378	88	144	56	27	....
5	195	1,270	345	87	166	45	27	....
6	210	1,220	312	58	167	45	27	....
7	268	1,160	312	428	122	45	27	....
8	325	930	283	650	84	45	27	....
9	332	885	253	880	84	45	27	....
10	340	840	253	1,100	56	45	27	....
11	280	985	228	495	56	45	27	....
12	280	835	202	300	70	45	27	....
13	250	830	202	88	84	45	27	....
14	280	825	177	450	84	45	30	....
15	340	784	177	800	68	45	36	....
16	334	857	166	1,160	56	45	36	....
17	328	980	155	1,200	56	45	36	....
18	322	738	180	1,200	60	45	56	....
19	380	602	204	1,100	64	45	36	....
20	448	671	154	480	68	45	56	....
21	510	740	154	1,200	68	45	55	....
22	624	700	154	92	56	45	54	....
23	546	780	154	98	56	45	53	....
24	468	860	147	80	56	45	52	....
25	521	740	140	80	56	45	51	....
26	574	585	134	80	56	45	50	....
27	768	548	125	80	56	45	49	....
28	808	510	112	125	56	36	48	....
29	848	406	112	170	56	36	47	....
30	856	628	95	210	56	36	46	....
31	864	....	93	530	....	36	....	....

NOTE.—These discharges were obtained from rating curves applicable as follows: May 1 to Aug. 31, indirect method for shifting channels used; Sept. 1 to Dec. 3, not well defined. Discharges interpolated for days when gage was not read.

Monthly discharge of Cottonwood Creek near Orangeville, Utah, for 1909.

Month.	Discharge in second-feet.			Run-off (total in acre-feet).	Accuracy.
	Maximum.	Minimum.	Mean.		
May	....	....	416	25,600	B.
June	1,280	406	842	50,100	B.
July	513	93	218	13,400	B.
August	1,200	80	438	26,900	C.
September	500	56	104	6,190	C.
October	56	36	45.3	2,790	C.
November	....	....	38.7	2,300	C.
The period	....	....	....	* 127,000	

FERRON CREEK NEAR FERRON, UTAH.

This station, which was established April 28, 1909, is located near the mouth of the canyon, about 2½ miles above the town of Ferron, Utah, and is below all important tributaries.

Practically all the normal flow in the low-water season is diverted above the station by the North and South canals, only enough water passing to supply one or two small ditches that take out below.

Several gages were used during 1909, all located in the same section. All gage heights were referred to one datum until August 31, when a flood destroyed the gage and bench mark and greatly changed the section. From September 1 to December 31 all gage heights refer to a new gage which was installed September 18, 1909, at a new datum.

Discharge measurements are made from a footbridge about 10 feet above the gage. Shifting of the stream bed makes it difficult to obtain accurate discharge records. The stream is icebound during the winter.

Discharge measurements of Ferron Creek near Ferron, Utah, in 1909.

Date.	Hydrographer.	Width.	Area of section.	Gage height.	Dis-charge.
		Feet.	*Sq. ft.*	*Feet.*	*Sec.-ft.*
Apr. 28	E. A. Fuller...........................	20	15.5	2.10	30.6
June 12	.....do.......................................	31	58	4.20	496
July 26	.....do.......................................	17	7.6	1.40	· 5.7
Aug. 23	.....do.......................................	14	16.9	2.15	36.5
Sept. 19	.....do.......................................	18.5	13.3	1.75	28.4
Dec. 4	.....do.......................................	....	....	1.40	6.5

NOTE.—Gage heights Apr. 28 to Aug. 23 are referred to same datum. Gage heights Sept. 19 and Dec. 4 are referred to new gage installed Sept. 18, 1909.

Daily gage height, in feet, of Ferron Creek near Ferron, Utah, for 1909.

[James Westenskaw, observer.]

Day.	Apr.	May.	June.	July.	Aug.	Sept.	Oct.	Nov.	Dec.
1..........	....	2.3	3.75	2.65	....	12.0	1.8	1.6	1.5
2..........	....	2.3	4.15	2.6	....	1.8	1.8	1.5	1.2
3..........	....	2.5	4.55	2.85	....	1.8	1.8	1.5	1.1
4..........	....	2.6	4.65	2.7	....	1.8	1.7	1.5	1.1
5..........	....	2.65	5.05	2.6	....	1.8	1.6	1.5	1.5
6..........	....	2.95	4.45	2.25	....	1.8	1.5	1.5	1.5
7..........	....	2.95	4.26	1.95	....	1.8	1.4	1.5	1.5
8..........	....	2.95	3.95	2.05	....	1.8	1.4	1.5	1.6
9..........	....	2.75	4.20	2.05	....	· 1.8	1.4	1.5	1.8
10..........	....	2.90	4.10	2.05	....	1.8	1.4	1.4	2.0
11..........	....	2.75	4.05	2.05	....	1.8	1.6	1.4	2.0
12..........	....	2.40	4.00	1.5	....	1.8	1.6	1.4	2.0
13..........	....	2.3	3.65	1.5	....	1.8	1.6	1.4	2.0
14..........	....	2.2	3.65	1.5	....	1.8	1.6	1.5	2.0
15..........	....	2.3	3.65	....	....	1.8	1.6	1.6	2.0
16..........	....	2.45	3.8	....	....	1.8	1.6	1.8	2.2
17..........	....	2.45	3.75	....	....	1.8	1.6	1.8	2.3
18..........	....	2.8	3.45	....	....	1.8	1.6	1.9	2.4
19..........	....	3.05	3.45	....	....	1.8	1.6	2.1	2.8
20..........	....	3.1	3.45	....	....	1.8	1.6	2.2	2.6
21..........	....	3.2	3.5	....	....	1.8	1.6	2.2	2.6
22..........	....	3.1	3.35	....	....	1.8	1.6	1.4	2.6
23..........	....	3.1	3.5	....	2.15	1.8	1.6	1.8	2.5
24..........	....	2.75	3.45	....	2.0	1.8	1.6	1.7	2.4
25..........	....	2.9	3.4	....	1.9	1.8	1.6	1.6	2.0
26..........	....	3.45	3.35	1.4	1.9	1.8	1.6	1.5	2.4
27..........	....	3.5	3.05	....	1.9	1.8	1.6	1.4	2.6
28..........	2.1	3.45	2.95	....	1.9	1.8	1.6	1.9	2.7
29..........	1.8	3.15	2.95	....	1.8	1.8	1.6	2.0	2.6
30..........	1.5	3.0	2.80	....	1.8	1.8	1.6	1.8	2.8
31..........	....	3.3	....	....	8.0	....	1.6	....	2.8

NOTE.—After Sept. 1 gage heights referred to new datum. Ice conditions after Nov. 13.

Daily discharge, in second-feet, of Ferron Creek near Ferron, Utah, for 1909.

Day.	Apr.	May.	June.	July.	Aug.	Sept.	Oct.	Nov.
1		51	367	105	....	3,000	32	20
2		51	485	96	....	32	32	16
3		78	606	143	....	32	32	16
4		96	638	114	....	32	25	16
5		105	766	96	....	32	20	16
6		164	575	46	....	32	16	16
7		164	515	22	....	32	13	16
8		164	425	28	....	32	13	16
9		124	500	28	....	32	13	16
10		153	470	28	....	32	13	13
11		124	455	28	....	32	20	13
12		63	440	7	....	32	20	13
13		51	339	7	....	32	20	13
14		41	339	7	....	32	20	13
15		51	339	6	....	32	20	13
16		70	381	6	....	32	20	13
17		70	367	6	....	32	20	13
18		133	284	6	....	32	20	12
19		185	284	6	....	32	20	12
20		196	284	6	....	32	20	12
21		220	298	6	....	32	20	11
22		196	258	6	....	32	20	11
23		196	298	6	36	32	20	10
24		124	284	6	25	32	20	10
25		153	271	6	20	32	20	10
26		284	258	6	20	32	20	9
27		298	185	6	20	32	20	9
28	32	284	164	6	20	32	20	8
29	16	208	164	6	16	32	20	8
30	7	174	133	6	16	32	20	8
31		245	....	6	1,690	....	20	....

NOTE.—These discharges are based on rating curves applicable as follows: Apr. 28 to Aug. 31, fairly well defined below discharge of 100 second-feet. Sept. 1 to Nov. 13, not well defined. Discharge estimated, July 15 to 25, 27 to 30, and Nov. 14 to 30.

Monthly discharge of Ferron Creek near Ferron, Utah, for 1909.

Month.	Discharge in second-feet.			Run-off (total in acre-feet).	Accuracy.
	Maximum.	Minimum.	Mean.		
May	298	41	146	8,980	B.
June	766	133	372	22,100	B.
July	143	6	27.6	1,700	C.
Aug. 23–31	1,690	16	207	3,700	C.
September	3,000	32	131	7,800	D.
October	32	13	20.3	1,250	D.
November	....	....	a 12.7	756	D.
December	....	....	b 4.0	246	D.
The period	....	....	....	46,500	

a Partly estimated.   b Estimated.

HUNTINGTON CREEK NEAR HUNTINGTON, UTAH.

This station, which was established May 3, 1909, is located at Cunha's ranch, in the canyon about 6 miles northwest of Huntington, Utah, and is below all important tributaries.

The ditch for the Cunha ranch diverts a small amount of water a short distance above the station; practically all the normal low-water flow is diverted for irrigation by canals heading near Huntington. A storage reservoir above the station controls the distribution of the discharge to a considerable extent.

The vertical staff gage is in two sections. The low-water part is nailed to an old bridge abutment on the right bank about 3 feet from the cable; the high-water section is nailed to the west face of a cottonwood tree near the low-water section. Discharge measurements are made from a cable.

The gage datum has remained unchanged since the station was established.

The flow at the station is not seriously affected by ice. The shifting of the stream bed during the spring high water and summer floods impairs the reliability of the records.

Discharge measurements of Huntington Creek near Huntington, Utah, in 1909.

Date.	Hydrographer.	Width.	Area of section.	Gage height.	Dis- charge.
		Feet.	*Sq. ft.*	*Feet.*	*Sec.-ft.*
June 17	E. A. Porter	44	133	4.80	730
July 29	.....do...	35	78.3	3.38	138
Aug. 21	.....do...	30.5	33.3	2.65	78.8
21	.....do...	32	30.1	2.75	65.5
Nov. 6	.....do...	31	24.5	2.55	45.7
30 *a*	.....do...	31	27.6	2.65	65.3
Dec. 5 *b*	.....do...	31	21.9	2.51	40.4

a Ice 0.25 foot thick at gauge.   *b* Ice 6 inches thick at gage.

Daily gage height, in feet, of Huntington Creek near Huntington, Utah, for 1909.

[Joseph Cunha, observer.]

Day.	May.	June.	July.	Aug.	Sept.	Oct.	Nov.	Dec.
1.....................		4.5	3.5	3.4	5.0	....	2.5	....
2.....................	....	4.7	3.5	3.3	4.0	2.55	2.5	....
3.....................	3.0	5.05	3.5	3.3	3.0	2.8	2.45	....
4.....................	3.2	5.15	3.5	3.4	3.0	2.5	2.5	....
5.....................	3.45	5.4	3.5	3.3	2.9	2.5	2.5	....
6.....................	3.55	5.45	3.5	3.3	2.9	2.5	2.4	2.60
7.....................	3.7	5.25	3.5	3.4	2.9	2.5	2.4	....
8.....................	3.85	5.1	3.4	2.9	2.8	2.5	2.5	....
9.....................	3.9	5.0	3.4	2.8	2.8	2.5	2.5	....
10....................	3.85	5.0	3.4	2.8	2.8	2.5	2.45	2.7
11....................	4.0	5.0	3.4	2.7	2.8	2.4	2.5	....
12....................	3.85	5.0	3.4	2.7	2.7	2.4	2.5	....
13....................	3.7	4.9	3.4	2.6	2.7	2.4	2.45	....
14....................	3.75	4.9	3.4	2.6	2.6	2.4	2.5	2.5
15....................	3.8	4.9	3.3	2.6	2.6	2.4	2.5	....
16....................	3.7	4.7	3.3	2.5	2.6	2.4	2.4	2.6
17....................	3.7	4.85	3.5	2.9	2.6	2.4	2.5	....
18....................	3.9	4.75	3.9	2.9	2.6	2.4	2.5	....
19....................	4.1	4.8	3.8	2.95	2.6	2.4	2.55	2.5
20....................	4.25	4.65	3.7	2.95	2.5	2.4	2.5	....
21....................	4.4	4.7	3.7	2.95	2.5	2.35	....	....
22....................	4.35	4.75	3.6	2.9	2.6	2.4	....	....
23....................	4.3	4.6	3.5	2.8	2.6	2.4	....	2.5
24....................	4.35	4.6	3.5	3.0	2.7	2.5	....	....
25....................	4.05	4.6	3.5	2.9	2.6	2.5	2.6	....
26....................	4.55	4.65	3.4	2.7	2.6	2.5	....	....
27....................	4.55	4.45	3.4	2.8	2.7	2.45	....	2.4
28....................	4.55	4.45	3.4	2.9	2.6	2.45	....	....
29....................	4.55	4.45	3.4	2.8	2.6	2.5	....	....
30....................	4.3	4.45	3.4	2.8	2.6	2.45	2.7	2.5
31....................	4.4	....	3.4	2.8	....	2.5	....	....

NOTE.—Ice prevailed from about Nov. 20 to Dec. 31; Nov. 30, ice 3 inches thick.

Daily discharge, in second-feet, of Huntington Creek near Huntington, Utah, for 1909.

Day.	May.	June.	July.	Aug.	Sept.	Oct.	Nov.
1	....	655	225	152	800	42	42
2	....	732	224	134	408	74	42
3	160	875	222	160	124	38	33
4	212	914	219	163	122	38	42
5	284	920	216	147	104	38	42
6	314	1,040	213	152	103	38	29
7	362	948	210	177	102	38	31
8	412	885	183	82	85	38	42
9	430	838	180	70	84	38	42
10	412	835	178	75	82	38	35
11	468	826	175	62	80	29	39
12	412	824	173	64	66	29	38
13	362	780	170	53	64	29	32
14	378	775	168	54	53	29	38
15	392	772	144	58	52	29	38
16	362	694	143	46	51	29	32
17	362	750	189	106	50	29	41
18	430	704	308	110	49	29	41
19	503	726	268	122	48	29	45
20	560	662	238	124	38	29	41
21	615	670	235	127	37	26	....
22	598	688	207	120	48	30	....
23	578	622	180	128	48	30	....
24	580	621	178	140	60	38	....
25	484	620	175	118	52	38	....
26	672	634	152	84	48	38	....
27	672	556	150	98	59	33	....
28	672	552	148	114	48	33	....
29	672	551	147	96	48	38	....
30	578	550	142	94	48	33	....
31	615	....	147	93	....	38	....

NOTE.—These discharges, except for the ice period, were obtained by the indirect method for shifting channels.

Monthly discharge of Huntington Creek near Huntington, Utah, for 1909.

Month.	Discharge in second-feet.			Run-off (total in acre-feet).	Accuracy.
	Maximum.	Minimum.	Mean.		
May 3-31	672	160	467	26,900	C.
June	1,040	550	741	44,100	C.
July	308	142	191	11,700	C.
August	177	46	107	6,580	C.
September	800	37	102	6,070	B.
October	74	26	35.0	2,150	B.
November	42	29	38.8	2,310	C.
December	....	....	a 34.8	2,140	D.
The period	....	....	215	102,000	

a Estimated.

NOTE.—Nov. 21 to 30 and Dec. 1 to 15, daily discharge estimated as 40 second-feet; Dec. 16 to 31, estimated as 30 second-feet, because of ice conditions.

GRAND RIVER BASIN.

DESCRIPTION.

Grand River and its tributaries drain an area comprising approximately 26,000 square miles, of which 22,290 are in Colorado and the rest in eastern Utah. On the east and southeast the basin is bounded by the high ranges of the Continental Divide, which separate it from the basins of Platte and Arkansas rivers; on the north it is limited by the White River and Book Cliffs Plateau, on the west by the canyon district of the Colorado.

Rising among the high peaks of the Rocky Mountains in the north-central portion of Colorado, the Grand flows southwestward to its junction with Green River, traversing approximately 350 miles. Its tributaries include Fraser, Blue, Eagle, Williams and Roaring forks, Gunnison and Dolores rivers, all of which enter from the south.

In most respects the Grand is a typical mountain stream, flowing throughout its course in a succession of deep canyons, whose precipitous or even perpendicular walls range in height up to 3,000 feet above the water's edge, alternating with long, narrow fertile valleys. The headwater region, comprising approximately 50 per cent of the basin, is extremely rugged, elevations ranging from 7,000 to 14,000 feet above sea level. Stream channels are numerous, tributaries are rapid, and gradients are steep, the fall ranging from 20 to 150 feet to the mile. The intermediate or middle portion of the basin—that portion immediately east and west of the Colorado State line—is a dry, broken, much-eroded region.

The rocks of the basin include all varieties, from the granites and masses of igneous origin on the crest of the Continental Divide to the younger and less resistant sedimentary rocks of the plateau region. The soils of the upper basin, though shallow, generally contain considerable organic matter; those of the intermediate basin are largely decomposed and disintegrated sedimentary rocks, which grade imperceptibly from one to the other. The scant vegetation of the lower basin renders soil erosion large.

The precipitation ranges from 5 to 10 inches in the lower basin, 10 to 20 inches in the intermediate region, and 20 to 30 inches in the headwater region. By far the greater part of the precipitation is in the form of snow.

The forestation of the mountainous part of the basin, except in a few localities, is good—the equal of any in Colorado. The forests consist of spruce, quaking asps, cedars, and piñon. The intermediate basin is fairly well forested with quaking asp, cedar, and piñon. The lower basin supports only scattered pines, cedars, and piñons, the prevailing vegetation being sagebrush, chico, and cactus pads.

The greater part of the timbered area in the Grand River basin above the Gunnison is included in the Arapahoe and Holy Cross national forests. These reserves in the Grand drainage basin include about 1,400 square miles of merchantable timberland, 900 square miles of woodland, and about 800 square miles of burned area.

In the middle basin, from the lower end of Gore Canyon to about Rifle, 30,000 to 35,000 acres will be irrigated under half a dozen small projects now contemplated. In the lower basin the Reclamation

Service has under way the Grand Valley project, to cover an irrigable area of 60,000 to 70,000 acres. Under other schemes 40,000 to 50,000 acres more will be irrigated. The Uncompahgre Valley project, which diverts water from the Gunnison, has finished structures capable of irrigating about 50,000 acres. The completed project will serve about 150,000 acres.

Natural storage within the basin is restricted to a few high mountain lakes, of which Grand Lake is the largest. There are, however, reservoirs sites along the Grand and its tributaries which if utilized would make possible a development of 1,000,000 horsepower. The Kremmling reservoir site is by far the best in the drainage basin. It is located near the upper end of Gore Canyon and with a 230-foot dam would impound about 2,200,000 acre-feet of water. A standard-gauge railroad now runs through this site.

Until recently the splendid power resources of this drainage basin have remained practically untouched. The estimated available power, including that on Dolores and Gunnison rivers, is as follows:

```
Minimum horsepower........................................   540,000
Minimum horsepower (6 high months).......................  1,000,000
Horsepower from storage (6 months' period)...............  1,600,000
```

Of this amount less than 40,000 horsepower has so far been developed.

Hot sulphur springs are located along Grand River at two points—Hot Sulphur Springs and Glenwood Springs, Colo.—and in both localities they increase the temperature of the river water, but probably all these springs together add less than 20 second-feet to the flow of the river. The years of maximum run-off in this drainage basin were 1897 and 1907; the year of maximum run-off since records were begun was 1902.

Fraser, Eagle, Williams and Roaring forks and Gunnison River are described in connection with gaging stations now maintained. The importance of Dolores River entitles it to the following brief description, although no station is now maintained in its basin.

The Dolores rises in the La Plata and San Miguel mountains, whose highest peak, Mount Wilson, attains an elevation of over 14,000 feet. Its course is southwesterly for about 50 miles, when it turns and flows almost due north for nearly 100 miles, when it again turns to the west and enters Grand River about 15 miles west of the Colorado-Utah line. For the greater part of its course the river flows through deep canyons, and along the stream itself comparatively little irrigation is practiced. In the vicinity of Dolores, however, the valley broadens, and for about 40 miles has a width of half a mile to a mile. A considerable part of this area is cultivated. In Paradox Valley also considerable land is cultivated, chiefly from small tributaries running into the main stream. By far the greater part of the Dolores River water is used for irrigation in the San Juan drainage basin, to which it is diverted by means of a tunnel and a great cut into the Montezuma Valley.

San Miguel River, the most important tributary of the Dolores, which drains an area immediately west of the headwaters of the Uncompahgre River, rises in San Miguel County, Colo., and enters the Dolores about 12 miles east of the Colorado-Utah line at an elevation of about 5,000 feet. In general the stream and its tributaries flow northeasterly. Considerable land along the San Miguel is irrigated.

The mean annual run-off of Dolores River above the mouth of the San Miguel is nearly 400,000 acre-feet, and the San Miguel furnishes at least half that amount.

Probably 600 square miles of the Dolores River basin is covered with merchantable timber and as much more is woodland. The total area of this basin is about 4,500 square miles.

The basin contains several small storage reservoir sites, a few of which have been developed, both for power and irrigation. Theoretically, by utilizing storage it would be possible to develop from 75,000 to 100,000 horsepower in the Dolores drainage basin. The river has an average fall of over 20 feet per mile throughout almost its whole course, and a great stretch of the San Miguel has an average fall of more than 50 feet to the mile. Several water-power plants are in operation along the upper San Miguel and its tributaries, the development aggregating nearly 10,000 horsepower, of which about 7,500 horsepower is developed at the Ames, Howard Fork, and Illium plants of the Telluride Power Co. One plant on Bridal Veil Creek is utilizing a head of 2,000 feet to develop 1,200 horsepower.

NORTH FORK OF GRAND RIVER NEAR GRAND LAKE, COLO.

This station, which was established July 29, 1904, is located at the highway bridge on the road between Grand Lake and Grandby, and is about 3 miles southwest of Grand Lake post office, Colo., and about 2 miles above Grand Lake Outlet, which is the most important tributary of this fork of the Grand. The nearest railroad is the Denver, Northwestern & Pacific, at Granby, Colo., distant about 12 miles.

One large ditch above the station diverts water into the headwaters of the Cache la Poudre, in the South Platte drainage basin.

Winter records at this station are more satisfactory than at the other stations on the headwaters of the Grand, as near-by springs tend to keep the stream at the gaging station more or less open.

The location and datum of the staff gage, which is at the bridge, have remained unchanged during the maintenance of the station.

Fairly good results have been obtained at this station, though low-stage measurements, because of sluggish current, are not entirely satisfactory.

The station was discontinued September 30, 1909.

Discharge measurements of North Fork of Grand River near Grand Lake, Colo., in 1909.

Date.	Hydrographer.	Width.	Area of section.	Gage height.	Discharge.
		Feet.	*Sq. ft.*	*Feet.*	*Sec.-ft.*
May 3 *a*	C. L. Chatfield	40	59	3. 60	45
Aug. 9 *b*	W. B. Freeman	44	42	3. 97	88

a Channel practically open.   *b* Made by wading at different sections.

Daily gage height, in feet, of North Fork of Grand River near Grand Lake, Colo., for 1909.

[Harry W. Carr, observer.]

Day.	Jan.	Feb.	Mar.	Apr.	May.	June.	July.	Aug.	Sept.
1		3.35	3.3	3.35		4.7	5.7	4.15	3.9
2	3.35			3.35		4.8	5.8	4.15	3.9
3	3.35	3.3	3.35	3.35		5.0	5.8	4.1	3.85
4				3.35		5.25	6.1	4.05	3.85
5	3.35	3.3	3.35			5.35	6.1	4.05	3.85
6				3.35	4.1	5.8	5.8	4.0	3.85
7	3.35	3.3	3.35			5.8	5.6	4.0	3.85
8				3.35		5.8	5.5	4.0	4.2
9	3.35	3.35	3.35		4.25	5.8	5.4	3.95	4.15
10			3.4	3.35		5.7	5.0	3.95	4.0
11		3.35		3.35		5.5	4.8	3.95	3.95
12	4.0		3.4			5.45	4.8	4.0	3.9
13		3.3		3.35		5.45	4.7	4.0	3.8
14	3.4		3.35		4.5	5.5	4.65	3.95	3.8
15		3.3		3.35	4.45	5.5	4.6	3.9	3.8
16	3.35		3.35	3.4	4.6	5.5	4.55	4.0	3.8
17	3.35	3.35			4.5	5.8	4.5	4.1	3.8
18	3.35		3.35	3.4	4.65	6.1	4.5	4.4	3.8
19		3.35			4.8	6.5	4.5	4.1	3.8
20	3.3		3.35	3.35	5.0	6.6	4.5	4.0	3.75
21		3.35	3.35		5.1	6.2	4.5	3.95	3.75
22	3.35			3.35	5.0	6.1	4.45	3.9	3.75
23	3.35	3.3	3.35		5.0	6.1	4.45	3.9	3.7
24				3.4	4.85	6.2	4.6	3.85	3.7
25	3.35	3.3	3.55		4.7	6.1	4.4	3.85	3.7
26				3.4	4.6	6.1	4.35	3.85	3.7
27	3.6	3.35	3.3		4.65	5.9	4.3	3.85	3.65
28		3.4	3.3	3.6	4.75	5.9	4.2	3.8	3.65
29	3.6			3.6	5.0	5.7	4.2	3.8	3.65
30			3.35	3.7	4.75	5.7	4.15	3.8	3.65
31	3.4				4.5		4.15	3.9	

NOTE.—Slight ice conditions from Jan. 1 to Apr. 26. Ice affected gage heights on Jan. 12, 27, and 29.

Daily discharge, in second-feet, of North Fork of Grand River near Grand Lake, Colo., for 1909.

Day.	Jan.	Feb.	Mar.	Apr.	May.	June.	July.	Aug.	Sept.
1	15	15	12	15	65	347	902	145	85
2	15	14	14	15	80	390	970	145	85
3	15	12	15	15	90	479	970	130	77
4	15	12	15	15	105	611	1,190	118	77
5	15	12	15	15	120	671	1,190	118	77
6	15	12	15	15	130	970	970	105	77
7	15	12	15	15	145	970	834	105	77
8	15	14	15	15	161	970	767	105	160
9	15	15	15	15	177	970	702	95	145
10	16	15	19	15	195	902	479	95	105
11	17	15	19	15	213	767	390	95	95
12	18	14	19	15	231	734	390	105	85
13	19	12	17	15	249	734	347	105	69
14	19	12	15	15	267	767	326	95	69
15	17	12	15	15	248	767	306	85	69
16	15	14	15	19	306	767	286	105	69
17	15	15	15	19	267	970	267	130	69
18	15	15	15	19	326	1,190	467	230	69
19	14	15	15	17	390	1,530	267	130	69
20	12	15	15	15	479	1,620	267	105	62
21	14	15	15	15	528	1,270	267	95	62
22	15	14	15	15	479	1,190	248	85	62
23	15	12	15	17	479	1,190	248	85	55
24	15	12	22	19	412	1,270	306	77	55
25	15	12	30	19	347	1,190	230	77	55
26	15	14	21	19	306	1,190	212	77	55
27	15	15	12	31	326	1,040	194	77	49
28	15	19	12	43	368	1,040	160	69	49
29	17	....	14	43	479	902	160	69	49
30	17	....	15	55	368	902	145	69	49
31	19	....	15	....	347	....	145	85	....

NOTE.—These discharges are based on rating curves applicable as follows: Jan. 1 to Apr. 26, an ice curve, not defined; Apr. 28 to Sept. 30, well defined below 1,000 second-feet. Discharges for days on which gage was not read estimated. Discharges for Jan. 12, 27, and 29 estimated because of serious ice conditions.

Monthly discharge of North Fork of Grand River near Grand Lake, Colo., for 1909.

Month.	Discharge in second-feet.			Run-off (total in acre-feet).	Accu-racy.
	Maximum.	Minimum.	Mean.		
January	19	12	15.6	969	D.
February	19	12	13.7	761	D.
March	30	12	16.0	984	C.
April	55	15	19.7	1,170	C.
May	528	65	280	17,200	B.
June	1,620	347	944	56,200	A.
July	1,190	145	465	28,600	A.
August	230	69	104	6,400	B.
September	160	49	74.3	4,420	B.
The period	....	....	....	117,000	

GRAND RIVER NEAR GRANBY, COLO.

This station, which was established June 19, 1908, is located at a highway bridge that crosses the river about 4 miles from Granby on the road to Grand Lake.

The station is about 4 miles below the junction of North and South forks, about the same distance above the mouth of Fraser River, and is above the mouth of Willow Creek. The drainage area is about 500 square miles.

No important diversions are made on the South Fork or on the main stream above the station. This basin affords some excellent storage sites. . Several filings for power development have been made above this station, but additional opportunities for filing no doubt exist. A small power plant is located on a tributary of the South Fork.

Measurements of discharge are made from a cable 300 feet downstream from the bridge.

Thick ice covers the river for about four months each year and anchor ice also occurs.

The location and datum of the gage have remained unchanged during the maintenance of the station.

Discharge measurements of Grand River near Granby, Colo., in 1909.

Date.	Hydrographer.	Width.	Area of section.	Gage height.	Discharge.
		Feet.	*Sq. ft.*	*Feet.*	*Sec.-ft.*
Mar. 10a....	C. L. Chatfield..	25	30	2.40	33
May 4........	.....do..	95	145	2.10	259
7........	.....do..	98	153	2.55	386
8........	.....do..	100	144	2.46	343
June 17.....	.....do..	111	411	4.65	2,590
Aug. 9.......	W. B. Freeman..	90	170	2.41	366
Oct. 14......	R. H. Woolsey..	85	104	1.67	122

a Ice conditions.

Daily gage height, in feet, of Grand River near Granby, Colo., for 1909.

[J. P. Switzer, observer.]

Day.	Jan.	Feb.	Mar.	Apr.	May.	June.	July.	Aug.	Sept.	Oct.	Nov.	Dec.
1...............	2.25	2.5	2.5	2.9	1.8	3.15	5.05	2.55	2.2	1,7	1.35	1,3
2...............	2.2	2.5	2.35	2.9	1.8	3.2	4.95	2.55	2.25	1`7	1.35	1`3
3...............	2.15	2.5	2.5	2.95	1.85	3.4	5.05	2.55	2.20	1 7	1.35	1 3
4...............	2.25	2.5	2.8	2.9	2.05	3.7	4.95	2.5	2.20	1 7	1.35	1 3
5...............	2.25	2.5	2.6	2.9	2.30	4.2	4.70	2.4	2.15	1 7	1.35	1 3
6...............	2.4	2.4	2.7	2.6	2.5	4.5	4.55	2.45	2.2	1 7	1.35	1 3
7...............	2.35	2.35	3.05	2.8	2.55	4.75	4.25	2.45	2.55	1 7	1.35	1 3
8...............	2.45	2.4	2.55	2.6	2.65	4.85	4.15	2.5	2.55	1 6	1.35	1 3
9...............	2.5	2.4	2.6	2.8	2.6	4.8	4.05	2.4	2.5	1 6	1.35	1 3
10...............	2.5	2.4	2.85	2.85	2.6	4.6	3.95	2.45	2.45	1 6	1.35	1 3
11...............	2.45	2.35	2.85	2.75	2.75	4.4	3.75	2.4	2.35	1.6	1.35	1.3
12...............	2.45	2.4	2.8	2.65	2.8	4.3	3.6	2.55	2.3	1.6	1.35	1.3
13...............	2.5	2.4	2.7	2.7	2.85	4.35	3.55	2.6	2.35	1.6	1.4	1.3
14...............	2.5	2.4	2.6	2.85	2.75	4.3	3.35	2.55	2.3	1.6	1.4	1.75
15...............	2.5	2.35	2.8	2.9	2.8	4.3	3.3	2.5	2.2	1.6	1.4	1.8
16...............	2.5	2.4	2.85	2.9	2.8	4.35	3.3	2.6	2.1	1.6	1.4	1.9
17...............	2.55	2.45	2.95	3.1	2.9	4.55	3.25	2.55	2.05	1.6	1.4	1.85
18...............	2.5	2.45	3.15	3.0	2.95	4.9	3.2	2.8	2.0	1.6	1.4	1.55
19...............	2.5	2.3	3.05	2.95	3.05	5.3	3.2	2.65	2.0	1.6	1.4	1.85
20...............	2.55	2.6	3.05	2.6	3.3	5.45	3.2	2.55	2.0	1.5	1.4	1.85
21...............	2.5	2.65	3.05	1.55	3.4	5.05	3.2	2.55	1.9	1.5	1.4	1.8
22...............	2.5	2.95	3.1	1.4	3.55	5.05	3.1	2.5	1.8	1.5	1.4	1.8
23...............	2.5	2.7	3.05	1.65	3.7	5.05	3.0	2..5	1.8	1.5	1.4	1.8
24...............	2.45	2.7	2.95	1.55	3.5	5.05	3.1	2.35	1.8	1.5	1.4	1.8
25...............	2.55	2.6	2.95	1.55	3.3	5.15	3.05	2.3	1.8	1.45	1.4	1.8
26...............	2.65	2.55	2.95	1.65	3.1	4.95	3.05	2.3	1.7	1.45	1.4	1.8
27...............	2.65	2.45	3.05	1.7	3.0	5.0	3.05	2.2	1.7	1.45	1.4	1.8
28...............	2.7	2.5	3.05	1.75	3.1	4.95	2.85	2.2	1.7	1.45	1.4	1.75
29...............	2.6	....	3.15	1.85	3.45	4.75	2.75	2.2	1.7	1.4	1.4	1.7
30...............	2.6	....	3.05	1.9	3.25	5.05	2.7	2.2	1.7	1.4	1.4	1.8
31...............	2.6	....	3.05	....	3.1	....	2.55	2.2	....	1.4	....	1.8

NOTE.—Ice conditions from Jan. 1 to Apr. 20 and from. Dec. 14 to 31.

Daily discharge, in second-feet, of Grand River near Granby, Colo., for 1909.

Day.	Apr.	May.	June.	July.	Aug.	Sept.	Oct.	Nov.	Dec.
1	....	157	802	3,250	430	280	132	62	54
2	....	157	840	3,070	430	298	132	62	54
3	....	170	1,000	3,250	430	280	132	62	54
4	....	230	1,290	3,070	405	280	132	62	54
5	....	317	1,870	2,630	359	263	132	62	54
6	....	405	2,310	2,390	382	280	132	62	54
7	....	430	2,720	1,940	382	430	132	62	54
8	....	482	2,890	1,800	405	430	109	62	54
9	....	455	2,800	1,680	359	405	109	62	54
10	....	455	2,470	1,560	382	382	109	62	54
11	....	540	2,160	1,340	359	338	109	62	54
12	....	570	2,010	1,190	430	317	109	62	54
13	....	600	2,080	1,140	455	338	109	70	54
14	....	540	2,010	962	430	317	109	70	....
15	....	570	2,010	920	405	280	109	70	....
16	....	570	2,080	920	455	246	109	70	....
17	....	630	2,390	880	430	230	109	70	....
18	....	662	2,980	840	570	214	109	70	....
19	....	730	3,720	840	482	214	109	70	....
20	....	920	4,000	840	430	214	88	70	....
21	98	1,000	3,250	840	430	184	88	70	....
22	70	1,140	3,250	765	405	157	88	70	....
23	120	1,290	3,250	695	382	157	88	70	....
24	98	1,100	3,250	765	338	157	88	70	....
25	98	920	3,440	730	317	157	79	70	....
26	120	765	3,070	730	317	132	79	70	....
27	132	695	3,160	730	280	132	79	70	....
28	144	765	3,070	600	280	132	79	70	....
29	170	1,050	2,720	540	280	132	70	70	....
30	184	880	3,250	510	280	132	70	70	....
31	....	765	....	430	280	....	70	....	....

NOTE.—These discharges are based on a rating curve that is well defined.
Daily discharges from Jan. 1 to Apr. 20 probably average from 33 to 50 second-feet. Mean discharge from Dec. 14 to 31 estimated as 40 second-feet.

Monthly discharge of Grand River near Granby, Colo., for 1909.

Month.	Discharge in second-feet.			Run-off (total in acre-feet).	Accuracy.
	Maximum.	Minimum.	Mean.		
May	1,290	157	644	39,600	B.
June	4,000	802	2,540	151,000	B.
July	3,250	430	1,350	83,000	A.
August	570	280	387	23,800	A.
September	430	132	250	14,900	A.
October	132	70	103	6,330	B.
November	70	62	67	3,990	B.
December	....	....	45.9	2,820	C.
The period	....	....	....	325,000	

GRAND RIVER AT SULPHUR SPRINGS, COLO.

This station was originally established July 27, 1904, at the highway bridge one-eighth mile below Sulphur Springs. On April 17, 1906, it was moved to a new highway bridge, about 1,000 feet above the old location, and a standard chain gage was installed. This gage has no determined relation to the old gage. The station was discontinued September 30, 1909. The data obtained are used to check up the results at other stations on Grand River and its tributaries and to afford a basis for estimating the flow of some of the smaller tributaries.

The Grand is joined by Fraser River about 10 miles above Sulphur Springs and by Williams Fork a few miles below. The drainage area at the station is about 950 square miles.

A number of small private ditches divert water for meadow irrigation along the principal tributaries above the station and along the Grand. A number of large diversion ditches are located in the lower drainage basin. Filings for power development are numerous from source to mouth, but it is probable that a large amount of this water would still be available for appropriation for irrigation.

The river at the station freezes across for about four months each year, the ice sometimes reaching a depth of 2 feet. No artificial control is used.

On account of unfavorable measuring conditions during the winter months at the regular section, temporary gages have been maintained in a canyon one-fourth mile below, where the river is open and where measurements can be made by wading. A temporary gage was used from January 25 to March 17, 1908, and another, 200 feet nearer the regular station, from November 25 to December 20, 1908. From April 1 to 16, 1908, an old staff gage was read by the observer instead of the chain gage. Gage heights for this period were adjusted to refer to the chain-gage datum. Beginning December 22, 1908, readings were resumed at the chain gage by measuring through a hole cut in the ice.

The accuracy of the results obtained at the old location of the station was affected to a certain extent by a bend in the river above the station, by the bridge pier, and by ice.

Discharge measurements of Grand River at Sulphur Springs, Colo., in 1909.

Date.	Hydrographer.	Width.	Area of section.	Gage height.	Discharge.
		Feet.	*Sq. ft.*	*Feet.*	*Sec.-ft.*
Mar. 10a	C. L. Chatfield	56	72	b 2.55	102
Apr. 29	.....do	103	194	2.40	543
May 8	.....do	83	278	3.20	1,080
June 17c	.....do	303	902	6.17	5,140
Aug. 8	.....do	110	226	2.70	608

a Made by wading in mouth of canyon.
b Gage height distorted by ice conditions.
c Stay line used.

Daily gage height, in feet, of Grand River at Sulphur Springs, Colo., for 1909.

[E. L. Chatfield, jr., observer.]

Day.	Jan.	Feb.	Mar.	Apr.	May.	June.	July.	Aug.	Sept.
1	2.6	2.8	2.7	2.6	2.15	4.2	6.3	3.2	2.6
2	2.55	2.75	2.7	2.3	2.2	4.3	6.3	3.2	2.6
3	2.6	2.75	2.75	2.45	2.3	4.7	6.2	3.1	2.6
4	2.6	2.75	2.7	2.45	2.45	5.2	6.1	3.05	2.6
5	2.6	2.7	2.7	2.2	2.5	5.5	6.15	3.0	2.5
6	2.65	2.7	2.65	2.4	3.1	5.9	6.0	2.9	2.5
7	2.6	2.8	2.65	2.35	3.05	6.5	5.7	2.8	2.5
8	2.55	2.8	2.65	2.25	2.95	6.6	5.35	2.8	2.5
9	2.55	2.75	2.7	2.3	3.0	6.65	5.05	2.65	2.5
10	2.6	2.8	2.65	2.25	3.4	6.6	4.85	2.7	2.4
11	2.65	2.75	2.6	2.1	3.55	6.35	4.8	2.8	2.4
12	2.75	2.75	2.6	2.0	3.6	6.2	4.75	2.9	2.5
13	3.0	2.75	2.6	1.9	3.4	6.0	4.8	2.9	2.4
14	3.05	2.8	2.6	1.95	3.4	5.8	4.85	2.9	2.4
15	3.1	2.8	2.65	2.0	3.4	5.55	3.75	2.85	2.35
16	3.1	2.75	2.65	2.0	3.4	5.55	3.7	2.95	2.4
17	3.1	2.75	2.65	2.1	3.5	6.25	3.65	3.05	2.4
18	3.1	2.75	2.65	2.1	3.7	6.35	3.65	3.2	2.4
19	3.05	2.75	2.7	2.05	3.9	6.45	3.7	3.3	2.35
20	3.05	2.8	2.75	2.2	4.2	6.45	3.6	3.2	2.4
21	3.05	2.8	2.8	2.2	4.55	6.6	3.65	3.1	2.35
22	3.05	2.75	2.9	2.2	4.55	6.5	3.6	3.0	2.3
23	3.0	2.8	3.0	2.35	4.7	6.3	3.6	2.8	2.2
24	3.0	2.8	3.0	2.35	4.8	6.35	3.55	2.7	2.25
25	2.9	2.75	3.05	2.5	4.55	6.55	3.45	2.6	2.25
26	2.9	2.75	2.95	2.5	4.0	6.9	3.4	2.6	2.2
27	2.9	2.7	2.9	2.7	3.9	6.65	3.4	2.6	2.2
28	2.9	2.7	2.75	2.7	4.1	6.5	3.4	2.65	2.2
29	2.85	....	2.7	2.5	4.7	6.2	3.4	2.45	2.2
30	2.8	....	2.65	2.6	4.4	6.25	3.3	2.45	2.3
31	2.8	....	2.65	....	4.3	....	3.3	2.7	....

NOTE.—Ice conditions from Jan. 1 to Apr. 1.

Daily discharge, in second-feet, of Grand River at Sulphur Springs, Colo., for 1909.

May.	Apr.	May.	June.	July.	Aug.	Sept.	May.	Apr.	May.	June.	July.	Aug.	Sept.
1	a 485	401	2,100	5,350	980	555	16	320	1,260	4,090	1,530	780	445
2	485	429	1,230	5,350	935	555	17	373	1,340	5,260	1,470	850	445
3	573	485	2,780	5,180	890	555	18	373	1,530	5,440	1,460	960	445
4	573	573	3,530	5,010	850	555	19	346	1,740	5,610	1,500	1,040	418
5	429	603	4,010	5,100	810	500	20	429	2,100	5,610	1,390	960	445
6	543	1,020	4,670	4,840	745	500	21	429	2,570	5,880	1,430	885	418
7	514	978	5,700	3,770	680	500	22	429	2,570	5,700	1,370	815	390
8	457	902	5,880	3,770	680	500	23	514	2,780	5,350	1,360	680	340
9	485	940	5,970	3,300	585	500	24	514	2,930	5,440	1,310	615	365
10	457	1,260	5,880	3,000	615	445	25	603	2,570	5,790	1,220	555	365
11	373	1,380	5,440	2,930	680	445	26	603	1,860	6,420	1,170	555	340
12	320	1,430	5,180	2,860	745	500	27	729	1,740	5,970	1,170	555	340
13	271	1,260	4,840	2,930	745	445	28	729	1,980	5,700	1,160	585	340
14	296	1,260	4,500	3,000	745	445	29	603	2,780	5,180	1,150	472	340
15	320	1,260	4,090	1,580	712	418	30	665	2,360	5,260	1,070	472	390
							31	....	1,230	....	1,060	615	......

a Estimated.

NOTE.—These discharges are based on rating curves applicable as follows: Apr. 2 to July 14, fairly well defined; July 15 to Aug. 5, indirect method for shifting channels used; Aug. 6 to Sept. 30, fairly well defined.

Monthly discharge of Grand River at Sulphur Springs, Colo., for 1909.

Month.	Discharge in second-feet.			Run-off (total in acre-feet).	Accuracy.
	Maximum.	Minimum.	Mean.		
April...	729	271	475	28,300	A.
May...	2,930	401	1,530	94,100	A.
June...	6,420	1,230	4,950	295,000	A.
July...	5,350	1,060	2,560	157,000	B.
August...	1,040	472	735	45,200	A.
September...	555	340	- 441	26,200	A.
The period.................................	....	....	....	646,000	

GRAND RIVER NEAR KREMMLING, COLO.

This station, which was established July 24, 1904, is located at the upper end of Gore Canyon, about 3 miles southwest of Kremmling, Colo., near the Kremmling reservoir dam site, which is the largest in Colorado.

The records obtained at this station show the water available for storage and power development and are used also to estimate probable run-off of some of the smaller tributaries between Kremmling and Sulphur Springs.

Blue River, the largest tributary above this station, empties into the Grand about 2 miles above. Other important tributaries between Kremmling and Sulphur Springs are Williams Fork and Troublesome and Muddy rivers.

A number of private ditches divert water for meadow irrigation from both the main stream and its tributaries between Sulphur Springs and this station.

The river is frozen completely across at the station for about four months each year. During this period the records are affected by the surface ice and also by anchor ice forming in the riffle just below the gage.

On October 18, 1906, the present slope gage was established on the opposite side of the river from the old chain gage. The zero of the slope gage is about 0.70 foot above the zero of the old gage. Measurements of discharge are made from a cable a few feet downstream from the gage.

Scouring during high stages and silting during low stages affect the accuracy of results. Data obtained are good at high and medium stages, but at low stages are not so accurate.

95620°—wsp 269—11——7

Discharge measurements of Grand River near Kremmling, Colo., in 1909.

Date.	Hydrographer.	Width.	Area of section.	Gage. height.	Dis-charge.
		Feet.	*Sq. ft.*	*Feet.*	*Sec.-ft.*
Mar. 12a	C. L. Chatfield	100	138	0.80	260
Apr. 29b	.....do	115	658	4.20	1,320
May 10	.....do	133	1,250	8.00	3,290
June 18	.....do	160	3,090	16.22	c13,000
19	.....do	160	3,180	16.90	c13,800
Aug. 7	W. B. Freeman	121	1,240	5.20	1,840
Oct. 10	R. H. Woolsey	105	549	2.50	749
Nov. 23	.....do	103	305	1.80	609

a Made through the ice.
b Some floating ice in river.
c Subsurface velocity method used. Coefficient taken as 95 per cent.

Daily gage height, in feet, of Grand River near Kremmling, Colo., for 1909.

[H. A. Howe, observer.]

Day.	Jan.	Feb.	Mar.	Apr.	May.	June.	July.	Aug.	Sept.	Oct.	Nov.	Dec.
1	1.05	0.75	1.1	1.4	3.35	9.85	14.8	5.25	5.25	2.8	1.5	1.5
2	1.05	.9	1.15	1.45	3.15	9.95	14.65	5.25	5.1	2.75	1.55	1.55
3	.9	.95	1.15	2.3	3.05	10.65	14.6	5.25	4.55	2.7	1.8	1.1
4	.75	.95	1.05	2.05	4.1	12.0	14.85	5.0	4.35	2.7	1.9	1.15
5	.8	1.3	1.0	1.95	5.65	13.5	15.15	4.8	4.35	2.6	1.7	.65
6	.85	1.3	1.0	1.7	6.65	14.95	14.65	4.65	4.8	2.65	1.6	.8
7	.9	1.25	.95	1.4	7.4	16.0	13.35	5.1	5.75	2.65	1.6	1.1
8	.9	1.2	.9	1.4	7.15	16.8	12.1	4.95	6.05	2.85	1.4	1.1
9	.9	1.2	.85	1.4	7.25	17.2	11.4	5.9	5.5	2.7	1.3	1.4
10	.8	1.1	.75	1.6	7.9	16.95	10.9	5.4	5.15	2.65	1.65	1.35
11	.8	1.0	.8	1.6	8.5	16.3	10.1	5.45	4.8	2.65	1.5	1.45
12	.9	1.1	.8	1.55	8.95	15.55	9.5	5.75	4.35	2.55	1.7	1.3
13	.9	1.15	.8	1.85	8.35	15.05	8.95	5.85	4.35	2.5	1.5	1.05
14	.9	1.3	.8	1.7	8.05	15.05	8.55	5.6	4.35	2.5	1.3	1.3
15	1.0	1.2	.8	2.15	8.05	14.9	8.25	5.1	4.2	2.5	1.3	1.25
16	1.05	1.15	.85	3.2	8.35	14.6	8.0	5.1	4.15	2.4	1.4	.95
17	1.05	1.15	.9	4.35	8.4	14.85	7.6	5.0	4.0	2.3	1.3	.95
18	.85	1.1	.9	4.65	8.8	15.8	7.45	7.4	3.9	2.3	.9	.9
19	.8	1.05	.9	3.35	9.2	16.95	7.5	7.45	3.6	2.25	1.15	1.0
20	.8	1.0	.85	3.25	10.0	18.0	7.6	6.2	3.6	2.2	1.75	1.3
21	.8	1.0	.9	2.85	10.8	18.0	7.7	5.6	3.45	2.1	1.95	1.8
22	.8	.9	.9	2.35	11.15	17.2	7.2	5.4	3.5	1.95	1.95	2.15
23	.9	1.0	1.0	2.25	11.65	16.7	7.3	4.95	3.5	2.0	1.8	2.1
24	.8	1.0	1.05	2.7	11.6	16.45	7.75	4.9	3.4	1.85	2.1	2.05
25	.8	1.0	1.1	2.7	10.9	16.65	8.1	5.05	3.3	1.7	1.8	2.4
26	.75	1.05	1.0	3.4	9.7	16.55	7.15	4.7	3.3	1.75	1.75	2.35
27	.8	1.1	1.0	4.7	9.5	16.0	6.9	4.25	3.15	1.8	1.65	2.45
28	.7	1.1	1.4	4.7	10.15	15.6	6.6	4.2	3.0	1.75	1.35	2.45
29	.7	....	1.0	4.85	10.75	15.15	6.05	4.3	2.9	1.75	1.45	2.5
30	.7	....	1.05	4.05	11.2	14.9	5.8	4.15	2.85	1.75	1.45	2.6
31	.6	....	1.4	....	10.0	....	5.55	4.7	....	1.55	....	2.2

NOTE.—Ice conditions from Jan. 1 to Mar. 31 and from Dec. 20 to 31.

Daily discharge, in second-feet, of Grand River near Kremmling, Colo., for 1909.

Day.	Apr.	May.	June.	July.	Aug.	Sept.	Oct.	Nov.	Dec.
1............................	490	1,080	4,960	11,000	1,860	1,860	890	515	515
2............................	502	1,010	5,050	10,800	1,860	1,800	875	528	528
3............................	740	978	5,740	10,700	1,860	1,550	860	590	415
4............................	665	1,360	7,280	11,000	1,750	1,460	860	620	428
5............................	535	2,060	9,220	11,500	1,660	1,460	830	565	302
6............................	565	2,620	11,200	10,800	1,590	1,660	845	540	340
7............................	490	3,080	12,600	9,020	1,800	2,120	845	540	415
8............................	490	2,920	13,700	7,400	1,730	2,280	908	490	415
9............................	490	2,990	14,200	6,570	2,200	1,990	860	465	490
10............................	540	3,410	13,900	6,010	1,940	1,820	845	552	478
11............................	540	3,840	13,000	5,190	1,960	1,660	845	515	502
12............................	528	4,180	12,000	4,640	2,120	1,460	815	565	465
13............................	605	3,720	11,300	4,180	2,170	1,460	800	515	402
14............................	565	3,510	11,300	3,870	2,040	1,460	800	465	465
15............................	695	3,510	11,100	3,650	1,800	1,400	800	465	452
16............................	1,030	3,720	10,700	3,480	1,800	1,380	770	490	378
17............................	1,460	3,760	11,000	3,220	1,750	1,320	740	465	378
18............................	1,590	4,060	12,300	3,120	3,080	1,280	740	365	365
19............................	1,080	4,380	13,900	3,150	3,120	1,260	725	428	390
20............................	1,050	5,100	15,300	3,220	2,360	1,170	710	578	465
21............................	908	5,900	15,300	3,280	2,040	1,120	680	635	450
22............................	755	6,280	14,200	2,960	1,940	1,140	635	635	450
23............................	725	6,850	13,500	3,020	1,730	1,140	650	590	450
24............................	800	6,800	13,200	3,310	1,700	1,100	605	680	450
25............................	860	6,010	13,500	3,540	1,770	1,060	565	590	450
26............................	1,100	4,820	13,300	2,920	1,620	1,060	578	578	450
27............................	1,320	4,640	12,600	2,760	1,420	1,010	590	552	450
28............................	1,620	5,240	12,100	2,580	1,400	960	578	478	450
29............................	1,680	5,850	11,500	2,280	1,440	925	578	502	450
30............................	1,340	6,340	11,100	2,140	1,380	908	552	502	450
31............................	....	5,100	....	2,020	1,620	....	528	....	450

NOTE.—These discharges are based on a rating curve that is well defined above 500 second-feet. Daily discharge for ice period from Dec. 20 to 31, estimated.

Monthly discharge of Grand River near Kremmling, Colo., for 1909.

[Drainage area, 2,380 square miles.]

Month.	Discharge in second-feet.				Run-off.		Accuracy.
	Maximum.	Minimum.	Mean.	Per square mile.	Depth in inches on drainage area.	Total in acre-feet.	
April.......................	1,680	490	864	0.363	0.40	51,400	B.
May........................	6,860	978	4,040	1.70	1.96	248,000	A.
June.......................	15,300	4,960	11,700	4.92	5.49	696,000	A.
July.......................	11,500	2,020	5,270	2.21	2.55	324,000	A.
August....................	3,120	1,380	1,890	.794	.92	116,000	A.
September.................	2,280	908	1,410	.592	.66	83,900	A.
October...................	908	528	739	.311	.36	45,400	A.
November.................	680	365	533	.224	.25	31,700	A.
December.................	515	302	437	.184	.21	27,000	B.
The period.............	....	....	....	....	....	1,620,000	

GRAND RIVER AT GLENWOOD SPRINGS, COLO.

This station, which is located at Glenwood Springs, was established May 12, 1899, discontinued July 17, and reestablished January 7, 1900. The position and datum of the gage have remained unchanged. A float gage is placed on the right bank of the river about one-fourth of a mile above the State bridge. Discharge measurements are made from a cable underneath this bridge.

This may be considered a base station, as the records show the entire run-off of the basin above the mouth of Roaring Fork. Ice never forms at this station, as the hot water from the near-by springs keeps the water above the freezing point even in the most severe weather. The winter records are, therefore, of especial value, as they furnish a basis for estimating approximately the discharge of the streams at other stations in the basin during the ice period.

The station is about one-fourth mile above Roaring Fork, which is the third largest tributary of the Grand.

A few minor irrigation ditches are taken out between this and the Kremmling station, but do not affect the discharge to any appreciable extent. The Shoshone plant of the Central Colorado Power Co., having a head of 170 feet, was practically completed in 1908. The tail water from this plant is returned to the river above the gaging station.

Results at this station are satisfactory. The channel has always been permanent except during 1907 and 1908, when the débris from the Shoshone plant was deposited in the river bed, thereby changing the rating curve of the stream.

Discharge measurements of Grand River at Glenwood Springs, Colo., in 1909.

Date.	Hydrographer.	Width.	Area of section.	Gage height.	Dis-charge.
		Feet.	*Sq. ft.*	*Feet.*	*Sec.-ft.*
Jan. 13	C. L. Chatfield..................................	180	407	3.78	628
Mar. 16	.....do......................................	185	464	4.04	788
Apr. 20	.....do......................................	195	710	5.28	1,920
June 21a	Chatfield and Matthes..........................	190	2,230	12.00	27,300
30	W.B. Freeman.................................	222	1,900	9.45	18,500
July 13	W. H. Suelson, jr.............................	220	1,300	7.12	7,530
24	W. B. Freeman...............................	210	1,080	6.50	5,050
Aug. 5	W. H. Snelson, jr.............................	205	843	5.10	2,640
Oct. 9	G. H. Russell................................	190	649	4.20	1,630

a Measurement from foot bridge a few feet upstream from the cable section.

Daily gage height, in feet, of Grand River at Glenwood Springs, Colo., for 1909.

[W. H. Richardson, observer.]

Day.	Jan.	Feb.	Mar.	Apr.	May.	June.	July.	Aug.	Sepr.	Oct.	Nov.	Dec.
1	3.9	3.7	3.75	4.3	5.5	7.4	9.35	5.35	5.15	4.3	3.75	3.6
2	3.9	3.7	3.8	4.3	5.3	7.3	9.25	5.3	5.25	4.2	3.7	3.65
3	3.9	3.7	3.9	4.4	5.2	7.65	9.2	5.25	5.1	4.25	3.7	3.7
4	3.9	3.8	3.95	4.5	5.2	8.35	9.2	5.2	5.0	4.2	3.8	3.5
5	3.95	3.9	4.35	4.5	5.7	9.3	9.35	5.1	4.95	4.2	3.8	3.15
6	4.4	3.9	4.2	4.6	6.3	10.0	9.1	5.05	5.0	4.2	3.8	3.1
7	4.3	3.9	4.2	4.45	6.5	10.8	8.6	5.1	5.25	4.15	3.8	3.2
8	4.25	3.9	4.15	4.2	6.7	10.95	8.4	5.15	5.6	4.3	3.8	3.05
9	4.15	3.9	4.1	4.2	6.6	11.2	8.0	5.1	5.5	4.2	3.7	3.35
1).	4.2	3.9	4.05	4.2	6.75	11.2	7.8	5.5	5.3	4.2	3.7	3.6
11	4.1	3.9	4.0	4.2	7.0	10.75	7.45	5.3	5.1	4.15	3.75	3.4
1?	3.8	4.0	3.9	4.4	.7.0	10.25	7.2	5.35	5.1	4.2	3.8	3.4
13	3.8	4.0	3.9	4.4	7.0	10.4	7.1	5.45	4.95	4.1	3.8	3.4
14	3.9	4.0	3.9	4.3	6.9	10.4	6.9	5.45	4.95	4.1	3.8	3.4
15	4.35	4.0	4.0	4.3	6.9	10.3	6.8	5.3	4.9	4.1	3.6	3.35
16	4.25	3.9	4.0	4.6	6.95	10.1	6.7	5.15	4.9	4.1	3.6	3.2
17	4.1	3.95	4.1	5.0	7.0	10.25	6.6	5.15	4.85	4.05	3.4	3.15
18	4.05	3.95	4.15	5.5	7.0	10.7	6.5	5.6	4.85	4.0	3.4	3.05
19	4.05	3.95	4.15	5.7	7.25	11.3	6.5	6.1	4.7	4.0	3.5	2.9
20	4.1	3.85	4.2	5.5	7.5	11.85	6.55	6.05	4.75	4.0	3.7	2.9
21	4.1	3.85	4.25	5.3	7.7	12.0	6.5	5.55	4.7	4.0	3.8	2.9
22	4.1	3.8	4.25	5.1	8.0	11.65	6.5	5.4	4.6	4.0	3.85	3.2
23	4.3	3.8	4.4	5.0	8.3	11.3	6.4	5.2	4.55	3.9	3.8	3.25
24	4.05	3.75	4.3	4.9	8.3	10.65	6.55	5.05	4.5	3.9	3.85	3.3
25	3.9	3.8	4.3	4.95	8.0	10.55	6.7	5.15	4.5	3.9	3.8	3.3
26	3.9	3.8	4.3	5.05	7.5	10.4	6.55	5.1	4.45	3.9	3.8	3.25
27	3.8	3.7	4.3	5.3	7.5	10.25	6.25	4.9	4.4	3.75	3.7	3.3
2?	3.85	3.7	4.35	5.6	7.7	9.9	5.95	4.9	4.4	3.85	3.7	3.4
29	3.7	....	4.4	5.8	8.05	9.6	5.7	4.9	4.35	3.85	3.6	3.3
30	3.65	....	4.4	5.75	8.2	9.5	5.6	4.9	4.35	3.8	3.6	3.3
31	3.6	....	4.3	....	7.7	....	5.45	4.95	....	3.8	....	3.35

Daily discharge, in second-feet, of Grand River at Glenwood Springs, Colo., for 1909.

Day.	Jan.	Feb.	Mar.	Apr.	May.	June.	July.	Aug.	Sept.	Oct.	Nov.	Dec.
1	690	575	602	980	2,440	7,500	18,100	2,970	2,800	1,750	1,140	1,000
2	690	575	630	980	2,150	7,250	17,400	2,890	2,970	1,630	1,090	1,040
3	690	575	690	1,070	2,010	8,470	17,000	2,820	2,730	1,690	1,090	1,090
4	690	630	720	1,170	2,010	11,300	17,000	2,740	2,580	1,630	1,190	920
5	720	690	1,020	1,170	2,750	17,600	18,100	2,590	2,500	1,630	1,190	672
6	1,070	690	900	1,270	3,810	22,600	16,300	2,520	2,580	1,630	1,190	640
7	980	690	900	1,120	4,220	28,300	13,200	2,590	3,020	1,570	1,190	705
8	940	690	860	900	4,650	29,400	12,100	2,670	3,660	1,750	1,190	610
9	860	690	820	900	4,430	31,200	10,200	2,590	3,470	1,630	1,090	810
10	900	690	785	900	4,760	31,200	9,380	3,240	3,100	1,630	1,090	1,000
11	820	690	750	900	5,360	28,000	8,080	2,900	2,780	1,570	1,140	845
12	630	750	690	1,070	5,360	24,400	7,260	2,290	2,780	1,630	1,190	845
13	630	750	690	1,070	5,360	25,400	6,960	3,150	2,550	1,510	1,190	845
14	690	750	690	980	5,120	25,400	6,390	3,210	2,550	1,510	1,190	845
15	1,020	750	750	980	5,120	24,700	6,110	2,960	2,530	1,510	1,000	810
16	940	690	750	1,270	5,240	23,300	5,840	2,720	2,530	1,510	1,000	705
17	820	720	820	1,740	5,360	24,400	5,580	2,720	2,450	1,460	845	672
18	785	720	860	2,440	5,360	27,600	5,330	3,500	2,450	1,400	845	610
19	785	720	860	2,750	6,000	32,000	5,330	4,520	2,250	1,400	920	520
20	820	660	900	2,440	6,710	36,100	5,460	4,400	2,310	1,400	1,090	520
21	820	660	940	2,150	7,000	37,200	5,330	3,390	2,250	1,400	1,190	520
22	820	630	940	1,870	8,100	34,600	5,330	3,180	2,100	1,400	1,240	705
23	980	630	1,070	1,740	9,350	32,000	5,090	2,850	2,070	1,290	1,190	740
24	785	602	980	1,610	9,500	27,200	5,460	2,610	2,000	1,290	1,240	775
25	690	630	980	1,680	8,500	26,500	5,840	2,760	2,000	1,290	1,190	775
26	690	630	980	1,800	7,000	25,400	5,460	2,700	1,940	1,290	1,190	740
27	630	575	980	2,150	7,150	24,400	4,750	2,400	1,870	1,140	1,090	775
28	660	575	1,020	2,590	7,900	21,900	4,100	2,400	1,870	1,240	1,090	845
29	575	......	1,070	2,910	9,350	19,800	3,590	2,410	1,800	1,240	1,000	775
30	550	......	1,070	2,830	10,200	19,800	3,400	2,420	1,800	1,800	1,000	775
31	525	......	980	......	8,350	......	3,140	2,500	......	1,190	......	810

NOTE.—These discharges are based on rating curves applicable as follows: Jan. 1 to May 20, not well defined; May 21 to June 5 and Aug. 6 to Sept. 30, indirect method for shifting channels used; June 6 to Aug. 5, fairly well defined above discharge of 3,200 second-feet; Oct. 1 to Dec. 31, not well defined.

Monthly discharge of Grand River at Glenwood Springs, Colo., for 1909.

[Drainage area, 4,520 square miles.]

Month.	Discharge in second-feet.				Run-off.		Accu-racy.
	Maximum.	Minimum.	Mean.	Per square mile.	Depth in inches on drainage area.	Total in acre-feet.	
January...................	1,070	525	771	0.171	0.20	47,400	A.
February..................	750	575	665	.147	.15	36,900	A.
March....................	1,070	602	861	.190	.22	52,900	A.
April.....................	2,910	900	1,580	.350	.39	94,000	B.
May......................	10,200	2,010	5,830	1.29	1.49	358,000	B.
June.....................	37,200	7,250	24,500	5.42	6.05	1,460,000	C.
July.....................	18,100	3,140	8,470	1.87	2.16	521,000	B.
August...................	4,520	2,400	2,910	.644	.74	179,000	B.
September................	3,660	1,800	2,480	.549	.61	148,000	B.
October..................	1,750	1,140	1,460	.323	.37	89,800	B.
November................	1,240	845	1,110	.246	.27	66,000	B.
December................	1,090	520	772	.171	.20	47,500	C.
The year............	37,200	520	4,280	.948	12.85	3,100,000	

GRAND RIVER NEAR PALISADES, COLO.[1]

This station was established April 9, 1902, at a steel highway bridge 2 miles above Palisades, Colo., at a point where the river enters Grand Valley.

The station is below all important tributaries except Gunnison and Dolores rivers, and is above all the irrigating ditches supplying water for irrigation in Grand Valley, excepting a ditch for one pumping plant which diverts about 80 second-feet for irrigation one-fourth mile above the gage. The proposed high-line canal of the United States Reclamation Service will take its water about 7 miles above Palisades.

The gage has been permanently located at the highway bridge above Palisades and no change in datum has occurred. The original wire gage was replaced by a chain gage on April 5, 1904. The river usually freezes over a portion of the year, but except for the interference of slush ice and an occasional thin ice cover the winter records are good.

Measuring conditions at the gage are poor, especially at high water. Beginning September 27, 1905, the measurements were made from a suspension bridge in the town of Palisades. Measurements are now made from the new steel bridge opened in the spring of 1909. The measuring conditions at both of these bridges are about the same. The sections are permanent, but at flood stages the velocities are high and the interference of bridge piers vitiates the results somewhat. Flood measurements prior to 1906 made at the upper bridge where the gage is located are less reliable than those made at the lower bridges.

---

[1] Called "at Palisades" in Water-Supply Paper 249.

Discharge measurements of Grand River near Palisades, Colo., in 1909.

Date.	Hydrographer.	Width.	Area of section.	Gage height.	Dis-charge.
		Feet.	*Sq. ft.*	*Feet.*	*Sec.-ft.*
May 6	S. O. Harper...	315	1,530	15.65	7,750
8	....do...	321	1,850	16.45	9,690
11	D. L. Henderson.......................................	325	2,130	17.40	11,800
19 a	....do...	366	2,180	17.65	14,100
22 a	S. O. Harper...	381	2,560	18.75	18,300
June 1 a	D. L. Henderson.....................................	370	2,24ᴦ	18.05	15,400
5 a	....do...	390	3,030	20.15	26,600
8 a	S. O. Harper...	400	3,860	22.05	37,500
11 a	D. L. Henderson.....................................	380	4,150	22.35	35,600
21 a	....do...	380	4,390	23.15	44,800
29 a	Freeman, Harper, and Henderson......................	383	3,820	21.45	29,300
July 28 a	D. L. Henderson.....................................	330	1,820	15.70	7,300
Aug. 5 a	....do...	300	1,410	14.40	3,950
Sept. 9	S. O. Harper...	254	612	12.50	1,740
Oct. 13	G. H. Russell...	300	1,020	13.45	2,810

a Subsurface velocity method used. Coefficients used, from 8 to 87 per cent.

Daily gage height, in feet, of Grand River near Palisades, Colo., for 1909.

[J. J. Morrow and Alex. Gowdey, observers.]

Day.	Jan.	Feb.	Mar.	Apr.	May.	June.	July.	Aug.	Sept.	Oct.	Nov.	Dec.
1...............	14.7	13.15	12.05	12.45	14.35	18.1	20.95	14.85	14.4	13.6	12.9	12.9
2...............	14.3	13.1	12.3	12.45	14.1	17.85	20.65	14.65	14.75	13.55	12.9	12.8
3...............	14.4	13.1	12.45	12.55	14.0	18.25	20.65	14.7	14.65	13.5	12.9	12.95
4...............	14.45	13.1	12.8	12.7	14.1	19.2	20.85	14.6	14.5	13.4	12.9	12.75
5...............	14.45	13.1	13.1	12.95	14.7	20.35	20.8	14.4	14.8	13.4	12.9	12.7
6...............	14.25	13.1	13.05	12.9	15.75	21.5	20.65	14.3	15.25	13.4	12.9	12.7
7...............	13.9	12.95	13.05	12.65	16.2	22.05	20.25	14.3	15.2	13.5	12.9	12.6
8...............	13.85	12.85	12.55	12.55	16.5	22.65	19.4	14.4	15.6	13.5	14.9	12.6
9...............	13.4	12.7	12.3	12.4	16.0	22.7	18.95	14.45	15.65	13.6	12.9	12.6
10...............	13.45	12.7	12.3	12.4	16.8	22.55	18.3	14.7	15.2	13.55	12.9	12.6
11...............	13.2	12.4	12.25	12.45	17.4	22.2	17.95	14.75	15.0	13.4	12.9	12.6
12...............	13.2	12.35	12.15	12.55	17.5	21.85	17.6	14.75	15.1	13.4	12.9	12.6
13...............	12.7	12.55	12.15	12.55	17.5	21.75	17.35	14.75	14.85	13.45	12.9	12.6
14...............	13.1	12.25	12.35	12.55	17.15	21.65	16.9	14.8	14.8	13.45	12.9	12.6
15...............	13.5	12.2	12.4	12.55	16.95	21.5	16.7	14.65	14.65	13.3	12.9	12.5
16...............	13.8	12.15	12.45	12.75	16.95	21.4	16.65	14.5	14.5	13.35	12.9	12.5
17...............	13.9	12.25	12.55	13.2	17.0	21.6	16.5	14.6	14.55	13.3	12.9	12.5
18...............	13.0	12.2	12.55	13.85	17.3	21.8	16.2	14.8	14.5	13.3	12.9	12.5
19...............	13.2	12.2	12.55	14.7	17.85	22.55	16.1	15.5	14.4	13.25	12.9	12.5
20...............	13.15	12.2	12.6	14.9	18.3	23.15	16.2	15.55	14.25	13.05	12.9	12.5
21...............	13.05	12.15	12.6	14.2	18.55	23.05	16.1	15.6	14.15	13.15	12.9	12.5
22...............	14.45	12.1	12.5	13.85	18.95	23.05	15.95	15.1	14.0	13.2	13.0	12.5
23...............	13.55	12.1	12.6	13.55	19.15	22.65	15.85	14.8	13.95	13.15	13.0	12.5
24...............	12.6	12.1	12.65	13.45	19.3	22.5	16.05	14.6	13.9	13.1	13.0	12.5
25...............	12.25	12.05	12.5	13.4	18.95	22.3	16.3	14.55	13.9	13.0	13.0	12.5
26...............	12.15	12.05	12.45	13.6	18.55	22.2	16.2	14.55	13.8	13.0	13.3	12.5
27...............	12.15	12.05	12.5	14.05	18.5	21.8	16.0	14.4	13.8	13.0	13.3	12.5
28...............	12.15	12.05	12.5	14.4	18.8	21.8	15.7	14.25	13.75	13.0	13.0	12.5
29...............	12.1	....	12.55	14.65	19.3	21.2	15.5	14.3	13.7	13.0	13.05	12.5
30...............	12.15	....	12.6	14.5	19.1	20.9	15.25	14.25	13.6	13.0	13.05	12.5
31...............	12.7	....	12.6	....	18.6	....	14.95	14.5	....	13.0	....	12.5

NOTE.—Ice conditions from Jan. 1 to 24 and Jan. 31 to Feb. 15.

Daily discharge, in second-feet, of Grand River near Palisades, Colo., for 1909.

Day.	Jan.	Feb.	Mar.	Apr.	May.	June.	July.	Aug.	Sept.	Oct.	Nov.	Dec.
1	1,200	1,200	1,120	1,480	4,500	15,600	29,300	5,570	4,600	3,070	2,010	2,010
2	1,200	1,200	1,320	1,480	3,990	14,600	27,700	5,130	5,350	2,980	2,010	1,880
3	1,200	1,200	1,480	1,580	3,800	16,200	27,700	5,240	5,130	2,900	2,010	2,080
4	1,300	1,200	1,880	1,760	3,990	20,300	28,800	5,020	4,810	2,730	2,010	1,820
5	1,300	1,200	2,280	2,080	5,240	26,000	28,500	4,600	5,460	2,730	2,010	1,760
6	1,500	1,200	2,210	2,010	7,660	32,500	27,700	4,390	6,460	2,730	2,010	1,760
7	1,500	1,200	2,210	1,700	8,880	35,800	25,500	4,390	6,350	2,900	2,010	1,640
8	1,500	1,200	1,580	1,580	9,810	39,700	21,300	4,600	7,290	2,900	2,010	1,640
9	1,500	1,200	1,320	1,420	10,100	40,000	19,200	4,700	7,410	3,070	2,010	1,640
10	1,500	1,200	1,320	1,420	10,800	39,000	16,400	5,240	6,350	2,980	2,010	1,640
11	1,400	1,250	1,280	1,480	12,900	36,800	15,000	5,350	5,900	2,730	2,010	1,640
12	1,200	1,250	1,190	1,580	13,200	34,600	13,600	5,350	6,120	2,730	2,010	1,640
13	1,300	1,250	1,190	1,580	13,200	34,000	12,700	5,350	5,570	2,820	2,010	1,640
14	1,400	1,250	1,370	1,580	12,000	33,400	11,100	5,460	5,460	2,570	2,010	1,640
15	1,500	1,200	1,420	1,580	11,300	32,500	10,400	5,130	5,130	2,570	2,010	1,530
16	1,450	1,190	1,480	1,820	11,300	31,900	10,300	4,810	4,810	2,650	2,010	1,530
17	1,400	1,280	1,580	2,420	11,400	33,100	9,810	5,020	4,920	2,570	2,010	1,530
18	1,400	1,230	1,580	3,520	12,500	34,300	8,880	5,460	4,810	2,570	2,010	1,530
19	1,400	1,230	1,580	5,240	14,600	39,000	8,590	7,050	4,600	2,500	2,010	1,530
20	1,400	1,230	1,640	5,680	16,400	43,000	8,880	7,170	4,290	2,210	2,010	1,530
21	1,400	1,190	1,640	4,190	17,400	42,300	8,590	7,290	4,090	2,350	2,010	1,530
22	1,400	1,150	1,530	3,520	19,200	42,300	8,180	6,120	3,800	2,420	2,140	1,530
23	1,400	1,150	1,640	2,980	20,100	39,700	7,910	5,460	3,700	2,350	2,140	1,530
24	1,300	1,150	1,700	2,820	20,800	38,700	8,450	5,020	3,610	2,280	2,140	1,530
25	1,280	1,120	1,530	2,730	19,200	37,400	9,180	4,920	3,610	2,140	2,140	1,530
26	1,190	1,120	1,480	3,070	17,400	36,800	8,880	4,920	3,430	2,140	2,570	1,530
27	1,190	1,120	1,530	3,900	17,200	34,300	8,310	4,600	3,430	2,140	2,570	1,530
28	1,190	1,120	1,530	4,600	18,500	34,300	7,530	4,290	3,340	2,140	2,140	1,530
29	1,150	....	1,580	5,130	20,800	30,700	7,050	3,990	3,250	2,140	2,210	1,530
30	1,190	....	1,640	4,820	19,900	29,000	6,460	4,290	3,070	2,140	2,210	1,530
31	1,200	....	1,640	....	17,600	....	5,790	4,810	....	2,140	....	1,530

NOTE.—These discharges, except for the ice periods, are based on a rating curve that is well defined above 1,200 second-feet.

Daily discharges Jan. 1 to 24 and Jan. 31 to Feb. 15 estimated from a general comparison with Grand and Roaring Fork at Glenwood Springs, considering also the open-water periods of January and February.

Monthly discharge of Grand River near Palisades, Colo., for 1909.

[Drainage area, 8,550 square miles.]

Month.	Discharge in second-feet.				Run-off.		
	Maximum.	Minimum.	Mean.	Per square mile.	Depth in inches on drainage area.	Total in acre-feet.	Accuracy.
January	....	....	1,340	0.157	0.18	82,400	D.
February	....	....	1,200	.140	.15	66,600	D.
March	2,280	1,120	1,560	.182	.21	95,900	C.
April	5,680	1,420	2,690	.315	.35	160,000	C.
May	20,800	3,800	13,100	1.53	1.76	806,000	B.
June	43,000	14,600	33,300	3.89	4.34	1,980,000	B.
July	29,300	5,790	14,400	1.68	1.94	885,000	B.
August	7,290	3,990	5,190	.607	.70	319,000	B.
September	7,410	3,070	4,870	.570	.64	290,000	B.
October	3,070	2,140	2,570	.301	.35	158,000	A.
November	2,570	2,010	2,080	.243	.27	124,000	B.
December	2,080	1,530	1,630	.191	.22	100,000	C.
The year	....	....	7,000	.817	11.11	5,100,000	

NORTH INLET TO GRAND LAKE AT GRAND LAKE, COLO.

Two streams, known as the North and East Inlets, flow into Grand Lake, North Inlet being the larger.

The gaging station on the North Inlet, which was established August 3, 1905, is located at the footbridge which crosses the stream about 100 yards north of the mouth and 300 yards east of the Grand Lake post office.

The approximate elevation of this basin is 8,000 to 11,000 feet above sea level, and the fall of the stream is very great.

No important tributaries enter above the station, which therefore gives results for the whole drainage area, measuring 36.5 square miles.

No water is diverted above the station, but filings have already been made for power development.

The stream is covered with thick ice for about four months, and winter gage readings are therefore of little value.

No change has been made in the datum of the staff gage at the bridge during the maintenance of the station.

The accuracy of the records is affected by the roughness of the stream bed and by ice. The records are fragmentary, as the gage has not been read continuously. No gage records were obtained in 1909.

The station was discontinued September 30, 1909.

Discharge measurements of North Inlet to Grand Lake at Grand Lake, Colo., in 1909.

Date.	Hydrographer.	Width.	Area of section.	Gage height.	Dis- charge.
		Feet.	*Sq. ft.*	*Feet.*	*Sec.-ft.*
May 3a	C. L. Chatfield...	14	6.4	3.00	8.5
Aug.10b	W. B. Freeman...	50	....	2.51	69.

a Ice at station. Measurement made by wading 100 feet below footbridge.
b Made by wading about 30 feet upstream from gauge.

GRAND LAKE OUTLET AT GRAND LAKE, COLO.

This station, which was established July 31, 1904, and discontinued September 30, 1909, is located at a footbridge at the west end of Grand Lake, about half a mile south of Grand Lake post office, Colo., in sec. 6, T. 3 N., R. 75 W. Granby, about 15 miles distant, on the Denver, Northwestern & Pacific Railroad, is the nearest railroad point.

The records of the station show the available storage of Grand Lake and the total run-off from the North and East inlets, and are valuable for use in connection with power development. The drainage area at the station is 62 square miles.

Shore ice forms at the station for about four months, but the stream does not freeze over because of the higher temperature of the water coming out of Grand Lake.

The location and datum of the staff gage at the footbridge have remained unchanged during the maintenance of the station. Measurements have been taken at various stations, more usually at a ford one-fourth mile downstream from footbridge, where a tag wire has been stretched across the stream 50 feet above the ford.

During low stages the rough bottom and sluggish character of the stream affect the accuracy of the results to a considerable extent.

Discharge measurements of Grand Lake Outlet at Grand Lake, Colo., in 1909.

Date.	Hydrographer.	Width.	Area of section.	Gage height.	Discharge.
		Feet.	*Sq. ft.*	*Feet.*	*Sec.-ft.*
May 3a	C. L. Chatfield...	17	12	1.60	20
Aug. 10b	W. B. Freeman	95	118	2.20	120

a Measurement made by wading. Not at regular section.
b Measurement made by wading at regular section.

Daily gage height, in feet, of Grand Lake Outlet at Grand Lake, Colo., for 1909.

[M. Wescott, observer.]

Day.	Jan.	Feb.	Mar.	Apr.	May.	June.	July.	Aug.	Sept.
1....................	1.35	1.4	1.3	1.35	1.5	2.4	4.3	2.35	2.1
2....................	1.35	1.4	1.3	1.35	1.55	2.4	4.3	2.3	2.1
3....................	1.35	1.4	1.3	1.35	1.55	2.55	4.3	2.3	2.05
4....................	1.35	1.4	1.3	1.35	1.55	3.0	4.25	2.3	2.05
5....................	1.35	1.4	1.3	1.35	1.6	3.4	4.45	2.25	2.05
6....................	1.35	1.35	1.3	1.35	1.7	3.8	4.1	2.2	2.05
7....................	1.4	1.35	1.3	1.35	1.8	3.9	3.85	2.2	2.1
8....................	1.4	1.35	1.3	1.35	1.85	4.05	3.7	2.2	2.35
9....................	1.4	1.35	1.3	1.35	1.9	4.0	3.5	2.2	2.4
10....................	1.4	1.35	1.25	1.35	2.0	3.9	3.4	2.2	2.3
11....................	1.35	1.35	1.25	1.35	2.0	3.75	3.3	2.3	2.2
12....................	1.35	1.35	1.25	1.35	2.05	3.5	3.2	2.3	2.1
13....................	1.35	1.35	1.25	1.35	2.05	3.5	3.1	2.35	2.1'
14....................	1.35	1.35	1.25	1.35	2.05	3.6	2.95	2.35	2.05
15....................	1.4	1.4	1.25	1.35	2.05	3.6	2.9	2.3	2.05
16....................	1.4	1.4	1.25	1.35	2.0	3.7	2.8	2.2	2.05
17....................	1.4	1.4	1.25	1.4	2.0	3.95	2.8	2.2	2.0
18....................	1.4	1.4	1.25	1.4	2.1	4.3	2.8	2.4	1.95
19....................	1.4	1.4	1.25	1.45	2.2	4.6	2.8	2.4	1.95
20....................	1.4	1.4	1.25	1.45	2.4	4.6	2.8	2.3	1.9
21....................	1.4	1.4	1.25	1.45	2.65	4.25	2.8	2.2	1.85
22....................	1.45	1.35	1.25	1.45	2.8	4.2	2.8	2.2	1.85
23....................	1.45	1.35	1.3	1.45	2.8	4.15	2.75	2.1	1.85
24....................	1.45	1.35	1.35	1.45	2.8	4.3	2.8	2.2	1.8
25....................	1.45	1.35	1.35	1.45	2.6	4.4	2.8	2.2	1.8
26....................	1.45	1.35	1.35	1.45	2.45	4.25	2.75	2.2	1.8
27....................	1.45	1.35	1.35	1.45	2.4	4.3	2.7	2.1	1.75
28....................	1.4	1.35	1.35	1.5	2.4	4.2	2.6	2.1	1.75
29....................	1.4	....	1.35	1.5	2.6	4.3	2.5	2.1	1.75
30....................	1.4	....	1.35	1.5	2.6	4.2	2.45	2.05	1.75
31....................	1.4	....	1.35	....	2.5	....	2.4	2.05	....

NOTE.—Although Grand Lake was frozen over it is probable that the above gage heights during the winter months are not materially affected by ice.

Daily discharge, in second-feet, of Grand Lake Outlet at Grand Lake, Colo., for 1909.

Day.	Jan.	Feb.	Mar.	Apr.	May.	June.	July.	Aug.	Sept.
1	9	11	7.5	9	16	170	1,050	156	97
2	9	11	7.5	9	19	170	1,050	143	97
3	9	11	7.5	9	19	214	1,050	143	87
4	9	11	7.5	9	19	372	1,020	143	87
5	9	11	7.5	9	22	550	1,140	131	87
6	9	9	7.5	9	32	762	931	119	87
7	11	9	7.5	9	44	818	790	119	97
8	11	9	7.5	9	52	902	706	119	156
9	11	9	7.5	9	60	874	600	119	170
10	11	9	6.0	9	77	818	550	119	143
11	9	9	6.0	9	77	734	502	143	119
12	9	9	6.0	9	87	600	456	143	97
13	9	9	6.0	9	87	600	413	156	97
14	9	9	6.0	9	87	652	352	156	87
15	11	11	6.0	9	87	652	333	143	87
16	11	11	6.0	9	77	706	297	119	87
17	11	11	6.0	11	77	846	297	119	77
18	11	11	6.0	11	97	1,050	297	170	68
19	11	11	6.0	13	119	1,220	297	170	68
20	11	11	6.0	13	170	1,220	297	143	60
21	11	11	6.0	13	246	1,020	297	119	52
22	13	9	6.0	13	297	989	297	119	52
23	13	9	7.5	13	297	960	280	97	52
24	13	9	9	13	297	1,050	297	119	44
25	13	9	9	13	230	1,100	297	119	44
26	13	9	9	13	184	1,020	280	119	44
27	13	9	9	13	170	1,050	263	97	38
28	11	9	9	16	170	989	230	97	38
29	11		9	16	230	1,050	199	97	38
30	11		9	16	230	989	184	87	38
31	11		9		199		170	87	

NOTE.—These discharges are based on a rating curve that is well defined below a discharge of 400 second-feet.

Monthly discharge of Grand Lake Outlet at Grand Lake, Colo., for 1909.

Month.	Discharge in second-feet.			Run-off (total in acre-feet).	Accuracy.
	Maximum.	Minimum.	Mean.		
January	13	9	10.7	658	B.
February	11	9	9.86	548	B.
March	9	6	7.26	446	B.
April	16	9	11.0	655	B.
May	297	16	92.7	5,700	A.
June	1,220	170	805	47,900	A.
July	1,140	170	491	30,200	A.
August	156	87	127	7,810	A.
September	170	38	79.8	4,750	A.
The period				98,000	

SOUTH FORK OF GRAND RIVER NEAR LEHMAN, COLO.

This station is located at a footbridge near Lehman's ranch house, about a mile above the junction with the North Fork and about 2 miles from Lehman post office, Colo.

From September 25, 1907, to June 10, 1908, the records of the flow at this station were kept by the Central Colorado Power Co. The engineer in charge of the work used, in general, United States Geological Survey methods, but the results furnished by the company for publication have not been verified by engineers of the survey.

The drainage area at the station is about 79 square miles. The vertical staff gage is located at the footbridge.

Practically no water is diverted above the station other than that used for meadow irrigation.

Discharge measurements of South Fork of Grand River near Lehman, Colo., 1907–8.

Date.	Hydrographer.	Width.	Area of section.	Gage height.	Discharge.
1907.		*Feet.*	*Sq. ft.*	*Feet.*	*Sec.-ft.*
Sept. 11	R. I. Meeker..	58	88	4.30	47
Oct. 16	C. L. Chatfield..	45	42	4.16	25
Nov. 21	R. I. Meeker..	22	23	4.02	20
Dec. 10	.....do...	13	10.2	3.93	9.8
1908.					
Jan. 14	R. I. Meeker..	23	25.2	3.76	8.7
Feb. 21 a	C. L. Chatfield..	28.5	15.8	4.08	8.4
Mar. 17	.....do...	27	17	4.00	9.2
Apr. 16	.....do...	68	62	4.65	90
May 20	.....do...	79	12.5	5.40	258
Sept. 16	.....do...	45	49	4.33	44
Oct. 21	.....do...	44	42	4.15	18
Nov. 24 b	.....do...	35	22	4.05	12.7
Dec. 20 a	.....do...	24	16.8	4.15	10.9

a Made through ice.   b Frozen at gage. Ice 1.5 feet thick.

NOTE.—Measurements in 1908 made by wading at various sections.

Daily gage height, in feet, of South Fork of Grand River near Lehman, Colo., for 1907–8.

Day.	Sept.	Oct.	Nov.	Dec.	Day.	Sept.	Oct.	Nov.	Dec.	Day.	Sept.	Oct.	Nov.	Dec.
1907.					**1907.**					**1907.**				
1....	......	4.3	4.1	3.95	11....	......	4.35	4.1	3.95	21....	......	4.25	4.05	4.1
2....	......	4.25	4.1	4.0	12....	......	4.3	4.1	4.0	22....	......	4.2	4.05	4.1
3....	......	4.25	4.0	4.0	13....	......	4.2	4.2	4.0	23....	......	4.2	4.05	4.1
4....	......	4.25	4.15	3.95	14....	......	4.3	4.3	4.0	24....	......	4.2	4.0	4.1
5....	......	4.25	4.1	3.95	15....	......	4.25	4.35	4.0	25....	4.25	4.25	4.15	4.1
6....	......	4.20	4.1	4.0	16....	......	4.25	4.3	4.0	26....	4.25	4.25	4.15	4.0	4.1
7....	......	4.35	4.1	4.0	17....	......	4.2	4.3	4.0	27....	4.25	4.25	4.1	4.05	4.1
8....	......	4.45	4.1	4.0	18....	......	4.2	4.3	4.0	28....	4.25	4.25	4.15	4.0	4.1
9....	......	4.4	4.05	4.0	19....	......	4.2	4.1	4.1	29....	4.20	4.20	4.15	4.0	4.1
10....	......	4.35	4.0	4.0	20....	......	4.1	4.1	4.1	30....	4.25	4.25	4.15	3.95	4.1
										31....	......	4.15	......	4.1

[Ed. W. Lehman, observer.]

Day.	Jan.	Feb.	Mar.	Apr.	May.	June.	Day.	Jan.	Feb.	Mar.	Apr.	May.	June.
1908.							**1908.**						
1..........	4.1	4.10	4.00	4.00	4.35	5.00	16..........	4.3	4.00	4.00	4.65	4.55	......
2..........	4.1	4.10	4.00	4.00	4.40	5.02	17..........	4.3	4.00	4.00	4.70	4.82	......
3..........	4.2	4.10	4.00	4.00	4.52	5.25	18..........	4.2	4.10	4.00	4.65	4.95	......
4..........	4.2	4.10	4.00	4.02	4.60	5.60	19..........	4.2	4.10	4.00	4.60	5.05	......
5..........	4.2	4.10	4.00	4.08	4.52	5.60	20..........	4.2	4.10	4.05	4.60	5.38	......
6..........	4.2	4.10	4.00	4.05	4.50	5.55	21..........	4.2	4.10	4.00	4.65	5.02	......
7..........	4.2	4.10	4.00	4.05	4.50	5.40	22..........	4.2	4.08	4.00	4.75	4.82	......
8..........	4.2	4.10	4.00	4.10	4.62	5.25	23..........	4.2	4.00	4.00	4.90	4.78	......
9..........	4.2	4.10	4.00	4.10	4.82	5.25	24..........	4.2	4.00	4.00	4.78	4.85	......
10..........	4.3	4.10	4.00	4.10	4.78	5.65	25..........	4.2	4.02	4.05	4.62	4.95	......
11..........	4.3	4.00	4.00	4.10	4.65	......	26..........	4.2	4.00	4.05	4.50	4.88	......
12..........	4.3	4.00	4.00	4.18	4.60	......	27..........	4.2	4.05	4.02	4.48	4.80	......
13..........	4.3	4.00	4.00	4.08	4.58	......	28..........	4.2	4.00	4.00	4.42	4.75	......
14..........	4.3	4.00	4.00	4.22	4.52	......	29..........	4.2	4.00	4.00	4.35	4.68	......
15..........	4.3	4.00	4.00	4.55	4.52	......	30..........	4.2	......	4.00	4.32	4.70	......
							31..........	4.2	......	4.00	......	4.72	......

NOTE.—Ice conditions from Jan. 1 to Mar. 10, 1908.

Daily discharge, in second-feet, of South Fork of Grand River near Lehman, Colo., for 1907–8.

Day.	Sept.	Oct.	Nov.	Dec.	Day.	Sept.	Oct.	Nov.	Dec.	Day.	Sept.	Oct.	Nov.	Dec.
1907.					1907.					1907.				
1....	......	47	22	10	11....	......	55	22	10	21....	......	40	17.5	9
2....	......	40	22	13.5	12....	......	47	22	10	22....	......	33	17.5	9
3....	......	40	13.5	13.5	13....	......	33	33	10	23....	......	33	17.5	9
4....	......	40	27.5	10	14....	......	47	47	10	24....	......	33	13.5	9
5....	......	40	22	10	15....	......	40	55	10	25....	40	28	22	9
6....	......	33	22	13.5	16....	......	40	47	10	26....	40	28	13.5	9
7....	......	55	22	13.5	17....	......	33	47	10	27....	40	22	17.5	9
8....	......	73	22	13.5	18....	......	33	47	10	28....	40	28	13.5	9
9....	......	64	17.5	13.5	19....	......	33	22	10	29....	33	28	13.5	9
10....	......	55	13.5	13.5	20....	......	22	22	10	30....	40	28	10	9
										31....	......	28	......	9

Day.	Mar.	Apr.	May.	June.	Day.	Mar.	Apr.	May.	June.	Day.	Mar.	Apr.	May.	June.
1908.					1908.					1908.				
1....	......	9	44	161	11....	9	16	91	......	21....	9	91	166	......
2....	......	9	50	166	12....	9	24	82	......	22....	9	109	122	......
3....	......	9	69	220	13....	9	14	79	......	23....	9	138	115	......
4....	......	10	82	313	14....	9	28	69	......	24....	9	115	128	......
5....	......	14	69	313	15....	9	74	69	......	25....	12	86	150	......
6....	......	12	66	299	16....	9	91	74	......	26....	12	66	134	......
7....	......	12	66	259	17....	9	100	122	......	27....	10	62	118	......
8....	......	16	86	220	18....	9	91	150	......	28....	9	54	109	......
9....	......	16	122	220	19....	9	82	172	......	29....	9	44	96	......
10....	9	16	115	327	20....	12	82	254	......	30....	9	40	100	......
										31....	9	......	104	......

NOTE.—These discharges are based on rating curves furnished by the Central Colorado Power Co. applicable as follows: Sept. 25 to Dec. 31, 1907, and Mar. 10 to June 11, 1908. The daily discharges have been changed slightly to conform to the computation rules used by the United States Geological Survey.

Monthly discharge of South Fork of Grand River near Lehman, Colo., for 1907–8.

Month.	Discharge in second-feet.			Run-off (total in acre-feet).
	Maximum.	Minimum.	Mean.	
1907.				
Sept. 25–30........	40	33	38.8	462
October................	73	22	38.6	2,370
November..............	55	10	24.3	1,450
December..............	14	9	10.5	646
1908.				
January...............	9	8.5	8.7	535
February..............	8.5	8.5	8.5	489
March.................	12	8.5	9.5	584
April.................	138	9	51	3,150
May...................	254	44	106	6,490
June 1–10.............	327	161	250	4,960

NOTE.—The values have been changed slightly to conform to the computation rules used by the United States Geological Survey.

FRASER RIVER BASIN.

FRASER RIVER AT GRANBY, COLO.

Fraser River rises among the peaks of the Front Range in southeastern Grand County and flows in a general northwesterly direction to its junction with Grand River in the east-central part of Middle Park.

The station, which was established July 28, 1904, to obtain data for use in determining the availability of the stream for power, storage, and irrigation, is located at the wagon bridge three-quarters of a mile southwest of Granby, about 4 miles above the mouth of Fraser River, in sec. 9, T. 1 N., R. 76 W., and is below all tributaries. The station was discontinued September 30, 1909.

The drainage area is about 220 square miles.

Other than small irrigation ditches, there are no important diversions above the station. A small canal is taken out a few feet downstream from the measuring section. It is proposed to divert water by means of a tunnel from the headwaters of this river into the headwaters of South Boulder Creek, in the Platte drainage basin.

Thick ice covers the stream during about four months of the year; anchor ice also occurs. Each year in the low-water season a small temporary dam is generally constructed about 50 feet below the station to divert the water up on the gage, thus affecting the conditions of free flow. It was taken out May 5, 1909.

Neither the location or the datum of the staff gage at the highway bridge was changed during the maintenance of the station.

Measuring conditions are rather poor. During high stages the flow is affected by backwater and a boiling effect caused by the crib piers of the bridge. During low stages ice and the temporary diversion dam interfere with good results.

Discharge measurements of Fraser River at Granby, Colo., in 1909.

Date.	Hydrographer.	Width.	Area of section.	Gage height.	Discharge.
		Feet.	*Sq. ft.*	*Feet.*	*Sec.-ft.*
Mar. 9a	C. L. Chatfield	44	29	5.08	44
Apr. 28b	.....do	45	61	5.30	226
May 6	.....do	69	148	5.54	462
June 16	Chatfield, Matthes, and Snelson	61	218	6.35	1,080
Aug. 11	Freeman and Woolsey	56	125	5.15	257

a Made by wading 250 feet above gage; 3.5 feet of ice at gage.
b Made by wading 200 feet above pumping station opposite Granby.

NOTE.—Measurements beginning May 6 were made after the temporary dam was removed.

Daily gag height, in feet, of Fraser River at Granby, Colo., for 1909.

[J. N. Ostrander, observer.]

Day.	Jan.	Feb.	Mar.	Apr.	May.	June.	July.	Aug.	Sept.
1	5.5	5.0	5.1	5.1	4.95	5.7	6.8	5.1	5.0
2	5.7	4.95	5.1	5.1	4.85	5.7	6.8	5.1	5.0
3	5.7	4.95	5.1	5.1	4.95	5.85	6.8	5.1	5.0
4	5.6	5.0	5.1	5.25	5.1	6.0	6.7	5.1	5.0
5	5.6	5.1	5.1	5.1	5.6	6.35	6.55	5.1	5.1
6	5.6	5.1	5.1	5.0	5.55	6.5	6.45	5.1	5.1
7	5.6	5.1	5.3	4.9	5.4	6.6	6.35	5.1	5.1
8	5.6	5.1	5.4	4.9	5.4	6.7	6.2	5.1	5.0
9	5.6	5.1	5.05	4.8	5.5	6.7	6.0	5.1	5.0
10	5.6	5.1	5.05	4.8	5.45	6.7	5.85	5.1	5.0
11	5.15	5.1	5.15	4.8	5.55	6.7	5.65	5.15	5.0
12	5.3	5.1	5.0	4.8	5.55	6.6	5.6	5.1	5.0
13	5.3	5.1	5.0	4.8	5.5	6.45	5.6	5.1	5.0
14	5.2	5.1	5.0	4.7	5.4	6.45	5.5	5.1	5.0
15	5.4	5.0	5.1	4.7	5.35	6.5	5.5	5.1	5.0
16	5.3	5.0	5.1	4.7	5.4	6.5	5.5	5.1	4.9
17	5.2	5.0	5.1	4.7	5.4	6.65	5.5	5.1	4.9
18	5.1	5.0	5.15	4.7	5.5	6.85	5.4	5.15	4.9
19	5.1	5.0	5.25	4.95	5.75	6.95	5.4	5.4	4.9
20	5.15	5.0	5.3	4.9	5.85	7.15	5.4	5.25	4.9
21	5.1	5.0	5.3	4.95	5.9	7.15	5.4	5.2	4.9
22	5.1	5.0	5.45	5.0	5.95	7.1	5.4	5.15	4.9
23	5.1	5.0	5.5	5.0	6.0	7.1	5.5	5.1	4.9
24	5.1	5.0	5.5	5.0	5.85	7.0	5.5	5.1	4.9
25	5.3	5.0	5.5	5.0	5.75	7.0	5.4	5.1	4.9
26	5.4	5.0	5.5	5.1	5.55	6.9	5.3	5.1	4.8
27	5.2	5.0	5.35	5.1	5.55	7.0	5.3	5.0	4.8
28	5.2	5.1	5.3	5.3	5.75	6.95	5.2	5.0	4.8
29	5.1	....	5.2	5.25	5.85	6.9	5.2	5.0	4.8
30	5.0	....	5.1	5.0	5.7	6.8	5.2	5.0	4.8
31	5.0	....	5.1	....	5.6	....	5.1	5.0	....

NOTE.—Gage heights probably affected by ice from Jan. 1 to Apr. 16.

Daily discharge, in second-feet, of Fraser River at Granby, Colo., for 1909.

Day.	Apr.	May.	June.	July.	Aug.	Sept.	Day.	Apr.	May.	June.	July.	Aug.	Sept.
1	......	107	581	1,510	226	185	16	......	380	1,240	444	226	150
2	......	83	581	1,510	226	185	17	......	380	1,380	444	226	150
3	......	107	692	1,510	226	185	18	52	444	1,560	380	249	150
4	......	150	810	1,420	226	185	19	107	618	1,650	380	380	150
5	......	511	1,110	1,280	226	226	20	94	692	1,840	380	298	150
6	......	478	1,240	1,200	226	226	21	107	730	1,840	380	272	150
7	......	380	1,330	1,110	226	226	22	120	770	1,790	380	249	150
8	......	380	1,420	976	226	185	23	120	810	1,790	444	226	150
9	......	444	1,420	810	226	185	24	120	692	1,700	444	226	150
10	......	412	1,420	692	226	185	25	120	618	1,700	380	226	150
11	......	478	1,420	546	249	185	26	150	478	1,610	323	226	120
12	......	478	1,330	511	226	185	27	150	478	1,700	323	185	120
13	......	444	1,200	511	226	185	28	226	618	1,650	272	185	120
14	......	380	1,200	444	226	185	29	206	692	1,610	272	185	120
15	......	352	1,240	444	226	185	30	120	581	1,510	272	185	120
							31	......	511	......	226	185	.....

NOTE.—These discharges are based upon rating curves applicable as follows: Apr. 17 to May 4, not well defined; May 5 to Sept. 30, fairly well defined.

Monthly discharge of Fraser River at Granby, Colo., for 1909.

Month.	Discharge in second-feet.			Run-off (total in acre-feet).	Accu-racy.
	Maximum.	Minimum.	Mean.		
January....................	....	....	a 45	2,770	D.
February....................	....	....	a 45	2,500	D.
March....................	....	....	a 45	2,770	D.
April....................	226	....	b 83.6	4,970	D.
May....................	810	83	473	29,100	B.
June....................	1,840	581	1,390	82,700	B.
July....................	1,510	226	653	40,200	B.
August....................	380	185	230	14,100	B.
September....................	226	120	167	9,940	B.
The period....................	....	....	....	189,000	

a Estimated.   b Apr 1-17 estimated as equivalent to 48 second-feet per day.

COOPERATIVE DATA FOR STATIONS IN FRASER RIVER BASIN.

DESCRIPTION.

The data for the following stations in the Fraser River drainage are published practically without change as received from cooperating engineers. The computations have in general been made to conform with the computation rules used by the United States Geological Survey. On all the streams except Big Jim, Little Jim, and Elk Creek, on each of which the station was established at the mouth, there is an upper and a lower station. At all the upper stations a Lallie automatic gage was used to get the fluctuations in the daily gage height, which on the upper reaches of mountain streams are considerable.

Each upper station is at an altitude of about 10,000 feet, and several of the streams head in the vicinity of Corona, on the "Moffat road" (Denver, Northwestern & Pacific Railway). The United States Weather Bureau has for some time maintained a station near Corona at an altitude of 11,660 feet above sea level, and it is believed that the records of precipitation at this station will represent closely the average for the drainage areas of the upper gaging stations.

Unfortunately no published map shows accurately the drainage areas and location of these streams, so that it is rather difficult to define the exact location of the gaging stations.

Little use is made of the water except along the lower courses of the streams, where some water is diverted for meadow irrigation during the summer months. A scheme is contemplated to divert most of the water by tunnel through the continental divide to the head waters of the South Boulder Creek.

Most of these stations were established in June, 1907, and some measurements were made, but some of the gages were not established until the spring of 1908. All were discontinued in 1909 except the upper station on the Fraser, which is still being maintained. The flow of the streams is little affected by ice.

FRASER RIVER AT UPPER STATION NEAR FRASER, COLO.

This station, which was established May 1, 1908, is located about 10 miles above Fraser post office, Colo., at an elevation of approximately 10,000 feet above sea level. The drainage area is about 9 square miles. Currant Creek is the only important tributary above.

The location and datum of the Lallie automatic gage have remained constant since station was established.

Discharge measurements of Fraser River at upper station near Fraser, Colo., in 1908–9.

Date.	Hydrographer.	Gage height.	Discharge.
1908.		*Feet.*	*Sec.-ft.*
May 2	Stanley Krajicek	0.60	7.5
17	....do.	1.00	28.4
28	....do.	.80	15.0
June 7	....do.	1.20	54.7
9	....do.	1.60	120
10	....do.	1.50	115
21	....do.	1.30	70
24	....do.	1.25	56.5
July 5	....do.	.95	28.5
13	....do.	.70	12.7
25	....do.	.75	19.0
30	....do.	.80	19.1
Aug. 28	....do.	.60	9.8
Sept. 16	....do.	.55	6.9
1909.			
July 7	N. O'Daniels.	1.15	106
12	....do.	.90	48
27	....do.	.80	35
30	....do.	.75	31
Aug. 5	....do.	.65	24
Sept. 4	....do.	.60	21
Oct. 14	....do.	.40	10
28	....do.	.55	12

Daily gage height, in feet, of Fraser River at upper station near Fraser, Colo., for 1908–9.

Day.	May.	June.	July.	Aug.	Sept.	Oct.	Day.	May.	June.	July.	Aug.	Sept.	Oct.
1908.							1908.						
1	0.56	1.12	0.93	0.75	0.55	0.55	16	0.85	1.30	0.75	0.72	0.55	......
2	.60	1.16	.97	.75	.55	.50	17	.90	1.35	.72	.73	.55	......
3	.70	1.30	.95	.75	.60	.50	18	.95	1.30	.80	.72	.55	......
4	.70	1.38	.95	.72	.60	.50	19	1.07	1.23	.83	.73	.55	......
5	.70	1.37	.95	.75	.60	.50	20	1.00	1.30	.80	.70	.55	......
6	.70	1.27	.85	.85	.60	.50	21	.95	1.27	.80	.70	.55	......
7	.70	1.33	.85	.75	.60	.50	22	.97	1.38	.80	.70	.55	......
8	.70	1.17	.97	.75	.60	.50	23	1.00	1.25	.88	.70	.50	......
9	.70	1.40	.95	.75	.60	.50	24	.95	1.23	.75	.70	.50	......
10	.70	1.55	.80	.72	.60	.50	25	.85	1.20	.75	.68	.50	......
11	.70	1.50	.80	.73	.60	.50	26	.93	1.10	.88	.65	.50	......
12	.70	1.60	.80	.72	.60	.50	27	.95	1.25	.75	.63	.50	......
13	.70	1.40	.78	.72	.60	.50	28	.85	1.10	.78	.60	.50	......
14	.70	1.25	.78	.72	.60	.50	29	.87	1.10	.75	.60	.50	......
15	.80	1.30	.80	.73	.60	......	30	.80	.93	.85	.60	.50	......
							31	.95	......	.83	.60	......	......

Daily gage height, in feet, of Fraser River at upper station near Fraser, Colo., for 1908-9—Continued.

Day.	July.	Aug.	Sept.	Oct.	Nov.	Day.	July.	Aug.	Sept.	Oct.	Nov.
1909.						1909.					
1	1.30	0.75	0.60	0.60	0.50	16	0.85	0.60	0.70	0.40	....
2	1.30	.75	.60	.60	.50	17	.85	.90	.70	.40	....
3	1.25	.70	.60	.50	.50	18	.85	.70	.70	.35	....
4	1.25	.70	.60	.50	.50	19	.90	.70	.70	.35	....
5	1.20	.65	.60	.50	.50	20	.90	.70	.70	.30	....
6	1.20	.70	.60	.50	.60	21	.85	.70	.70	.25	....
7	1.15	.70	.65	.50	.55	22	.85	.70	.70	.50	....
8	1.10	.70	.65	.50	.55	23	.85	.70	.70	.50	....
9	1.05	.80	.65	.50	.50	24	.85	.70	.65	.50	....
10	1.00	.80	.70	.50	.45	25	.80	.70	.65	.50	....
11	.95	.80	.70	.45	.30	26	.80	.70	.65	.50	....
12	.90	.80	.70	.45	.40	27	.80	.70	.65	.50	....
13	.90	.70	.70	.40	.40	28	.80	.70	.65	.55	....
14	.90	.70	.70	.40	.40	29	.80	.70	.65	.50	....
15	.85	.70	.70	.40	....	30	.75	.70	.65	.50	....
						31	.75	.60	....	.50	....

Daily discharge, in second-feet, of Fraser River at upper station, near Fraser, Colo., for 1908-9.

Day.	May.	June.	July.	Aug.	Sept.	Oct.	Day.	May.	June.	July.	Aug.	Sept.	Oct.
1908.							1908.						
1	7	46	27	15	7	7	16	22	70	15	14	7	......
2	8	51	30	15	7	5	17	25	77	14	14	7	......
3	13	70	34	15	8	5	18	29	70	18	14	7	......
4	13	82	34	14	8	5	19	41	60	20	14	7	......
5	13	81	34	15	8	5	20	34	70	18	13	7	......
6	13	65	21	22	8	5	21	29	65	18	13	7	......
7	13	73	21	15	8	5	22	31	82	18	13	7	......
8	13	52	30	15	8	5	23	34	62	24	13	5	......
9	13	85	34	15	8	5	24	29	60	15	13	5	......
10	13	115	18	14	8	5	25	29	56	15	12	5	......
11	13	105	18	14	8	5	26	27	44	24	11	5	......
12	13	125	18	14	8	5	27	29	60	15	10	5	......
13	13	85	17	14	8	5	28	22	44	17	8	5	......
14	13	62	17	14	8	5	29	23	44	15	8	5	......
15	18	70	18	14	8	......	30	25	27	22	8	5	......
							31	29	......	20	8	......	......

Day.	July.	Aug.	Sept.	Oct.	Nov.	Day.	July.	Aug.	Sept.	Oct.	Nov.
1909.						1909.					
1	140	31	21	21	10.5	16	40	21	27	10	....
2	140	31	21	21	10.5	17	40	48	27	10	....
3	128	27	21	15	10.5	18	40	27	27	8	....
4	128	27	21	15	10.5	19	48	27	27	8	....
5	114	24	21	15	10.5	20	48	27	27	6	....
6	114	27	21	15	15	21	40	27	27	5	....
7	106	27	24	15	12	22	40	27	27	10.5	....
8	92	27	24	15	12	23	40	27	27	10.5	....
9	81	35	24	15	10.5	24	40	27	24	10.5	....
10	66	35	27	15	9.5	25	35	27	24	10.5	....
11	56	35	27	13	8	26	35	27	24	10.5	....
12	48	35	27	13	9	27	35	27	24	10.5	....
13	48	27	27	10	9	28	35	27	24	12	....
14	48	27	27	10	9	29	35	27	24	10.5	....
15	40	27	27	10	....	30	31	27	24	10.5	....
						31	31	21	....	10.5	....

Monthly discharge of Fraser River at upper station near Fraser, Colo., for 1908–9.

Month.	Discharge in second-feet.			Run-off (total in acre-feet).
	Maximum.	Minimum.	Mean.	
1908.				
May	41	7	20.9	1,290
June	125	27	68.6	4,080
July	34	14	21.3	1,300
August	22	8	13.3	818
September	8	5	6.9	411
Oct. 1–14	7	5	5.1	142
1909.				
July	140	31	63.3	3,890
August	48	21	28.5	1,750
September	27	21	24.8	1,480
October	21	5	12.0	738
Nov. 1–14	15	8	10.5	292

FRASER RIVER AT LOWER STATION NEAR FRASER, COLO.

This station, which was established June 1, 1907, and discontinued October 30, 1909, is located about 6 miles above Fraser post office, Colo., one-fourth mile below the mouth of Jim Creek and about 3 miles above the mouth of Vasquez River.

The drainage area is approximately 24 square miles. The location and datum of the rod gage remained constant while the station was being maintained.

Discharge measurements of Fraser River at lower station near Fraser, Colo., in 1907–1909.

Date.	Hydrographer.	Gage. height.	Discharge.
1907.		*Feet.*	*Sec.-ft.*
June 6	G. M. Bull	1.60	100
12	.....do.	1.70	105
July 2	.....do.	2.55	196
Aug. 2	.....do.	1.40	50
15	.....do.	1.05	30
23	.....do.	.80	23
1908.			
Apr. 28	Stanley Krajicek	.30	21
May 14	.....do.	.85	43
18	.....do.	1.20	60
26	.....do.	1.30	71
June 11	.....do.	2.50	250
21	.....do.	1.60	103
July 4	.....do.	1.60	99
19	.....do.	1.10	56
30	.....do.	1.00	56
Aug. 11	.....do.	.80	48
22	.....do.	.60	41
Sept. 3	.....do.	.40	33
21	.....do.	.25	24
1909.			
July 6	N. O'Daniels	2.00	320
12	.....do.	1.45	145
19	.....do.	1.25	115
30	.....do.	.90	72
Aug. 5	.....do.	.75	57
12	.....do.	.75	57
20	.....do.	1.00	83
24	.....do.	.80	62
Sept. 3	.....do.	.60	42
Oct. 18	.....do.	.35	23

Daily gage height, in feet, of Fraser River at lower station near Fraser, Colo., for 1907–1909.

Day.	June.	July.	Aug.	Day.	June.	July.	Aug.	Day.	June.	July.	Aug.
1907.				**1907.**				**1907.**			
1	0.54	.....	1.40	11	.....	.....	1.08	21	.....	2.45	.....	0.90
2	.....	2.55	1.40	12	1.70	.....	1.05	22	.....	2.70	.....	.....
3	.....	.....	1.40	13	2.05	2.45	1.00	23	.....	.....	.80
4	.....	2.50	1.30	14	.....	.....	.....	24	.....	.....	.....
5	.....	2.55	1.28	15	2.80	2.00	1.05	25	.....	3.00	.....
6	1.60	.....	1.25	16	2.55	.....	.98	26	.....	.....	.....
7	1.60	2.45	1.20	17	2.80	1.90	.95	27	.....	.....	.....
8	1.70	.....	1.18	18	2.85	.....	.....	28	.....	2.80	1.75	.....
9	.....	.....	1.18	19	2.70	1.80	.....	29	.....	.....	.....
10	.....	.....	1.15	20	2.60	.....	.....	30	.....	.....	1.60	.....
								31	.....	1.50	.....

Day.	May.	June.	July.	Aug.	Sept.	Oct.	Day.	May.	June.	July.	Aug.	Sept.	Oct.
1908.							**1908.**						
1	0.40	1.40	.....	1.00	.....	0.25	16	1.00	2.40	1.22	0.80	0.40	.....
2	.40	1.60	1.20	.98	.....	.25	17	1.10	2.35	1.14	.79	.40	.....
3	.40	1.70	1.60	.95	0.40	.25	18	1.20	2.30	1.10	.78	.37	.....
4	.40	1.80	.....	.92	.40	.25	19	1.35	2.20	1.10	.77	.34	.....
5	.40	1.80	1.52	.89	.40	.25	20	1.50	2.20	1.10	.76	.31	.....
6	.40	1.90	1.46	.86	.40	.25	21	1.49	2.0	1.10	.75	.25	.....
7	.40	1.85	1.40	.83	.40	.25	22	1.48	2.3	1.10	.74	.25	.....
8	.40	1.80	1.45	.80	.40	.25	23	1.47	2.3	1.10	.73	.25	.....
9	.40	1.75	1.49	.80	.40	.25	24	1.44	2.9	1.10	.71	.25	.....
10	.30	2.15	1.40	.80	.40	.25	25	1.40	2.4	1.10	.70	.25	.....
11	.50	2.50	1.35	.80	.40	.25	26	1.30	2.50	1.10	.68	.25	.....
12	.60	2.50	1.30	.80	.40	.25	27	1.20	2.30	1.10	.66	.25	.....
13	.75	2.45	1.30	.80	.40	.25	28	1.10	1.85	1.10	.63	.25	.....
14	.85	2.45	1.30	.80	.40	.25	29	1.10	1.57	1.05	.60	.25	.....
15	.95	2.40	1.30	.80	.40	.....	30	1.12	1.29	1.00	.59	.25	.....
							31	1.15	.....	1.00	.59	.....	.....

Day.	July.	Aug.	Sept.	Oct.	Day.	July.	Aug.	Sept.	Oct.
1909.					**1909.**				
1	.....	.....	0.65	0.65	16	.....	.....	.....	.....
2	.....	0.85	.....	.....	17	1.25	0.75	.....	.....
3	.....	.....	.65	.....	18	.....	1.25	.....	0.35
4	2.10	.....	.....	.50	19	1.25	1.20	.....	.....
5	.....	.75	.....	.....	20	.....	1.00	.....	.....
6	2.0	.75	.65	.....	21	.....	.90	0.75	.....
7	.....	.80	.....	.....	22	1.20	.....	.75	.....
8	1.80	.....	.80	.50	23	1.20	.....	.75	.40
9	1.75	.....	.90	.50	24	.....	.80	.75	.....
10	.....	.80	.85	.....	25	.....	.75	.75	.40
11	.....	.....	.....	.50	26	.....	.70	.....	.....
12	1.45	.75	.....	.50	27	.....	.70	.70	.40
13	1.40	.75	.....	.45	28	.....	.....	.....	.....
14	1.35	.75	.....	.....	29	.....	.....	.65	.....
15	.....	.....	.....	.....	30	.90	.65	.70	.35
					31	.90	.....	.....	.....

Daily discharge, in second-feet, of Fraser River at lower station near Fraser, Colo., for 1907–1909,

Day.	June.	July.	Aug.	Day.	June.	July.	Aug.	Day.	June.	July.	Aug.
1907.				**1907.**				**1907.**			
1	35	200	49	11	105	180	34	21	180	105	26
2	48	196	49	12	105	180	30	22	205	105	24
3	61	192	49	13	126	180	28	23	220	105	23
4	74	185	45	14	185	140	29	24	235	105	22
5	87	196	44	15	210	105	30	25	250	105	21
6	100	188	42	16	196	100	28	26	238	105	19
7	100	180	40	17	210	100	28	27	224	105	18
8	105	180	38	18	225	108	28	28	210	105	17
9	105	180	38	19	205	105	27	29	207	104	16
10	105	180	36	20	200	105	27	30	203	105	15
								31		75	15

Day.	May.	June.	July.	Aug.	Sept.	Oct.	Day.	May.	June.	July.	Aug.	Sept.	Oct.
1908.							**1908.**						
1	25	85	72	55	25	20	16	55	217	71	43	25	.....
2	25	108	69	54	25	20	17	62	208	64	42	25	.....
3	25	118	108	52	25	20	18	69	200	62	42	24	.....
4	25	130	104	50	25	20	19	81	185	62	41	23	.....
5	25	130	100	48	25	20	20	95	185	62	41	22	.....
6	25	142	90	47	25	20	21	93	155	62	40	20	.....
7	25	135	85	45	25	20	22	92	200	62	40	20	.....
8	25	130	81	43	25	20	23	91	210	62	40	20	.....
9	25	125	93	43	25	20	24	89	220	62	39	20	.....
10	22	175	85	43	25	20	25	85	230	62	38	20	.....
11	29	235	81	43	25	20	26	77	238	62	37	20	.....
12	33	235	77	43	25	20	27	69	200	62	35	20	.....
13	41	225	77	43	25	20	28	62	135	62	35	20	.....
14	46	225	77	43	25	20	29	62	103	58	33	20	.....
15	52	217	77	43	25	.....	30	63	76	55	33	20	.....
							31	66	.....	55	33	.....	.....

Day.	July.	Aug.	Sept.	Oct.	Nov.	Day.	July.	Aug.	Sept.	Oct.	Nov.
1909.						**1909.**					
1	420	68	47	47	25	16	121	57	62	25	....
2	420	68	47	42	25	17	115	57	57	23	....
3	375	68	47	38	25	18	115	115	57	23	....
4	375	57	47	34	25	19	115	108	57	23	....
5	320	57	47	34	23	20	115	83	57	23	....
6	320	57	47	34	23	21	108	72	57	23	....
7	265	62	62	34	23	22	108	72	57	25	....
8	225	62	62	34	23	23	108	68	57	25	....
9	210	62	72	34	23	24	108	62	57	25	....
10	185	62	68	34	23	25	102	57	57	25	....
11	165	57	68	34	23	26	95	52	57	25	....
12	145	57	68	34	23	27	90	52	52	25	....
13	137	57	68	30	23	28	83	52	52	25	....
14	128	57	62	30	23	29	77	47	52	25	....
15	121	57	62	25	....	30	72	47	52	23	....
						31	72	47	....	23	

NOTE.—Discharges estimated for days on which gage was not read.

Monthly discharge of Fraser River at lower station near Fraser, Colo., for 1907–1909.

Month.	Discharge in second-feet.			Run-off (total in acre-feet).
	Maximum.	Minimum.	Mean.	
1907.				
June.....................	250	35	160.0	9,590
July.....................	200	75	100.0	8,600
August...................	49	15	30.0	1,870
1908.				
May.....................	95	25	53.5	3,290
June....................	238	76	173.0	10,300
July....................	108	55	72.9	4,480
August.................	55	33	42.2	2,600
September..............	25	20	23.1	1,370
Oct. 1-14..............	20	20	20.0	555
1909.				
July....................	420	72	175.0	10,800
August.................	115	47	63.1	3,880
September..............	72	47	57.1	3,400
October................	47	23	29.2	1,800
Nov. 1-14..............	25	23	23.6	655

BIG (EAST) JIM CREEK AT MOUTH, NEAR FRASER, COLO.

Jim Creek drains the west slope of James Peak, and the two branches, East, or Big Jim, and the West, or Little Jim, unite a short distance above the point where they enter the Fraser River, some 6 miles above Fraser post office, Colo. The drainage area of the two creeks is about 7½ square miles.

This station, which was established June 1, 1907, and discontinued November 11, 1909, is located just above the junction of the Big and Little Jim, about 1 mile above Idlewild ranger station.

The location and datum of the rod gage remained constant while the station was being maintained.

Discharge measurements of Big Jim Creek at mouth, near Fraser, Colo., 1907–1909.

Date.	Hydrographer.	Gage height.	Discharge.
1907.		*Feet.*	*Sec.-ft.*
June 1	G. M. Bull..................	0.30	5.6
12	.....do..................	.60	15.0
July 2	.....do..................	.90	36.0
Aug. 6	.....do..................	.45	8.3
15	.....do..................	.35	6.2
1908.			
May 20	Stanley Krajicek...........	.40	10.7
26	.....do..................	.50	12.4
June 2	.....do..................	.50	13.4
11	.....do..................	.95	44.7
22	.....do..................	.90	45.4
July 4	.....do..................	.65	25.7
19	.....do..................	.40	13.7
23	.....do..................	.40	12.5
Aug. 11	.....do..................	.34	9.8
29	.....do..................	.27	7.4
Sept. 3	.....do..................	.20	5.4
21	.....do..................	.10	3.5
1909.			
July 10	N. O'Daniels...............	.95	38.5
12	.....do..................	.85	34.0
30	.....do..................	.60	18.5
Aug. 5	.....do..................	.50	14.5
12	.....do..................	.50	14.5
20	.....do..................	.70	24.5
Sept. 3	.....do..................	.45	12.5
Oct. 18	.....do..................	.30	7.0

Daily gage height, in feet, of Big Jim Creek at mouth, near Fraser, Colo., for 1907–1909.

Day.	June.	July.	Aug.	Day.	June.	July.	Aug.	Day.	June.	July.	Aug.
1907.				**1907.**				**1907.**			
1	0.30	......	0.50	11	......	......	0.38	21	0.70	......	0.30
2	......	0.90	.50	12	0.60	......	.40	22	.85	......	......
3	......	......	.50	13	.70	0.85	.35	23	......	......	.30
4	......	.85	.50	14	......	......	......	24	......	......	......
5	......	.85	.50	15	.90	.65	.35	25	.95	......	......
6	.55	......	.45	16	.75	......	.30	26	......	......	......
7	.55	.86	.48	17	.90	.65	......	27	......	......	......
7	.60	......	.40	18	.90	......	......	28	.90	0.55	......
9	......	......	.40	19	.80	.65	......	29	......	......	......
10	......	......	.40	20	.80	......	......	30	......	.58	......
								31	......	.53	......

Day.	Apr.	May.	June.	July.	Aug.	Sept.	Oct.	Day.	Apr.	May.	June.	July.	Aug.	Sept.	Oct.
1908.								**1908.**							
1	......	0.20	0.30	0.55	0.40	0.20	0.18	16	......	0.40	0.80	0.40	0.18	0.18	......
2	......	.20	.40	.55	.40	.20	.18	17	......	.40	.85	.40	.16	.18	......
3	......	.20	.40	.55	.39	.20	.18	18	......	.40	.85	.40	.15	.18	......
4	......	.25	.50	.55	.39	.20	.18	19	......	.40	.90	.40	.14	.18	......
5	......	.25	.50	.55	.37	.20	.18	20	......	.40	.90	.40	.12	.18	......
6	......	.30	.60	.55	.36	.20	.18	21	......	.40	.90	.40	.10	.18	......
7	......	.30	.60	.55	.35	.20	.18	22	......	.40	.90	.40	.13	.18	......
8	......	.30	.55	.55	.34	.20	.18	23	......	.40	.90	.40	.18	.18	......
9	......	.30	.55	.55	.34	.19	.18	24	......	.40	.90	.40	.18	.18	......
10	......	.30	.75	.55	.34	.19	.18	25	......	.40	.85	.40	.20	.18	......
11	......	.30	.95	.53	.34	.19	.18	26	......	.40	.85	.40	.21	.18	......
12	......	.30	.80	.50	.32	.19	.18	27	......	.35	.77	.40	.22	.18	......
13	......	.35	.80	.47	.28	.19	.18	28	0.30	.30	.70	.40	.24	.18	......
14	......	.35	.80	.44	.24	.19	.18	29	.25	.30	.70	.40	.27	.18	......
15	......	.40	.80	.40	.20	.19	......	30	.25	.30	.68	.40	.27	.18	......
								31	......	.30	......	.40	......	......	......

Day.	July.	Aug.	Sept.	Oct.	Nov.	Day.	July.	Aug.	Sept.	Oct.	Nov.
1909.						**1909.**					
1	......	......	0.45	0.40	......	16	......	0.50	......	......	......
2	......	......	......	......	......	17	0.75	.50	......	......	......
3	......	......	.45	......	......	18	.70	.70	......	0.30	......
4	......	......	......	.40	0.30	19	.70	.80	......	......	......
5	......	0.50	......	......	......	20	......	.70	0.50	......	......
6	1.20	......	.50	......	.30	21	......	.60	......	......	......
7	......	......	.60	......	......	22	......	.60	.50	......	......
8	1.10	......	.80	.40	......	23	.70	.50	.50	.30	......
9	1.05	......	.65	.40	.30	24	......	.50	......	......	......
10	.95	.50	.65	......	......	25	......	.50	......	.35	......
11	......	......	......	.40	.30	26	......	......	......	......	......
12	.85	.50	......	.40	......	27	......	.50	......	......	......
13	.80	......	......	.40	......	28	......	......	......	......	......
14	.80	......	......	......	......	29	......	......	......	......	......
15	......	.50	......	......	......	30	.60	.50	......	.30	......
						31	......	......	......	......	......

Daily discharge, in second-feet, of Big Jim Creek at mouth, near Fraser, Colo., for 1907–1909.

Day.	June.	July.	Aug.	Day.	June.	July.	Aug.	Day.	June.	July.	Aug.
1907.				**1907.**				**1907.**			
1	5.6	36	11	11	15	30	7	21	20	15	6
2	7	36	11	12	15	30	8	22	31	15	6
3	8	34	11	13	20	31	6	23	34	15	6
4	10	31	11	14	28	25	6	24	37	14	6
5	12	31	11	15	36	18	6	25	40	14	6
6	14	29	8.3	16	24	18	6	26	39	14	5
7	14	27	10	17	36	18	6	27	37	14	5
8	15	28	8	18	36	17	6	28	36	14	5
9	15	28	8	19	27	15	6	29	36	14	4
10	15	29	8	20	27	15	6	30	36	15	4
								31		14	4

Day.	May.	June.	July.	Aug.	Sept.	Oct.	Day.	May.	June.	July.	Aug.	Sept.	Oct.
1908.							**1908.**						
1	5	9	20	13	5	5	16	13	35	13	5	5	
2	5	13	20	13	5	5	17	13	39	13	4	5	
3	5	13	20	12	5	5	18	13	39	13	4	5	
4	7	17	20	11	5	5	19	13	43	13	4	5	
5	7	17	20	11	5	5	20	13	43	13	3	5	
6	9	22	20	10	5	5	21	13	43	13	3	5	
7	9	22	20	10	5	5	22	13	43	13	4	5	
8	9	20	20	10	5	5	23	13	43	13	4	5	
9	9	20	20	10	5	5	24	13	43	13	5	5	
10	9	32	20	10	5	5	25	13	39	13	5	5	
11	9	47	19	10	5	5	26	13	39	13	6	5	
12	9	35	17	9	5	5	27	10	33	13	6	5	
13	10	35	16	8	5	5	28	9	29	13	7	5	
14	10	35	14	7	5	5	29	9	29	13	8	5	
15	13	35	13	5	5		30	9	27	13	8	5	
							31	9		13			

Day.	July.	Aug.	Sept.	Oct.	Nov.	Day.	July.	Aug.	Sept.	Oct.	Nov.
1909.						**1909.**					
1	70	16.5	12.5	10.5	10.5	16	28	14.5	18.5	8.5	
2	70	16.5	12.5	10.5	10.5	17	28	14.5	16.5	8.5	
3	65	16.5	12.5	10.5	10.5	18	24.5	24.5	16.5	7	
4	60	14.5	12.5	10.5	7	19	24.5	31	16.5	7	
5	55	14.5	14.5	10.5	7	20	24.5	24.5	14.5	7	
6	50	14.5	14.5	10.5	7	21	24.5	18.5	14.5	7	
7	47	14.5	18.5	10.5	7	22	24.5	18.5	14.5	7	
8	45	14.5	31	10.5	7	23	24.5	14.5	14.5	7	
9	43	14.5	21	10.5	7	24	24.5*	14.5	14.5	8.5	
10	38.5	14.5	21	10.5	7	25	24.5	14.5	14.5	8.5	
11	36	14.5	21	10.5	7	26	24.5	14.5	14.5	8.5	
12	34	14.5	21	10.5	7	27	21	14.5	12.5	8.5	
13	31	14.5	18.5	10.5	7	28	21	14.5	12.5	7	
14	31	14.5	18.5	10.5	7	29	21	14.5	10.5	7	
15	31	14.5	18.5	10.5		30	18.5	14.5	10.5	7	
						31	18.5	14.5		7	

NOTE.—Discharges interpolated for days on which gage was not read.

Monthly discharge of Big Jim Creek at mouth, near Fraser, Colo., for 1907–1909.

Month.	Discharge in second-feet.			Run-off (total in acre-feet).
	Maximum.	Minimum.	Mean.	
1907.				
June..........	40	6	24	1,450
July..........	36	14	22	1,370
August........	11	4	7	434
1908.				
April 28–30....	9	7	7.7	45
May...........	13	5	10.1	621
June..........	47	9	31.3	1,860
July..........	20	13	15.7	965
August........	13	3	7.5	446
September.....	5	5	5.0	297
October 1–14...	5	5	5.0	139
1909.				
July..........	70	18.5	35.0	2,150
August........	31	14.5	16.1	990
September.....	31	10.5	16.1	958
October.......	10.5	7	8.98	552
November 1–14..	10.5	7	7.75	215

LITTLE (WEST) JIM CREEK AT MOUTH, NEAR FRASER, COLO.

This station, which was established June 1, 1907, and discontinued September 3, 1909, is located just above the junction of the Little Jim with the Big Jim, about 6 miles above the Fraser and at an elevation of about 9,000 feet above sea level.

The location and datum of the rod gage remained constant while the station was being maintained.

Discharge measurements of Little Jim Creek at mouth, near Fraser, Colo., in 1907–1909.

Date.	Hydrographer.	Gage height.	Discharge.
1907.		*Feet.*	*Sec.-ft.*
June 1	G. M. Bull..........	1.02	3.0
12	....do..........	1.55	9.6
July 2	....do..........	1.90	17.1
Aug. 6	....do..........	1.00	1.9
23	....do..........	.85	1.2
1908.			
May 20	Stanley Krajicek..........	.73	2.4
26	....do..........	.70	2.2
June 2	....do..........	.80	3.1
11	....do..........	1.70	14.2
July 4	....do..........	1.20	5.9
19	....do..........	.70	2.1
23	....do..........	.70	2.2
Aug. 11	....do..........	.60	1.2
29	....do..........	.58	.9
Sept. 3	....do..........	.58	.9
21	....do..........	.50	.5
1909.			
July 10	N. O'Daniels..........	1.50	13
12	....do..........	1.30	7.5
23	....do..........	1.00	4.0
30	....do..........	.90	3.0
Aug. 5	....do..........	.70	1.5
12	....do..........	.70	1.5
17	....do..........	.65	1.3
20	....do..........	.95	3.20
24	....do..........	.65	1.3
Sept. 3	....do..........	.50	.5

Daily gage height, in feet, of Little Jim Creek at mouth, near Fraser, Colo., for 1907–1909.

Day.	June.	July.	Aug.	Day.	June.	July.	Aug.	Day.	June.	July.	Aug.
1907.				**1907.**				**1907.**			
1	1.02	.....	1.10	11	.....	.....	0.95	21	1.75	.....	0.80
2	.....	1.90	1.05	12	1.55	.....	.90	22	1.85	.....	.....
3	.....	.....	.....	13	1.70	1.75	.90	23	.....	.....	.85
4	.....	1.70	1.08	14	.....	.....	.....	24	.....	.....	.....
5	.....	1.75	1.00	15	1.95	1.55	.90	25	1.95	.....	.....
6	1.45	.....	1.00	16	1.80	.....	.80	26	.....	.....	.....
7	1.45	1.70	1.00	17	1.90	1.50	.....	27	.....	.....	.....
8	1.50	.....	.98	18	1.90	.....	.....	28	1.90	1.35	.....
9	.....	.....	.98	19	1.85	1.35	.....	29	.....	.....	.....
10	.....	.....	.93	20	1.85	.....	.....	30	.....	1.20	.....
								31	.....	1.13	.....

Day.	Apr.	May.	June.	July.	Aug.	Sept.	Oct.	Day.	Apr.	May.	June.	July.	Aug.	Sept.	Oct.
1908.								**1908.**							
1	.....	0.50	0.80	1.20	0.70	0.58	0.50	16	.....	.75	1.50	0.81	0.50	0.56	.....
2	.....	.50	.80	1.20	.70	.58	.50	17	.....	.75	1.50	.80	.50	.56	.....
3	.....	.50	.85	1.20	.69	.58	.50	18	.....	.75	1.50	.75	.50	.56	.....
4	.....	.50	.90	1.20	.68	.58	.50	19	.....	.74	1.50	.70	.50	.56	.....
5	.....	.55	1.00	1.20	.66	.58	.50	20	.....	.73	1.50	.70	.50	.56	.....
6	.....	.55	1.10	1.20	.64	.58	.50	21	.....	.75	1.55	.70	.50	.56	.....
7	.....	.55	1.10	1.20	.62	.58	.50	22	.....	.75	1.60	.70	.50	.56	.....
8	.....	.60	1.10	1.13	.60	.58	.50	23	.....	.75	1.C0	.70	.50	.56	.....
9	.....	.60	1.10	1.07	.60	.58	.50	24	.....	.80	1.55	.70	.50	.56	.....
10	.....	.65	1.40	1.00	.60	.58	.50	25	.....	.80	1.55	.70	.50	.56	.....
11	.....	.65	1.70	.93	.60	.57	.50	26	.....	.70	1.50	.70	.52	.56	.....
12	.....	.65	1.60	.87	.58	.56	.50	27	.....	.65	1.45	.70	.54	.56	.....
13	.....	.70	1.55	.85	.56	.55	.50	28	0.70	.60	1.40	.70	.56	.56	.....
14	.....	.70	1.55	.84	.53	.55	.50	29	.65	.60	1.32	.70	.58	.56	.....
15	.....	.75	1.50	.82	.50	.55	.....	30	.60	.60	1.24	.70	.56	.56	.....
								31	.....	.60	.....	.70	.56	.56	.....

Day.	July.	Aug.	Day.	July.	Aug.	Day.	July.	Aug.
1909.			**1909.**			**1909.**		
1	.....	.....	11	.....	.....	21	.....	.....
2	.....	.....	12	1.30	.70	22	.....	.....
3	.....	.....	13	1.25	.....	23	1.00	.....
4	.....	.....	14	1.20	.....	24	.....	0.65
5	.....	0.70	15	.....	.....	25	.....	.....
6	.....	.....	16	.....	.....	26	.....	.....
7	.....	.....	17	1.15	.65	27	.....	.60
8	1.60	.....	18	.....	.....	28	.....	.....
9	1.60	.....	19	1.10	1.20	29	.....	.....
10	1.50	.65	20	.....	.95	30	.....	.90	.....
						31	.....	.....

Daily discharge, in second-feet, of Little Jim Creek at mouth, near Fraser, Colo., for 1907–1909.

Day.	June.	July.	Aug.	Day.	June.	July.	Aug.	Day.	June.	July.	Aug.
1907.				**1907.**				**1907.**			
1	3	17	3	11	9	14	2	21	14	6	1
2	4	17	3	12	9.6	14	2	22	16	6	1
3	5	15	3	13	13	14	2	23	17	6	1.2
4	6	13	3	14	16	12	2	24	18	6	1
5	7	14	2	15	19	10	2	25	19	6	1
6	8	14	1.9	16	15	10	1	26	19	6	1
7	8	13	2	17	17	9	1	27	18	6	1
8	9	13	2	18	17	7	1	28	17	6	1
9	9	13	2	19	16	6	1	29	17	5	1
10	9	13	2	20	16	6	1	30	17	4	1
								31	.....	4	1

Daily discharge, in second-feet, of Little Jim Creek at mouth, near Fraser, Colo., for 1907–1909—Continued.

Day.	Apr.	May.	June.	July.	Aug.	Sept.	Oct.	Day.	Apr.	May.	June.	July.	Aug.	Sept.	Oct.
1908.								1908.							
1....		1	2	6	2	1.4	1.1	16....			2	10	3	1	1.3	
2....		1	2	6	2	1.4	1.1	17....			2	10	3	1	1.3	
3....		1	3	6	2	1.4	1.1	18....			2	10	2	1	1.3	
4....		1	3	6	2	1.4	1.1	19....			2	10	2	1	1.3	
5....		1	4	6	2	1.4	1.1	20....			2	10	2	1	1.3	
6....		1	5	6	2	1.4	1.1	21....			2	11	2	1	1.3	
7....		1	5	6	2	1.4	1.1	22....			2	12	2	1	1.3	
8....		2	5	5	2	1.4	1.1	23....			2	12	2	1	1.3	
9....		2	5	5	2	1.4	1.1	24....			3	11	2	1	1.3	
10....		2	8	4	2	1.4	1.1	25....			3	11	2	1	1.3	
11....		2	14	3	2	1.4	1.1	26....			2	10	2	1	1.3	
12....		2	12	3	1	1.3	1.1	27....			2	9	2	1	1.3	
13....		2	11	3	1	1.3	1.1	28....			2	8	2	1	1.3	
14....		2	11	3	1	1.3	1.1	29....		2	2	7	2	1	1.3	
15....		2	10	3	1	1.3		30....		2	2	6	2	1	1.3	
								31....			2		2	1		

Day.	July.	Aug.	Sept.	Oct.	Day.	July.	Aug.	Sept.	Oct.	Day.	July.	Aug.	Sept.	Oct.
1909.					1909.					1909.				
1....	23	2.5	1.0	0.4	11....	10	1.5	1	0.4	21....	4.5	2.5	0.7	
2....	23	2.0	.5	.4	12....	7.5	1.5	1	.4	22....	4.5	1.7	.7	
3....	21.5	1.7	.5	.4	13....	7	1.3	1		23....	4	1.5	.7	
4....	19.5	1.5	.5	.4	14....	6	1.3	1		24....	4	1.3	.7	
5....	19.5	1.5	.7	.4	15....	6	1.5	.7		25....	4	1.3	.7	
6....	16.5	1.5	1	.4	16....	5.5	1.5	.7		26....	4	1	.5	
7....	16.5	1.5	1	.4	17....	5.5	1.3	.7		27....	3.5	1	.5	
8....	16.5	1.3	1	.4	18....	5.5	2	.7		28....	3.5	1	.4	
9....	16.5	1.3	1	.4	19....	5	6	.7		29....	3	1	.4	
10....	13	1.3	1	.4	20....	5	3.5	.7		30....	3	1	.4	
										31....	3	1		

NOTE.—Discharges interpolated for days when gage was not read. From Oct. 12 to Dec. 31, 1909, the flow was almost zero.

Monthly discharge of Little Jim Creek at mouth, near Fraser, Colo., for 1907–1909.

Month.	Discharge in second feet.			Run-off (total in acre-feet).
	Maximum.	Minimum.	Mean.	
1907.				
June............	19	3	12	756
July............	17	4	10	610
August............	3	1	2	100
1908.				
April 28–30............	2	2	2	12
May............	3	1	1.8	111
June............	14	2	8.2	488
July............	6	2	3.4	209
August............	2	1	1.4	86
September............	1.4	1.3	1.3	77
October 1–14............	1.1	1.1	1.10	30
1909.				
July............	23	3	9.34	574
August............	6	1	1.67	103
September............	1	.4	.74	44
October 1–12............	.4	.4	.40	10

VASQUEZ RIVER AT UPPER STATION, NEAR FRASER, COLO.

Vasquez River is tributary to Fraser River from the west.

This station, which was established May 1, 1908, and discontinued November 9, 1909, is located about 4 miles above the mouth of the stream, at an elevation of approximately 10,000 feet above sea level.

The drainage area is about 9 square miles.

The location and datum of the Lallie automatic gage remained constant while the station was maintained.

Discharge measurements of Vasquez River at upper station near Fraser, Colo., 1908-9.

Date.	Hydrographer.	Gage height.	Discharge.
1908.		*Feet.*	*Sec.-ft.*
May 1	Stanley Krajicek..	0.23	12.1
16	....do..	.33	18.2
22	....do..	.55	32.8
June 3	....do..	.83	50.0
11	....do..	1.15	63.5
17	....do..	1.35	103
30	....do..	.55	37.3
July 12	....do..	.75	40.3
17	....do..	.76	46.6
19	....do..	.72	43.4
28	....do..	.64	36.3
Sept. 19	....do..	.30	14.0
1909.			
July 4	N. O'Daniels..	1.90	208
14	....do..	1.30	114
17	....do..	1.25	107
30	....do..	1.00	74
Aug. 5	....do..	.85	58
12	....do..	90	63
Sept. 20	....do..	.70	42
Oct. 18	....do..	.45	22

Daily gage height, in feet, of Vasquez River at upper station near Fraser, Colo., for 1908-9.

Day.	June.	July.	Aug.	Sept.	Oct.	Day.	June.	July.	Aug.	Sept.	Oct.
1908.						1908.					
1........	0.48	0.53	0.38	0.34	0.34	16........	1.41	0.75	0.34	0.34	....
2........	.53	.53	.38	.34	.34	17........	1.51	.75	.34	.34	....
3........	.74	.53	.38	.34	.34	18........	1.41	.75	.34	.34	....
4........	1.12	.53	.38	.34	.34	19........	1.42	.75	.34	.34	....
5........	1.15	.53	.38	.34	.34	20........	1.38	.75	.34	:34	....
6........	1.29	.53	.38	.34	.34	21........	1.25	.70	.34	.34	....
7........	1.32	.41	.34	.34	.34	22........	1.05	.70	.34	.34	....
8........	1.41	.40	.34	.34	.34	23........	1.12	.70	.34	.34	....
9........	1.59	.48	.34	.34	.34	24........	1.53	.69	.34	.34	....
10........	1.50	.63	.34	.34	.34	25........	1.41	.69	.34	.34	....
11........	1.51	.75	.34	.34	.34	26........	1.36	.64	.34	.34	....
12........	1.42	.75	.34	.34	.34	27........	1.48	.64	.34	.34	....
13........	1.32	.75	.34	.34	.34	28........	1.48	.64	.34	.34	....
14........	1.32	.75	.34	.34	.34	29........	1.48	.54	.34	.34	....
15........	1.34	.75	.34	.34	....	30........	1.48	.48	.34	.34	....
						31........	....	.47	.34	....	....

Daily gage height, in feet, of Vasquez River at upper station near Fraser, Colo., for 1908-9—Continued.

Day.	July.	Aug.	Sept.	Oct.	Nov.	Day.	July.	Aug.	Sept.	Oct.	Nov.
1909.						**1909.**					
1		0.95	0.80	0.60	0.70	16		0.90			
2		.95	.80	.60		17	1.25	1.00	0.85		
3		.90	.80	.60		18	1.30	1.00	.90	0.45	
4	1.90	.90		.60		19	1.40	1.00			
5		.85		.60		20	1.25	.90			
6		.85		.60		21	1.20	.90	.75		
7		.90		.60		22	1.20	.85			
8		.90		.60		23	1.25	.85			
9		.90			.50	24	1.25	.85			
10		.85				25		.80			
11		.90				26		.80	.70	.50	
12		.85				27		.80	.65		
13		.85				28		.80	.65		
14	1.30	.85				29		.80	.65		
15		.85				30	1.00	.85	.60		
						31	1.00	.85			

Daily discharge, in second-feet, of Vasquez River at upper station near Fraser, Colo., for 1908-9.

Day.	June.	July.	Aug.	Sept.	Oct.	Day.	June.	July.	Aug.	Sept.	Oct.
1908.						**1908.**					
1	28	31	21	19	19	16	95	46	19	19	
2	31	31	21	19	19	17	103	46	19	19	
3	46	31	21	19	19	18	95	46	19	19	
4	73	31	21	19	19	19	98	46	19	19	
5	75	31	21	19	19	20	93	46	19	19	
6	86	31	21	19	19	21	83	43	19	19	
7	88	23	19	19	19	22	68	43	19	19	
8	95	23	19	19	19	23	73	43	19	19	
9	109	28	19	19	19	24	53	42	19	19	
10	102	38	19	19	19	25	41	42	19	19	
11	102	46	19	19	19	26	36	39	19	19	
12	96	46	19	19	19	27	28	39	19	19	
13	88	46	19	19	19	28	28	39	19	19	
14	89	46	19	19	19	29	28	32	19	19	
15	91	46	19	19		30	28	28	19	19	
						31		27	19		

Day.	July.	Aug.	Sept.	Oct.	Nov.	Day.	July.	Aug.	Sept.	Oct.	Nov.
1909.						**1909.**					
1	225	67	52	32	42	16	107	63	52	22	
2	225	67	52	32	42	17	107	74	58	22	
3	216	63	52	32	37	18	114	74	63	22	
4	208	63	52	32	37	19	128	74	58	22	
5	200	58	52	32	32	20	107	63	52	22	
6	192	58	52	32	32	21	101	63	47	22	
7	175	63	52	32	24	22	101	58	42	22	
8	167	63	52	32	24	23	107	58	42	22	
9	159	63	52	32	24	24	107	58	42	24	
10	143	58	47	27	24	25	107	52	42	24	
11	135	63	47	27	22	26	101	52	42	24	
12	128	58	47	24	22	27	95	52	37	22	
13	121	58	47	24	22	28	88	52	37	22	
14	114	58	42	24	22	29	82	52	37	22	
15	114	58	47	22		30	74	58	32	27	
						31	74	58		32	

NOTE.—Discharges in 1909 estimated for days on which gage was not read.

Monthly discharge of Vasquez River at upper station near Fraser, Colo., for 1908–9.

Month.	Discharge in second-feet.			Run-off (total in acre-feet).
	Maximum.	Minimum.	Mean.	
1908.				
June...	109	28	71.6	4,200
July...	46	23	37.9	2,330
August...	21	19	19.4	1,190
September..	19	19	19.0	1,130
October 1–14...	19	19	19.0	528
1909.				
July...	225	74	133	8,180
August...	74	52	60.6	3,730
September..	63	32	47.6	2,830
October..	32	22	26.1	1,600
November 1–14..	42	22	29.0	806

VASQUEZ RIVER AT LOWER STATION NEAR FRASER, COLO.

This station, which was established June 1, 1907 and discontinued November 9, 1909, is located about 1 mile above the mouth of the stream, at an elevation of probably 8,700 feet above sea level. The drainage area is about 12 square miles.

The location and datum of the rod gage remained constant while the station was being maintained.

Discharge measurements of Vasquez River at lower station near Fraser, Colo., in 1907–1909.

Date.	Hydrographer.	Gage height.	Discharge.
1907.		*Feet.*	*Sec.-ft.*
June 1	G. M. Bull..	0.80	46
7	.....do...	1.15	92
14	.....do...	1.75	170
27	.....do...	1.60	161
July 6	.....do...	1.35	133
30	.....do...	.95	72
Aug. 5	.....do...	.80	52
13	.....do...	.70	39
18	.....do...	.60	31
1908.			
Apr. 29	Stanley Krajicek...................................	.80	10.4
May 10	.....do...	.80	11.3
14	.....do...	1.50	23.0
23	.....do...	1.63	46.4
25	.....do...	1.80	66.0
June 2	.....do...	1.90	79.3
9	.....do...	2.00	93.2
11	.....do...	2.5	177
July 25	.....do...	1.72	52
Aug. 10	.....do...	1.70	49.1
17	.....do..:..	1.7	50.0
Sept. 22	.....do...	1.40	16
1909.			
July 6	N. O'Daniels.......................................	1.50	243
13	.....do...	1.10	125
30	.....do...	.80	73
Aug. 5	.....do...	.65	55
12	.....do...	.65	55
20	.....do...	.70	60
Sept. 20	.....do...	.60	49
Oct. 18	.....do...	.40	31

Daily gage height, in feet, of Vasquez River at lower station near Fraser, Colo., for 1907–1909.

Day.	June.	July.	Aug.	Day.	June.	July.	Aug.	Day.	June.	July.	Aug.
1907.				**1907.**				**1907.**			
1	0.80	1.55	0.93	11	1,25	1.25	0.75	21	1.42		0.68
2		1.55	.93	12	1 10	1.20	.73	22	1.40		.60
3		1.50	.83	13	1 10	1.30	.70	23	1.50		.60
4		1.45	.80	14	1 75	1.30	.73	24	1.50		
5	1.40	1.43	.80	15	1 65	1.15	.78	25	1.55	1.00	
6		1.35	.78	16	1,50	1.10	,73	26	1.60		
7	1.15	1.40	.78	17	1 40	1.10	68	27	1.60		.60
8	1.20	1.40	.75	18	1 60	1.05	60	28	1.50	.95	
9	1.00	1.35	.75	19	1 48	1.00	63	29	1.55	.95	
10	.90	1.30	.73	20	1 45	1.00	68	30	1.65	.93	
								31		.88	

Day.	June.	July.	Aug.	Sept.	Oct.	Day.	June.	July.	Aug.	Sept.	Oct.
1908.						**1908.**					
1	1.00	1.00	0.72	0.43	0.44	16	1.20	0.80	0.69	0.45	
2	1.02	.98	.72	.43	.44	17	1.20	.78	.68	.45	
3	1.04	.95	.72	.43	.43	18	1.16	.76	.67	.45	
4	1.06	.92	.71	.44	.43	19	1.12	.75	.66	.44	
5	1.08	.90	.71	.44	.43	20	1.08	.75	.65	.43	
6	1.10	.89	.71	.44	.42	21	1.04	.74	.64	.43	
7	1.13	.89	.70	.44	.42	22	1.00	.73	.63	.43	
8	1.15	.86	.70	.44	.42	23	1.00	.73	.62	.43	
9	1.20	.85	.70	.44	.44	24	1.00	.72	.61	.43	
10	1.20	.86	.70	.44	.46	25	1.00	.72	.60	.43	
11	1.20	.86	.70	.44	.40	26	1.00	.72	.56	.43	
12	1.20	.87	.70	,44	.40	27	1.00	.72	.52	.43	
13	1.20	.86	.70	.44	.40	28	1.00	.72	.48	.43	
14	1.20	.84	.70	.45	.40	29	1.00	.72	.45	.43	
15	1.20	.82	.70	.45		30	1.00	.72	.45	.43	
						31		.72	.45		

Day.	July.	Aug.	Sept.	Oct.	Nov.	Day.	July.	Aug.	Sept.	Oct.	Nov.
1909.						**1909.**					
1			0.60	0.50	0.50	16					
2						17		0.65			
3			.60	.50		18				0.40	
4	1.70					19		.80			
5		0.65				20		.70	0.70		
6	1.50				.50	21					
7					.40	22					
8			.60		.30	23	0.90		.60		
9					.30	24		.65			
10	1.00					25					
11						26				.30	
12		.65		.50		27		.60			
13	1.10					28		.60			
14						29		.60			
15						30	.80	.60			
						31		.60			

Daily discharge, in second-feet, of Vasquez River at lower station near Fraser, Colo., for 1907–1909.

Day.	June.	July.	Aug.	Day.	June.	July.	Aug.	Day.	June.	July.	Aug.
1907.				**1907.**				**1907.**			
1	46	148	65	11	108	108	42	21	130	76	38
2	66	148	65	12	90	103	42	22	128	76	31
3	86	142	53	13	90	115	39	23	142	76	31
4	106	135	52	14	170	115	42	24	142	76	31
5	128	135	52	15	164	92	48	25	148	76	31
6	110	133	48	16	142	90	42	26	160	75	31
7	92	128	48	17	128	90	38	27	161	74	31
8	103	128	45	18	160	84	31	28	142	72	31
9	76	133	45	19	140	76	32	29	148	72	31
10	64	115	42	20	135	76	38	30	164	65	31
								31	.....	62	31

Day.	June.	July.	Aug.	Sept.	Oct.	Day.	June.	July.	Aug.	Sept.	Oct.
1908.						**1908.**					
1	85	85	54	33	33	16	125	62	52	34	....
2	88	83	54	33	33	17	125	60	52	34	....
3	90	79	54	33	33	18	108	58	51	34	....
4	92	75	54	33	33	19	103	57	50	33	....
5	94	73	54	33	33	20	94	57	49	33	....
6	100	71	54	33	32	21	90	55	48	32	....
7	104	69	53	33	32	22	85	55	47	33	....
8	107	68	53	33	32	23	85	55	46	33	....
9	125	67	53	33	33	24	85	54	46	33	....
10	125	68	53	33	34	25	85	54	45	33	....
11	125	68	53	33	31	26	85	54	41	33	....
12	125	69	53	33	31	27	85	54	38	33	....
13	125	68	53	33	31	28	85	54	36	33	....
14	125	66	53	34	31	29	85	54	34	33	....
15	125	64	53	34	....	30	85	54	34	33	....
						31	....	54	34	....	....

Day.	July.	Aug.	Sept.	Oct.	Nov.	Day.	July.	Aug.	Sept.	Oct.	Nov.
1909.						**1909.**					
1	370	66	49	39	39	16	115	55	49	35	....
2	350	66	49	39	39	17	115	55	73	31	....
3	315	60	49	39	39	18	103	60	79	31	....
4	315	60	49	39	39	19	103	73	73	31	....
5	280	55	49	39	39	20	103	60	60	31	....
6	243	55	49	39	39	21	95	60	49	31	....
7	208	55	49	39	31	22	95	60	49	31	....
8	164	55	49	39	22	23	86	60	49	31	....
9	125	55	49	39	22	24	86	55	49	27	....
10	103	55	49	39	22	25	86	55	49	27	....
11	103	55	49	39	22	26	79	49	49	22	....
12	125	55	49	39	22	27	79	49	49	22	....
13	125	55	49	39	22	28	73	49	45	22	....
14	125	55	49	39	22	29	73	49	45	22	....
15	115	55	49	35	....	30	73	49	39	31	....
						31	73	49	....	35	....

NOTE.—Discharges interpolated for days on which gage was not read.

Monthly discharge of Vasquez River at lower station near Fraser, Colo., for 1907–1909.

Month.	Discharge in second-feet.			Run-off (total in acre-feet).
	Maximum.	Minimum.	Mean.	
1907.				
June	170	46	122	7,340
July	148	62	100	6,190
August	65	31	40	2,510
1908.				
June	125	85	102	6,070
July	85	54	63.4	3,900
August	54	34	48.5	2,980
September	34	32	33.2	1,980
October 1–14	34	31	32.3	897
1909.				
July	370	73	145	8,920
August	73	49	56.3	3,460
September	79	39	51.4	3,060
October	39	22	33.6	2,070
November 1–14	39	22	29.9	830

ELK CREEK AT MOUTH, NEAR FRASER, COLO.

Elk Creek, the next important stream tributary to the Fraser below the Vasquez River, enters about 1 mile above Fraser post office. The drainage area is about 8 square miles.

The only station on this stream was located a short distance above its mouth. The datum and location of the rod gage remained constant while the station was maintained. The station was established June 5, 1907, and discontinued November 9, 1909.

Discharge measurements of Elk Creek at mouth, near Fraser, Colo., in 1907–1909.

Date.	Hydrographer.	Gage height.	Discharge.
1907.		*Feet.*	*Sec.-ft.*
June 5	G. M. Bull	0.92	28
July 16	....do	.55	14
Aug. 9	....do	.25	2
1908.			
May 7	Stanley Krajicek	1.40	11.8
14	....do	1.30	7.0
25	....do	1.46	16.0
June 3	....do	1.60	24.2
July 4	....do	1.40	12.2
21	....do	1.30	7.7
Sept. 24	....do	1.00	1.5
1909.			
July 6	N. O'Daniels	.90	48
14	....do	.50	21
22	....do	.40	14
31	....do	.30	9
Aug. 7	....do	.20	5
16	....do	.20	5
25	....do	.20	5

Daily gage height, in feet, of Elk Creek at mouth, near Fraser, Colo., for 1907–1909.

Day.	June.	July.	Aug.	Day.	June.	July.	Aug.	Day.	June.	July.	Aug.
1907.				**1907.**				**1907.**			
1	......	0.85	0.30	11	0.80	0.80	0.00	21	......	1.10	0.50	......
2		.80	.30	12	1.00	.75	.00	22		.93	.50	0.20
3		.75	.25	13	1.05	.75		23		.90	.45	......
4	......	.75	.25	14	1.05	.70	.20	24		.90	.45	......
5	0.92	.70	.50	15	1.00	.70		25		1.00	.45	......
6		.70	.20	16	1.05	.55		26		.95	.40	......
7		.75	.00	17	1.00	.55	.20	27		1.00	.50	.15
8	.90	.80	.00	18	1.05	.50		28		.85	.35	......
9		.80	.25	19	1.05	.50		29		.85	.38	......
10	1.00	.80	.00	20	1.10	.50		30		.85	.35	......
								31			.33	......

Day.	May.	June.	July.	Aug.	Sept.	Oct.	Day.	May.	June.	July.	Aug.	Sept.	Oct.
1908.							**1908.**						
1	1.20	1 58	1.40	1 30	1.04	0.98	16	1,32	1,80	1.30	1.23	1.02	......
2	1.38	60	1.40	30	1.04	.98	17	1 34	1 70	1.30	1.20	1.02	......
3	1.				1.03	.98	18	1 36	1 80	1.30	1.18	1.02	......
4	1.				1.03	.98	19	1 57	1 60	1.30	L 16	1.02	......
5					1.02	.98	20	1 39	1 60	1.30	1.14	1.02	......
6	1.38	1 63	1.40	1 28	1.02	.98	21	1 41	1 60	1.30	1,12	1.02	......
7	1.36	66	1.40	28	1.02	.98	22	1 43	1 60	1.30	1 10	1.02	......
8					1.02	.98	23	1 45	1 60	1.30	1 08	1.02	......
9					1.02	.99	24	1 47	1 57	1.30	1 06	1.02	......
10					1.02	1.00	25	1 48	1 53	1.30	1 04	1.01	......
11	1.32	1 70	1.40	1 28	1.02	1.00	26	1.49	1.50	1.30	1 04	1.01	......
12	1.30	78	1.36	28	1.02	1.00	27	1.68	1.48	1.30	04	1.00	......
13					1.02	1.00	28	1.				1.00	......
14					1.02	1.00	29	1.			04	1.00	......
15					1.02		30	1.			04		......
							31	1.			04		......

Day.	July.	Aug.	Sept.	Oct.	Nov.	Day.	July.	Aug.	Sept.	Oct.	Nov.
1909.						**1909.**					
1			0.20			16		0.20			
2				0.10		17	0.40				
3				.10		18				0.10	
4				.10	0.10	19		.30			
5						20					
6	0.90		.20		.10	21		0.15			
7		0.20				22	.40				
8						23				.10	
9					.10	24					
10			.20			25		.20	.15		
11						26	.40				
12						27				.10	
13				.10		28	.20				
14	.50					29					
15						30				.10	
						31	.30				

Daily discharge, in second-feet, of Elk Creek at mouth, near Fraser, Colo., for 1907–1909.

Day.	June.	July.	Aug.	Day.	June.	July.	Aug.	Day.	June.	July.	Aug.
1907.				**1907.**				**1907.**			
1	26	26	4	11	24	24	2	21	37	12	2
2	26	24	4	12	33	22	2	22	28	12	2
3	27	22	2	13	35	22	2	23	27	10	2
4	27	22	2	14	35	19	2	24	27	10	2
5	28	19	12	15	33	19	2	25	33	10	2
6	27	19	2	16	35	14	2	26	30	8	2
7	27	22	1	17	33	14	2	27	33	12	2
8	27	24	1	18	35	12	2	28	26	6	1
9	30	24	2	19	35	12	2	29	26	7	1
10	33	24	2	20	37	12	2	30	26	6	1
								31	......	5	1

Day.	May.	June.	July.	Aug.	Sept.	Oct.	Day.	May.	June.	July.	Aug.	Sept.	Oct.
1908.							**1908.**						
1	4	24	12	7	1	1	16	8	49	7	5	1	......
2	5	25	12	7	1	1	17	9	35	7	4	1	......
3	6	25	12	7	1	1	18	10	49	7	4	1	......
4	7	26	12	7	1	1	19	10	25	7	3	1	......
5	8	27	12	7	1	1	20	11	25	7	3	1	......
6	11	28	12	7	1	1	21	12	25	7	3	1	......
7	12	29	12	7	1	1	22	13	25	7	2	1	......
8	11	31	12	7	1	1	23	14	25	7	2	1	......
9	10	34	12	7	1	1	24	16	23	7	2	1	......
10	9	35	12	7	1	1	25	17	20	7	2	1	......
11	8	35	12	7	1	1	26	17	18	7	2	1	......
12	8	38	12	7	1	1	27	17	17	7	2	1	......
13	8	39	7	7	1	1	28	18	17	7	2	1	......
14	7	42	9	6	1	1	29	19	16	7	2	1	......
15	8	47	9	6	1	......	30	20	14	7	2	1	......
							31	24	......	7	2		......

Day.	July.	Aug.	Sept.	Oct.	Nov.	Day.	July.	Aug.	Sept.	Oct.	Nov.
1909.						**1909.**					
1	54	9	5	2	2	16	17	5	5	2	....
2	54	9	5	2	2	17	14	5	2.5	2	....
3	51	9	5	2	2	18	14	7	2.5	2	....
4	51	7	5	2	2	19	14	9	2.5	2	....
5	48	7	5	2	2	20	14	9	2.5	2	....
6	48	5	5	2	2	21	14	7	2.5	2	....
7	44	5	5	2	2	22	14	5	2.5	2	....
8	41	5	5	2	2	23	14	5	2.5	2	....
9	38	5	5	2	2	24	14	5	2.5	2	....
10	35	5	5	2	2	25	14	5	2.5	2	....
11	31	5	5	2	2	26	14	5	2.5	2	....
12	28	5	5	2	2	27	14	5	2.5	2	....
13	24	5	5	2	2	28	14	5	2.5	2	....
14	21	5	5	2	2	29	11	5	2.5	2	....
15	21	5	5	2	....	30	11	5	2	2	....
						31	9	5	......	2	....

NOTE.—Discharges interpolated for days on which gage was not read.

Monthly discharge of Elk Creek at mouth, near Fraser, Colo., for 1907–1909.

Month.	Discharge in second-feet.			Run-off (total in acre-feet).
	Maximum.	Minimum.	Mean.	
1907.				
June......	37	24	30	1,810
July......	26	5	16	988
August......	12	1	2	140
1908.				
May......	24	4	11.5	707
June......	49	14	28.9	1,720
July......	12	7	9.1	560
August......	7	2	4.7	289
September......	1	1	1.0	60
Oct. 1-14......	1	1	1.0	28
1909.				
July......	54	9	26.0	1,600
August......	9	5	5.90	363
September......	5	2	3.82	227
October......	2	2.	2.0	123
Nov. 1-14......	2	2	2.0	56

ST. LOUIS CREEK AT UPPER STATION NEAR FRASER, COLO.

St. Louis Creek is tributary to the Fraser River from the west, entering about 1 mile below the town of Fraser.

The upper station is about 6 miles above the mouth of the stream at an elevation of approximately 10,000 feet above the sea level. The drainage area is about 15 square miles.

The location and datum of the Lallie automatic gage remained constant while the station was maintained. The station was established April 30, 1908, and discontinued November 9, 1909.

Discharge measurements of St. Louis Creek at upper station near Fraser, Colo., in 1908–1909.

Date.	Hydrographer.	Gage height.	Discharge.
1908.		*Feet.*	*Sec.-ft.*
Apr. 30	Stanley Krajicek......	0.90	11.8
May 9	.....do......	1.15	27.0
16	.....do......	1.20	28.6
22	.....do......	1.30	40.1
30	.....do......	.90	11.9
June 3	.....do......	1.50	64.1
11	.....do......	2.35	137
22	.....do......	3.75	294
July 3	.....do......	2.20	121
16	.....do......	1.95	91.8
30	.....do......	1.59	63.2
Aug. 8	.....do......	1.40	42.7
30	.....do......	1.20	30.4
Sept. 6	.....do......	1.15	33.2
6	.....do......	1.25	35.7
20	.....do......	1.10	27.8
Oct. 9	.....do......	1.15	26.5
1909.			
July 14	N. O'Daniels......	1.45	134
17	.....do......	1.35	122
22	.....do......	1.20	103
31	.....do......	.80	62
Aug. 7	.....do......	.65	50
16	.....do......	.50	40
19	.....do......	.70	54
28	.....do......	.55	43
Sept. 25	.....do......	.35	30
Oct. 13	.....do......	.20	23
20	.....do......	.20	23

Daily gage height, in feet, of St. Louis Creek at upper station near Fraser, Colo., for 1908-9.

Day.	June.	July.	Aug.	Sept.	Oct.	Day.	June.	July.	Aug.	Sept.	Oct.
1908.						1908.					
1......	1.41	2.43	1.72	1.20	1.08	16......	2.86	1.95	1.33	1.10	....
2......	1.47	2.34	1.64	1.20	1.05	17......	2.83	1.88	1.35	1.10	....
3......	1.90	2.17	1.58	1.20	1.05	18......	2.80	1.83	1.40	1.10	....
4......	2.06	2.27	1.50	1.20	1.05	19......	2.66	1.80	1.37	1.10	....
5......	2.03	2.23	1.45	1.20	1.04	20......	2.66	1.78	1.46	1.10	....
6......	1.98	2.19	1.45	1.20	1.03	21......	2.55	1.75	1.45	1.08	....
7......	1.77	2.14	1.45	1.15	1.02	22......	2.97	1.75	1.45	1.07	....
8......	1.70	2.10	1.35	1.15	.90	23......	2.97	1.77	1.50	1.08	....
9......	1.72	2.12	1.35	1.15	1.00	24......	2.75	1.71	1.45	1.07	....
10......	2.05	2.12	1.35	1.15	1.00	25......	2.91	1.65	1.45	1.08	....
11......	2.05	2.10	1.40	1.15	1.00	26......	2.91	1.65	1.45	1.08	....
12......	2.12	2.10	1.45	1.15	1.00	27......	2.88	1.65	1.47	1.08	....
13......	2.97	2.10	1.55	1.20	.90	28......	2.61	1.60	1.29	1.08	....
14......	2.79	2.02	1.40	1.15	.80	29......	2.61	....	1.29	1.05	....
15......	2.71	2.02	1.35	1.10	....	30......	2.35	1.58	1.24	1.05	....
						31......	....	1.70	1.20	....	....

Day.	July.	Aug.	Sept.	Oct.	Nov.	Day.	July.	Aug.	Sept.	Oct.	Nov.
1909.						1909.					
1......	....	....	0.50	....	....	16......	....	0.55	0.45	0.20	....
2......	....	....	.50	0.30	0.30	17......	1.35	.80	.45	.20	....
3......	2.60	....	.45	.30	....	18......	1.35	.80	....	.20	....
4......	....	....	.40	.30	....	19......	1.40	.70	....	.10	....
5......	....	....	....	.30	....	20......	1.35	.60	....	.10	....
6......	....	....	....	....	.20	21......	1.35	.55	.35	....	....
7......	....	....	....	....	....	22......	1.25	.55	.35	....	....
8......	....	....	....	....	....	23......	1.40	.55	.35	....	....
9......	1.95	....	....	.90	.35	24......	1.40	.60	.35	....	....
10......	....	....	.50	.40	....	25......	1.30	.50	.35	....	....
11......	....	....	.50	.40	....	26......	1.10	.50	.35	.20	....
12......	....	....	.50	.30	....	27......	....	.45	.30	....	....
13......	....	....	.50	.20	....	28......	....	.50	.30	....	....
14......	1.45	0.55	.50	.20	....	29......	....	.50	.30	....	....
15......	....	.50	.45	.20	....	30......	....	.55	....	.55	....
						31......	.80	.60	....	....	....

Daily discharge, in second-feet, of St. Louis Creek at upper station near Fraser, Colo., for 1908-9.

Day.	June.	July.	Aug.	Sept.	Oct.	Day.	June.	July.	Aug.	Sept.	Oct.
1908.						1908.					
1......	49	141	74	34	24	16......	187	94	42	27	....
2......	53	132	67	34	24	17......	186	88	44	27	....
3......	89	115	63	34	24	18......	180	83	48	27	....
4......	105	125	56	34	24	19......	166	81	46	27	....
5......	102	121	52	34	24	20......	166	79	53	27	....
6......	98	117	52	34	23	21......	153	76	52	26	....
7......	78	112	52	30	22	22......	200	76	52	25	....
8......	72	108	44	30	18	23......	200	78	56	26	....
9......	74	110	44	30	20	24......	175	73	52	25	....
10......	107	110	44	30	20	25......	193	68	52	26	....
11......	107	108	48	30	20	26......	193	68	52	26	....
12......	110	108	52	30	20	27......	188	68	53	26	....
13......	200	108	60	34	18	28......	160	64	40	26	....
14......	179	100	48	30	18	29......	160	63	40	24	....
15......	171	100	44	27	....	30......	133	62	37	24	....
						31......	....	72	34	....	....

Daily discharge, in second-feet, of St. Louis Creek at upper station near Fraser, Colo., for 1908-9—Continued.

Day.	July.	Aug.	Sept.	Oct.	Nov.	Day.	July.	Aug.	Sept.	Oct.	Nov.
1909.						1909.					
1.......	264	54	40	28	28	16......	129	43	36	23	....
2.......	254	54	40	28	28	17......	122	62	36	23	....
3.......	242	46	36	28	28	18......	122	62	36	23	....
4.......	232	43	33	28	25	19......	129	54	33	18	....
5.......	221	43	40	28	25	20......	122	46	33	18	....
6.......	212	43	33	33	23	21......	122	43	30	21	....
7.......	202	43	33	40	25	22......	110	43	30	23	....
8.......	194	40	33	54	28	23......	129	43	30	28	....
9.......	182	40	33	71	30	24......	129	46	30	28	....
10......	174	40	40	33	30	25......	115	40	30	25	....
11......	166	40	40	33	28	26......	92	40	30	23	....
12......	155	40	40	28	28	27......	86	36	28	28	....
13......	140	43	40	23	28	28......	81	40	28	33	....
14......	134	43	40	23	28	29......	76	40	28	40	....
15......	129	40	36	23	....	30......	71	43	28	43	....
						31......	62	46	....	43	....

NOTE.—Discharge interpolated for days on which gage was not read.

Monthly discharge of St. Louis Creek at upper station near Fraser, Colo., for 1908-9.

Month.	Discharge in second-feet.			Run-off (total in acre-feet).
	Maximum.	Minimum.	Mean.	
1908.				
June...	200	49	141	8,390
July..	141	62	93.8	5,770
August...	74	34	50.1	3,080
September..	34	24	28.8	1,710
Oct., 1-14...	24	18	21.4	594
1909.				
July..	264	62	148	9,100
August...	62	56	44.5	2,740
September..	40	28	34.1	2,030
October..	71	18	30.4	1,870
Nov. 1-14..	30	23	27.3	758

ST. LOUIS CREEK AT LOWER STATION NEAR FRASER, COLO.

This station, which was established June 3, 1907, and discontinued November 9, 1909, is located about 1 mile above the mouth of the stream at an elevation of about 8,600 feet above sea level. The drainage area is about 19 square miles.

The stream forks near its junction with the Fraser and enters that stream in two channels. This fact was overlooked when the station was first established in 1907, and so the discharge of a smaller channel was not included in the results obtained that year. These have not been published. In the spring of 1908 a new gage was established at a new datum and shows the entire flow. The datum of the gage was changed on September 21, 1909. Data collected in 1907 are not published, as they are not comparable with those for 1908 and 1909.

Discharge measurements of St. Louis Creek at lower station near Fraser, Colo., in 1908–9.

Date.	Hydrographer.	Gage height.	Discharge.
1908.		*Feet.*	*Sec.-feet.*
May 23	Stanley Krajicek...	0.42	52.3
31	....do..	.68	81.8
June 6	....do..	.73	97.1
11	....do..	1.03	148
17	....do..	1.03	148
July 3	....do..	.96	141
7	....do..	.8	110
20	....do..	.61	73.0
27	....do..	.50	51.1
Aug. 15	....do..	.37	41.0
Sept. 10	....do..	.20	30.8
18	....do..	.23	25.8
23	....do..	.20	24.2
1909.			
July 9	N. O'Daniels..	1.10	182
14	....do..	.90	114
17	....do..	.80	94
22	....do..	.75	87
31	....do..	.50	60
Aug. 7	....do..	.45	55
16	....do..	.40	49
Sept. 25	....do..	.40	39
Oct. 13	....do..	.25	33
20	....do..	.20	31

NOTE.—Gage heights for 1909 are not comparable with those for 1908. Beginning Sept. 25, 1909, the gage datum was changed again.

Mean daily gage height, in feet, of St. Louis Creek at lower station near Fraser, Colo., for 1908–9.

Day.	June.	July.	Aug.	Sept.	Oct.	Day.	June.	July.	Aug.	Sept.	Oct.
1908.						**1908.**					
1......	0.67	1.40	0.71	0.45	0.15	16......	1.03	0.66	0.48	0.26	....
2......	.68	1.20	.77	.45	.15	17......	1.03	.65	.48	.23	....
3......	.69	.96	.81	.45	.15	18......	1.10	.64	.48	.23	....
4......	.70	.97	.87	.45	.15	19......	1.18	.62	.48	.23	....
5......	.71	.98	.92	.45	.15	20......	1.25	.61	.47	.22	....
6......	.73	99	.97	45	.15	21......	1.32	.60	.47	.21	....
7......	.79	1.01	1.02	.45	.15	22......	1.4	.58	.47	.20	....
8......	.85	.96	.85	.46	.15	23......	1.4	.57	.47	.20	....
9......	.91	.90	.66	.46	.15	24......	1.4	.55	.47	.20	....
10......	.97	.85	.47	.46	.15	25......	1.4	.54	.47	.20	....
11......	1.08	.78	.47	.43	.15	26......	1.4	.52	.47	.21	....
12......	1.03	.72	.47	.40	.15	27......	1.4	.50	.46	.21	....
13......	1.03	.71	.47	.36	.15	28......	1.4	.57	.46	.21	....
14......	1.03	.69	.47	.33	.15	29......	1.4	.61	.45	.21	....
15......	1.03	.68	.48	.30	....	30......	1.4	.67	.45	.21	....
						31......	....	.71	.45	....	....

Day.	July.	Aug.	Sept.	Oct.	Nov.	Day.	July.	Aug.	Sept.	Oct.	Nov.
1909.						**1909.**					
1......	....	0.50	0.40	....	....	16......	....	0.40	....	....	....
2......	....	....	....	0.30	0.30	17......	....	....	....	....	....
3......	....	....	....	.30	....	18......	0.80	....	....	....	....
4......	....	....	....	....	....	19......	....	.50	....	....	....
5......	....	....	....	....	....	20......	....	....	....	0.20	....
6......	....	....	.40	....	.20	21......	....	....	0.40	....	....
7......	....	.45	....	....	....	22......	....	....	....	....	....
8......	1.10	....	....	....	....	23......	.70	....	....	.30	....
9	....	....	.50	....	.30	24	....	....	....	....	....
10......	....	....	....	....	....	25......	....	.40	.40	....	....
11......	....	....	....	....	....	26......	....	....	....	.20	....
12......	....	....	....	....	....	27......	....	....	....	....	....
13......	....	....	....	.25	....	28......	....	.40	.35	....	....
14......	....	.40	....	....	....	29......	....	....	....	....	....
15......	.85	....	....	....	....	30......	....	....	....	.40	....
						31......	.50	....	....	....	....

NOTE.—Gage heights for 1909 are not comparable with those for 1908. Beginning Sept. 25, a new gage datum used.

Daily discharge, in second-feet, of St. Louis Creek at lower station near Fraser, Colo., for 1908–9.

Day.	June.	July.	Aug.	Sept.	Oct.	Day.	June.	July.	Aug.	Sept.	Oct.
1908.						1908.					
1......	80	270	86	47	21	16......	150	76	52	29	....
2......	82	200	95	47	21	17......	150	74	52	26	....
3......	84	135	102	47	21	18......	170	72	52	26	....
4......	86	137	115	47	21	19......	195	70	52	26	....
5......	88	139	125	47	21	20......	220	68	49	25	....
6......	89	142	135	47	21	21......	245	66	49	25	....
7......	100	146	158	47	21	22......	270	64	49	24	....
8......	110	134	110	49	21	23......	270	62	49	24	....
9......	122	122	77	49	21	24......	270	60	49	24	....
10......	140	110	51	49	21	25......	270	58	49	24	....
11......	150	98	51	45	21	26......	270	56	49	25	....
12......	150	86	51	42	21	27......	270	53	49	25	....
13......	150	84	51	38	21	28......	270	63	48	25	....
14......	150	81	51	35	21	29......	270	68	47	25	....
15......	150	79	52	32	....	30......	270	77	47	25	....
						31......	....	84	47	....	....

Day.	July.	Aug.	Sept.	Oct.	Nov.	Day.	July.	Aug.	Sept.	Oct.	Nov.
1909.						1909.					
1......	250	60	49	35	35	16......	102	49	49	33	....
2......	250	60	49	35	35	17......	94	55	49	33	....
3......	236	60	49	35	35	18......	94	60	49	33	....
4......	218	55	49	35	35	19......	94	60	49	31	....
5......	218	55	49	37	33	20......	87	60	49	31	....
6......	218	55	49	37	21	21......	87	55	39	31	....
7......	200	55	49	39	33	22......	87	55	39	35	....
8......	182	55	49	39	35	23......	81	55	39	35	....
9......	182	55	49	44	35	24......	81	49	39	35	....
10......	164	49	49	39	35	25......	76	49	39	33	....
11......	146	49	49	37	35	26......	76	49	39	31	....
12......	128	49	49	35	35	27......	70	49	39	33	....
13......	114	49	49	33	35	28......	70	49	37	35	....
14......	114	49	49	33	35	29......	64	49	37	37	....
15......	102	49	49	33	....	30......	64	49	37	39	....
						31......	60	49	....	39	....

NOTE.—Discharges interpolated for days when gage was not read.

Monthly discharge of St. Louis Creek at lower station near Fraser, Colo., for 1908–9.

Month.	Discharge in second-feet.			Run-off (total in acre-feet).
	Maximum.	Minimum.	Mean.	
1908.				
June..	270	80	176	10,500
July...	270	53	97.9	6,020
August......................................	158	47	67.7	4,160
September...................................	49	25	34.9	2,080
Oct. 1–14...................................	21	21	21.0	583
1909.				
July...	250	60	129	7,930
August......................................	60	49	53.1	3,260
September...................................	49	37	45.5	2,710
October......................................	44	31	35.2	2,160
Nov. 1–14...................................	35	21	33.7	936

NORTH RANCH CREEK AT UPPER STATION NEAR ROLLINS PASS, COLO.

This station, which was established May 1, 1908, and discontinued November 9, 1909, is located at the foot of Rollins Pass, Colo., at an elevation of about 10,000 feet above sea level.

The approximate drainage area is 2 square miles. Records of gage height were obtained by the means of a Lallie float gage, the datum remaining constant during the maintenance of the station.

Discharge measurements of North Ranch Creek at upper station near Rollins Pass, Colo., in 1908-9.

Date.	Hydrographer.	Gage height.	Dis-charge.
1908.		*Feet.*	*Sec.-ft.*
May 8	Stanley Krajicek	1.54	2.4
17	....do	1.72	7.0
19	....do	1.86	12.2
June 8	....do	1.93	20.2
20	....do	2.20	26.3
July 7	....do	1.74	8.2
19	....do	1.74	8.1
27	....do	1.74	9.2
Aug. 31	....do	1.68	6.3
1909.			
July 13	N. O'Daniels	.70	10.5
Aug. 2	....do	.60	6
6	....do	.55	5
13	....do	.55	5
18	....do	.60	6
Sept. 2	....do	.50	4

Daily gage height, in feet, of North Ranch Creek at upper station near Rollins Pass, Colo., for 1908-9.

Day.	May.	June.	July.	Aug.	Sept.	Oct.	Day.	May.	June.	July.	Aug.	Sept.	Oct.
1908.							1908.						
1	1.54	1.64	1.87	1.50	1.50	1.50	16	1.54	1.89	1.50	1.50	1.50	......
2	1.54	1.79	1.82	1.50	1.50	1.50	17	1.69	1.86	1.50	1.50	1.50	......
3	1.54	1.84	1.77	1.50	1.50	1.50	18	1.69	1.86	1.50	1.50	1.50	......
4	1.55	1.84	1.72	1.50	1.50	1.50	19	1.70	1.86	1.50	1.50	1.50	......
5	1.56	1.86	1.67	1.50	1.50	1.50	20	1.66	1.84	1.50	1.50	1.50	......
6	1.57	1.94	1.62	1.50	1.50	1.51	21	1.84	1.85	1.50	1.50	1.50	......
7	1.58	1.84	1.50	1.50	1.50	1.51	22	1.82	1.85	1.50	1.51	1.56	......
8	1.50	1.74	1.50	1.50	1.50	1.51	23	1.74	1.85	1.50	1.51	1.50	......
9	1.60	1.94	1.50	1.50	1.50	1.51	24	1.74	1.86	1.50	1.51	1.50	......
10	1.56	1.84	1.50	1.50	1.50	1.51	25	1.74	1.86	1.50	1.51	1.50	......
11	1.54	1.86	1.50	1.50	1.50	1.51	26	1.74	1.86	1.50	1.50	1.49	......
12	1.54	1.84	1.50	1.50	1.50	1.51	27	1.69	1.87	1.50	1.50	1.49	......
13	1.54	1.84	1.50	1.50	1.50	1.51	28	1.69	1.87	1.50	1.50	1.49	......
14	1.54	1.94	1.50	1.50	1.50	1.51	29	1.71	1.87	1.50	1.50	1.49	......
15	1.54	......	1.50	1.50	1.50	......	30	1.74	1.87	1.50	1.50	1.49	......
							31	1.78	......	1.50	1.50		......

*Daily gage height, in feet, of North Ranch Creek at upper station near Rollins Pass, Colo., for 1908-9—*Continued.

Day.	July.	Aug.	Sept.	Oct.	Nov.	Day.	July.	Aug.	Sept.	Oct.	Nov.
1909.						1909.					
1......			0.50	0.60	....	16......	.65	.55	.60	....	....
2......		.60	.50	.60	....	17......	.65	.60	.60	....	....
3......		.65	.50	.55	....	18......	.65	.60	....	....	....
4......		.60	.50	.55	.40	19......	.65	.70	....	....	....
5......		.55	.50	.55	....	20......	.60	.55	.60	....	....
6......		.55	.55	.55	....	21......	.60	.55	.60	....	....
7......		.55	.60	.55	....	22......	.60	.60	.60	....	....
8......	.90	.55	.60	.50	.40	23......	....	.60	.60	....	....
9......	.90	.55	.60	....	....	24......		.60	.60	....	....
10......	.90	.60	.60	....	....	25......		.60	.60	.50	....
11......	.85	.60	60	....	....	26......		.55	.60	....	....
12......	.85	.60	.60	.50	....	27......		.55	.60	....	....
13......	.70	.55	.60	....	....	28......		.55	.60	....	....
14......	.70	.55	.60	....	....	29......		.55	.60	....	....
15......	.70	.55	.60	....	....	30......		.55	.60	....	....
						31......		.50		....	....

Daily discharge, in second-feet, of North Ranch Creek at upper station near Rollins Pass, Colo., for 1908-9.

Day.	May.	June.	July.	Aug.	Sept.	Oct.	Day.	May.	June.	July.	Aug.	Sept.	Oct.
1908.							1908.						
1..........	3	5	4	2	2	2	16..........	5	14	2	2	2	......
2..........	3	10	4	2	2	2	17..........	7	13	2	· 2	2	......
3..........	3	12	4	2	2	2	18..........	7	13	2	2	2	......
4..........	3	12	4	2	2	2	19..........	7	13	2	2	2	......
5..........	3	13	4	2	2	2	20..........	6	12	2	2	2	......
6..........	4	16	3	2	2	2	21..........	12	13	2	2	2	......
7..........	4	12	2	2	2	2	22..........	11	13	2	2	2	......
8..........	4	9	2	2	2	2	23..........	9	13	2	2	2	......
9..........	4	16	2	2	2	2	24..........	9	13	2	2	2	......
10..........	3	12	2	2	2	2	25..........	9	13	2	2	2	......
11..........	3	13	2	2	2	2	26..........	9	13	2	2	2	......
12..........	3	12	2	2	2	2	27..........	7	13	2	2	2	......
13..........	3	12	2	2	2	2	28..........	7	13	2	2	2	......
14..........	3	16	2	2	2	2	29..........	8	13	2	2	2	......
15..........	3	15	2	2	2	......	30..........	9	13	2	2	2	......
							31..........	10		2	2		......

Day.	July.	Aug.	Sept.	Oct.	Nov.	Day.	July.	Aug.	Sept.	Oct.	Nov.
1909.						1909.					
1......	52	6	4	6	2	16......	8	5	6	4	....
2......	52	6	4	6	2	17......	8	6	6	4	....
3......	52	8	4	5	2	18......	8	6	6	4	....
4......	46	6	4	5	2	19......	8	10.5	6	4	....
5......	46	5	4	5	2	20......	6	5	6	4	....
6......	46	5	5	5	2	21......	6	5	6	4	....
7......	37	5	6	5	2	22......	6	6	6	4	....
8......	37	5	6	4	2	23......	6	6	6	4	....
9......	37	5	6	4	2	24......	6	6	6	4	....
10......	37	6	6	4	2	25......	6	6	6	4	....
11......	27	6	6	4	2	26......	6	5	6	4	....
12......	27	6	6	4	2	27......	6	5	6	4	....
13......	10.5	5	6	4	2	28......	6	5	6	4	....
14......	10.5	5	6	4	2	29......	6	5	6	4	....
15......	10.5	5	6	4	....	30......	6	5	6	4	....
						31......	6	4	....	4	....

NOTE.—Discharges estimated for days in which gage was not read.

Monthly discharge of North Ranch Creek at upper station near Rollins Pass, Colo., for 1908-9.

Month.	Discharge in second-feet.			Run-off (total in acre-feet).
	Maximum.	Minimum.	Mean.	
1908.				
May..	12	3	5.8	357
June...	16	5	12.7	756
July...	4	2	2.4	148
August..	2	2	2.0	123
September...	2	2	2.0	119
Oct. 1-14	2	2	2.0	56
1909.				
July...	52	6	20.4	1,250
August..	10.5	4	5.6	344
September...	6	4	5.6	333
October...	6	4	4.3	264
November..	2	2	2.0	56

NORTH RANCH CREEK AT LOWER STATION NEAR ROLLINS PASS, COLO.

This station, which was established June 3, 1907, and discontinued November 9, 1909, is located near Rollins Pass, about 4 miles below the upper station, and just above its junction with Middle Ranch Creek. The drainage area is about 5 square miles.

The datum of the rod gage remained constant during the maintenance of the station.

Discharge measurements of North Ranch Creek at lower station near Rollins Pass, Colo., in 1907-9.

Date.	Hydrographer.	Gage height.	Discharge.
1907.		*Feet.*	*Sec.-ft.*
June 4	G. M. Bull..	0.80	23.3
9	.....do...	1.15	38.
14	.....do...	1.80	76.
26	.....do...	1.95	92.
July 3	.....do...	1.70	75.
14	.....do...	1.15	43.
Aug. 5	.....do...	.45	12.
13	.....do...	.20	8.2
20	.....do...	.10	4.8
1908.			
May 17	Stanley Krajicek..	1.30	8.4
June 8	.....do...	1.75	30.6
July 4	.....do...	1.57	20.9
11	.....do...	1.33	15.1
Aug. 12	.....do...	1.00	8.0
31	.....do...	1.77	5.0
Sept. 29	.....do...	.81	4.1
1909.			
July 8	N. O'Daniels..	1.40	58.
13	.....do...	1.00	39.
Aug. 2	.....do...	.40	13.
6	.....do...	.30	10.
13	.....do...	.30	10.

Daily gage height, in feet, of North Ranch Creek at lower station near Rollins Pass, Colo., for 1907-1909.

Day.	June.	July.	Aug.	Day.	June.	July.	Aug.	Day.	June.	July.	Aug.
1907.				**1907.**				**1907.**			
1			0.55	11	1.15			21		1.7	
2				12		1.1	0.2	22		1.8	
3	0.8	1.7	.5	13			.2	23			
4	.8			14	1.8	1.15		24			
5		1.5	.45	15	1.8			25			
6				16				26		1.95	
7			.4	17	1.7			27			
8	1.15			18		.95		28		1.7	
9	1.15	1.3		19	1.8			29			
10	1.15	1.2	.3	20		.8	.1	30			
								31			

Day.	May.	June.	July.	Aug.	Sept.	Oct.	Day.	May.	June.	July.	Aug.	Sept.	Oct.
1908.							**1908.**						
1	1.10	1.65	1.57	0.95	0.77	0.80	16	1.20	1.95	1.21	0.82	0.77	
2	1.10	1.80	1.57	.96	.77	.79	17	1.30	1.93	1.15	.75	.77	
3	1.10	1.85	1.57	.96	.77	.79	18	1.30	1.91	1.09	.75	.77	
4	1.10	1.87	1.57	.97	.77	.79	19	1.30	1.89	1.00	.75	.77	
5	1.10	1.87	1.55	.97	.77	.79	20	1.30	1.87	1.00	.75	.77	
6	1.10	1.95	1.53	.98	.77	.78	21	1.30	1.85	1.00	.75	.77	
7	1.10	1.85	1.51	.99	.77	.78	22	1.30	1.83	1.00	.75	.77	
8	1.10	1.75	1.49	.99	.77	.77	23	1.30	1.81	1.00	.75	.78	
9	1.10	1.85	1.47	1.00	.77	.76	24	1.31	1.79	1.00	.75	.78	
10	1.10	1.95	1.40	1.00	.77	.76	25	1.33	1.77	1.00	.76	.78	
11	1.10	2.05	1.33	1.00	.77	.75	26	1.34	1.75	1.00	.76	.79	
12	1.10	2.03	1.33	1.00	.77	.75	27	1.36	1.70	1.00	.76	.79	
13	1.10	2.01	1.33	1.00	.77	.75	28	1.38	1.66	.99	.77	.80	
14	1.10	1.99	1.33	1.00	.77	.75	29	1.40	1.62	.97	.77	.80	
15	1.15	1.97	1.27	.84	.77		30	1.43	1.57	.96	.77	.81	
							31	1.45		.95	.77		

Day.	July.	Aug.	Sept.	Oct.	Nov.	Day.	July.	Aug.	Sept.	Oct.	Nov.
1909.						**1909.**					
1						16					
2		0.40	0.30			17					
3				0.30		18		0.35			
4					0.10	19					
5						20					
6		.30				21		.40			
7			.40			22					
8					.10	23		.35	0.35		
9						24					
10			.30			25				0.20	
11						26		.30			
12				.25		27			.30		
13		.30				28					
14	1.00					29					
15						30		.25	.30		
						31	0.40				

Daily discharge, in second-feet, of North Ranch Creek at lower station near Rollins Pass, Colo., for 1907–1909.

Day.	June.	July.	Aug.	Day.	June.	July.	Aug.	Day.	June.	July.	Aug.
1907.				1907.				1907.			
1	21	75	16	11	38	40	9	21	75	22	4
2	22	75	15	12	51	36	8	22	75	22	4
3	23	75	14	13	64	39	8	23	76	21	4
4	23	67	13	14	76	43	8	24	81	21	3
5	27	58	12	15	76	39	8	25	86	20	3
6	31	55	12	16	76	36	7	26	92	20	3
7	35	52	11	17	75	33	7	27	84	19	3
8	38	49	11	18	75	30	6	28	75	19	2
9	38	47	10	19	76	27	5	29	75	18	2
10	38	43	10	20	76	23	5	30	75	18	2
								31		17	2

Day.	May.	June.	July.	Aug.	Sept.	Oct.	Day.	May.	June.	July.	Aug.	Sept.	Oct.
1908.													
1	10	25	23	7	5	5	16	12	37	12	5	5	......
2	10	31	23	7	5	5	17	15	36	11	4	5	......
3	10	33	23	7	5	5	18	15	35	11	4	5	......
4	10	33	23	8	5	5	19	15	35	8	4	5	......
5	10	34	22	8	5	5	20	15	34	8	4	.5	......
6	10	37	22	8	5	5	21	15	33	8	4	5	......
7	10	33	21	8	5	5	22	15	32	8	4	5	......
8	10	29	20	8	5	5	23	15	31	8	4	5	......
9	10	33	19	8	5	4	24	15	31	8	4	5	......
10	10	37	18	8	5	4	25	16	30	8	5	5	......
11	10	42	16	8	5	4	26	16	29	8	5	5	......
12	10	41	16	8	5	4	27	17	27	8	5	5	......
13	10	40	16	8	5	4	28	17	26	8	5	5	......
14	10	39	16	8	5	4	29	18	24	8	5	5	......
15	11	38	14	6	5	......	30	18	22	7	5	5	......
							31	19		7	5		......

Day.	July.	Aug.	Sept.	Oct.	Nov.	Day.	July.	Aug.	Sept.	Oct.	Nov.
1909.											
1	73	13	10	10	5.5	16	34	10	10	9	....
2	70.5	13	10	10	5.5	17	29.5	11.5	10	9	....
3	68	11.5	10	10	5.5	18	29.5	11.5	10	9	....
4	65.5	11.5	11.5	10	5.5	19	25	11.5	10	9	....
5	63	10	11.5	10	5.5	20	25	13	11.5	9	....
6	63	10	13	10	5.5	21	20.5	13	1..5	9	....
7	60.5	10	13	10	5.5	22	20.5	13	11.5	9	....
8	58	10	11.5	10	5.5	23	16.5	11.5	11.5	9	....
9	58	10	10	10	5.5	24	16.5	11.5	11.5	7.5	....
10	53.5	10	10	9	5.5	25	16.5	10	11.5	7.5	....
11	48.5	10	10	9	5.5	26	13	10	10	7.5	....
12	43.5	10	10	9	5.5	27	13	10	10	7.5	....
13	39	10	10	9	5.5	28	13	10	10	7.5	....
14	39	10	10	9	5.5	29	13	10	10	7.5	....
15	34	10	10	9	....	30	13	9	10	7.5	....
						31	13	9		7.5	....

NOTE.—Discharges estimated for days on which gage height was not read.

Monthly discharge of North Ranch Creek at lower station near Rollins Pass, Colo., for 1907–1909.

Month.	Discharge in second-feet.			Run-off (total in acre-feet).
	Maximum.	Minimum.	Mean.	
1907.				
June	92	21	59	3,550
July	75	17	37	2,320
August	16	2	7	454
1908.				
May	19	10	13.0	799
June	42	22	32.9	1,960
July	23	7	13.8	848
August	8	4	6.0	369
September	5	5	5.0	298
Oct. 1-14	5	4	4.6	128
1909.				
July	73	13	37.0	2,280
August	13	9	10.8	664
September	13	10	10.6	631
October	10	7.5	8.9	547
Nov. 1-14	5.5	5 5	5.5	153

MIDDLE RANCH CREEK AT UPPER STATION NEAR ARROW, COLO.

This station, which was established May 1, 1908, and discontinued November 8, 1909, is located 3 miles from Arrow, on the Colonial Wagon Road, at an elevation of about 10,000 feet above sea level. Its drainage area is about 2½ square miles, and there are no tributaries above, which could be designated.

The location and datum of the Lallie automatic gage remained the same while the station was maintained.

Discharge measurements of Middle Ranch Creek at upper station near Arrow, Colo., in 1908–1909.

Date.	Hydrographer.	Gage height.	Discharge.
1908.		*Feet.*	*Sec.-ft.*
May 7	Stanley Krajicek	0.73	2.6
19	....do	1.03	12.8
26	....do	.93	9.4
June 15	....do	1.93	46.9
20	....do	1.10	22.0
July 7	....do	.88	10.9
21	....do	.78	3.7
Aug. 31	....do	.50	.4
Sept. 22	....do	.55	.8
1909.			
July 19	N. O'Daniels	0.7	9.5
Aug. 2	....do	.60	6.5
6	....do	.55	5.5
13	....do	.50	5.0
18	....do	.55	5.5

Daily gage height, in feet, of Middle Ranch Creek at upper station near Arrow, Colo., for 1908–9.

Day.	May.	June.	July.	Aug.	Sept.	Oct.	Day.	May.	June.	July.	Aug.	Sept.	Oct.
1908.							**1908.**						
1	0.73	1.01	0.95	0.81	0.54	0.54	16	0.91	1.20	0.85	0.81	0.53	......
2	.73	1.02	.95	.81	.54	.54	17	.93	1.20	.85	.81	.52	......
3	.73	1.03	.95	.81	.54	.54	18	.95	1.20	.85	.83	.52	......
4	.73	1.14	.95	.81	.54	.54	19	.98	1.20	.85	.83	.51	......
5	.73	1.14	.95	.81	.54	.54	20	1.13	1.20	.85	.83	.51	......
6	.73	1.14	.93	.81	.54	.54	21	.96	.95	.74	.83	.50	......
7	.73	1.13	.93	.81	.54	.55	22	.99	.95	.74	.83	.51	......
8	.75	1.19	.85	.81	.54	.54	23	.93	.95	.74	.83	.51	......
9	.77	1.14	.85	.81	.54	.55	24	.93	.95	.74	.78	.51	......
10	.79	1.14	.85	.81	.54	.55	25	.93	.95	.74	.68	.52	......
11	.81	1.16	.85	.81	.54	.55	26	.93	.95	.74	.65	.52	......
12	.83	1.22	.85	.81	.55	.55	27	.93	.95	.74	.63	.53	......
13	.85	1.23	.85	.81	.55	.60	28	.88	.95	.81	.60	.53	......
14	.87	1.24	.85	.81	.55	.55	29	.88	.95	.81	.58	.54	......
15	.89	1.20	.85	.81	.53	......	30	.88	.95	.81	.58	.54	......
							31	1.01	......	.81	.58	......	......

Day.	July.	Aug.	Sept.	Oct.	Nov.	Day.	July.	Aug.	Sept.	Oct.	Nov.
1909.						**1909.**					
1	......	......	0.50	0.60	....	16	....	0.50	....	0.55	....
2	......	0.60	.50	.60	....	17	....	.50	....	.55	....
3	......	.60	.50	.55	....	18	....	.55	....	.55	....
4	......	.60	.50	.55	0.60	19	0.70	.60	....	.55	....
5	......	.55	.50	.55	.60	20	.70	.60	0.55	.55	....
6	......	.55	.55	.55	....	21	.70	.60	....	.55	....
7	......	.55	.55	.55	....	22	.65	.55	....	.55	....
8	......	....	.55	.55	.60	23	.65	.55	....	.55	....
9	......	....	.55	.55	....	24	.70	.55	....	....	....
10	......	....	.55	.55	....	25	.65	.55	....	....	....
11	......	....	.55	.55	....	26	.60	.55	.60	....	....
12	......	....	....	.55	....	27	.60	.50	.60	.60	....
13	......	.50	....	.55	....	28	....	.50	.60	....	....
14	......	....	....	.55	....	29	....	.50	.60	....	....
15	......	.50	....	.55	....	30	....	.50	.60	....	....
						31	....	.50	....	....	....

Daily discharge, in second-feet, of Middle Ranch Creek at upper station near Arrow, Colo., for 1908–9.

Day.	May.	June.	July.	Aug.	Sept.	Oct.	Day.	May.	June.	July.	Aug.	Sept.	Oct.
1908.							**1908.**						
1	3	13	10	5	1	1	16	8	31	6	5	1	......
2	3	14	10	5	1	1	17	9	31	6	5	1	......
3	3	15	10	5	1	1	18	10	31	6	5	1	......
4	3	24	10	5	1	1	19	10	31	6	5	1	......
5	3	24	10	5	1	1	20	23	31	6	5	1	......
6	3	24	9	5	1	1	21	10	10	3	5	1	......
7	3	24	9	5	1	1	22	10	10	3	5	1	......
8	3	24	6	5	1	1	23	9	10	3	5	1	......
9	3	24	6	5	1	1	24	9	10	3	4	1	......
10	4	27	6	5	1	1	25	9	10	3	2	1	......
11	5	34	6	5	1	1	26	9	10	3	2	1	......
12	5	34	6	5	1	1	27	9	10	3	1	1	......
13	6	37	6	5	1	1	28	7	10	5	1	1	......
14	7	31	6	5	1	1	29	7	10	5	1	1	......
15	7	31	6	5	1	......	30	7	10	5	1	1	......
							31	13	......	5	1	......	......

Daily discharge, in second-feet, of Middle Ranch Creek at upper station near Arrow, Colo.,
for 1908-9—Continued.

Day.	July.	Aug.	Sept.	Oct.	Nov.	Day.	July.	Aug.	Sept.	Oct.	Nov.
1909.						1909.					
1	25.0	6.5	5.0	6.5	6.5	16	11.0	5.0	5.5	5.5	....
2	25.0	6.5	5.0	6.5	6.5	17	10.5	5.0	5.5	5.5	....
3	22.5	6.5	5.0	5.5	6.5	18	10.0	5.5	5.5	5.5	....
4	22.5	6.5	5.0	5.5	6.5	19	9.5	6.5	5.5	5.5	....
5	22.5	5.5	5.0	5.5	6.5	20	9.5	6.5	5.5	5.5	....
6	21.0	5.5	5.5	5.5	6.5	21	9.5	6.5	5.5	5.5	....
7	21.0	5.5	5.5	5.5	6.5	22	7.5	5.5	5.5	5.5	....
8	21.0	5.5	5.5	5.5	6.5	23	7.5	5.5	5.5	5.5	....
9	19.0	5.5	5.5	5.5	6.5	24	9.5	5.5	5.5	6.5	....
10	17.0	5.5	5.5	5.5	6.5	25	7.5	5.5	5.5	6.5	....
11	16.0	5.5	5.5	5.5	6.5	26	6.5	5.5	6.5	6.5	....
12	15.0	5.0	5.5	5.5	6.5	27	6.5	5.0	6.5	6.5	....
13	13.5	5.0	5.5	5.5	6.5	28	6.5	5.0	6.5	6.5	....
14	13.5	5.0	5.5	5.5	6.5	29	6.5	5.0	6.5	6.5	....
15	12.5	5.0	5.5	5.5	....	30	6.5	5.0	6.5	6.5	....
						31	6.5	5.0	....	6.5	....

NOTE.—Discharges estimated for days on which gage height was not read.

Monthly discharge of Middle Ranch Creek at upper station near Arrow, Colo., for 1908-9.

Month.	Discharge in second-feet.			Run-off (total in acre-feet).
	Maximum.	Minimum.	Mean.	
1908.				
May	23	3	7.10	437
June	37	10	21.2	1,260
July	10	3	6.0	369
August	5	1	4.1	252
September	1	1	1.0	60
October 1-14	1	1	1.0	28
1909.				
July	25.0	6.5	13.5	830
August	6.5	5.0	5.55	341
September	6.5	5.0	5.58	332
October	6.5	5.5	5.76	354
November 1-14	6.5	6.5	6.5	180

MIDDLE RANCH CREEK AT LOWER STATION NEAR ARROW, COLO.

This station which was established June 4, 1907, and discontinued November 8, 1909, is situated about 3 miles below the upper station and just above the junction of the stream with North Ranch Creek. The drainage area is about 3½ square miles.

The datum and location of the rod gage remained constant while station was being maintained.

Discharge measurements of Middle Ranch Creek at lower station near Arrow, Colo., in 1907–1909.

Date.	Hydrographer.	Gage.	Discharge.
1907.		*Feet.*	*Sec. ft.*
June 4	G. M. Bull	1.35	18.5
9	....do	1.60	32
14	....do	2.00	64
July 10	....do	1.50	23
Aug. 1	....do	1.00	6.6
5	....do	1.00	6.1
13	....do	.95	4.3
20	....do	.90	3.6
1908.			
June 8	Stanley Krajicek	1.40	27.3
12	....do	1.65	33.7
July 4	....do	1.16	13.2
11	....do	1.05	9.8
27	....do	1.00	6.4
Aug. 12	....do	.90	5.3
Sept. 29	....do	.97	5.7
1909.			
July 13	N. O'Daniels	1.40	20
Aug. 2	....do	.95	9
6	....do	.90	8
13	....do	.90	8

Daily gage height, in feet, of Middle Ranch Creek at lower station near Arrow, Colo., for 1907–1909.

Day.	June.	July.	Aug.	Day.	June.	July.	Aug.	Day.	June.	July.	Aug.
1907.				**1907.**				**1907.**			
1			1.00	11	1.65			21	1.85		
2				12		1.40	0.95	22			
3	1.41	1.85	1.00	13			.95	23	1.95		
4	1.35			14	2.00	1.40		24			
5		1.70	1.00	15	2.00			25			
6				16				26	2.00		
7			1.00	17	1.90			27			
8	1.65	1.55		18		1.20		28	1.85		
9	1.60	1.55		19	1.95			29			
10	1.60	1.50	.95	20		1.15	.90	30			
								31			

Day.	May.	June.	July.	Aug.	Sept.	Oct.	Day.	May.	June.	July.	Aug.	Sept.	Oct.
1908.							**1908.**						
1	1.20	1.40	1.36	0.98	0.70	0.60	16	1.20	1.60	1.05	0.90	0.70	
2	1.20	1.40	1.32	.97	.70	.60	17	1.20	1.58	1.05	.90	.70	
3	1.20	1.40	1.24	.96	.70	.60	18	1.20	1.56	1.05	.90	.70	
4	1.20	1.40	1.16	.96	.70	.60	19	1.20	1.54	1.05	.90	.70	
5	1.20	1.40	1.14	.96	.70	.60	20	1.20	1.52	1.05	.90	.70	
6	1.20	1.40	1.12	.94	.70	.60	21	1.20	1.50	1.05	.90	.68	
7	1.20	1.40	1.10	.93	.70	.60	22	1.20	1.48	1.05	.86	.66	
8	1.20	1.40	1.07	.92	.70	.60	23	1.20	1.46	1.05	.82	.63	
9	1.20	1.40	1.06	.92	.70	.60	24	1.18	1.44	1.05	.78	.61	
10	1.20	1.40	1.05	.91	.70	.60	25	1.16	1.40	1.05	.74	.58	
11	1.20	1.40	1.05	.91	.70	.60	26	1.13	1.40	1.05	.70	.56	
12	1.20	1.40	1.05	.90	.70	.60	27	1.11	1.40	1.00	.70	.54	
13	1.20	1.65	1.05	.90	.70	.60	28	1.09	1.40	.99	.70	.52	
14	1.20	1.64	1.05	.90	.70	.60	29	1.06	1.40	.99	.70	.50	
15	1.20	1.62	1.05	.90	.70		30	1.03	1.39	.98	.70	.50	
							31	1.01		.98	.70		

Daily gage height, in feet, of Middle Ranch Creek at lower station near Arrow, Colo., for 1907–1909—Continued.

Day.	July.	Aug.	Sept.	Oct.	Nov.	Day.	July.	Aug.	Sept.	Oct.	Nov.
1909.						1909.					
1						16					
2		0.95	0.90			17					
3						18		0.90			
4					0.95	19					
5						20					
6		.90				21		.90			
7			.90			22					
8	1.65				1.10	23		.90	1.00		
9			.90			24					
10						25				0.95	
11						26		.90			
12				0.90		27			1.00		
13	1.40	.90				28					
14						29					
15						30		.90	.95		
						31					

Daily discharge, in second-feet, of Middle Ranch Creek at lower station near Arrow, Colo., for 1907–1909.

Day.	June.	July.	Aug.	Day.	June.	July.	Aug.	Day.	June.	July.	Aug.
1907.				1907.				1907.			
1	19	48	6.6	11	35	22	4	21	48	12	3
2	20	48	7	12	45	20	4	22	53	11	3
3	21	48	7	13	55	20	4.3	23	58	11	3
4	18.5	43	7	14	64	20	4	24	60	10	3
5	22	37	6.1	15	64	18	4	25	62	10	3
6	26	35	7	16	59	16	4	26	64	9	3
7	30	33	7	17	54	14	4	27	56	9	3
8	35	31	6	18	56	13	4	28	48	8	3
9	32	28	5	19	58	13	4	29	48	8	3
10	32	23	4	20	53	12	3.6	30	48	7	3
								31		7	2

Day.	May.	June.	July.	Aug.	Sept.	Oct.	Day.	May.	June.	July.	Aug.	Sept.	Oct.
1908.							1908.						
1	12	21	18	7	2.5	1.7	16	12	32	8	5	2.5	
2	12	21	16	6	2.5	1.7	17	12	31	8	5	2.5	
3	12	21	14	6	2.5	1.7	18	12	30	8	5	2.5	
4	12	21	17	6	2.5	1.7	19	12	29	8	5	2.5	
5	12	21	11	6	2.5	1.7	20	12	27	8	5	2.5	
6	12	21	10	6	2.5	1.7	21	12	26	8	5	2.4	
7	12	21	8	6	2.5	1.7	22	12	25	8	5	2.2	
8	12	21	8	5	2.5	1.7	23	12	25	8	4	2.0	
9	12	21	8	5	2.5	1.7	24	12	25	8	3	1.8	
10	12	21	8	5	2.5	1.7	25	11	21	8	3	1.6	
11	12	21	8	5	2.5	1.7	26	10	21	8	2	1.4	
12	12	21	8	5	2.5	1.7	27	10	21	8	2	1.3	
13	12	36	8	5	2.5	1.7	28	9	21	7	2	1.1	
14	12	35	8	5	2.5	1.7	29	8	21	7	2	1.1	
15	12	34	8	5	2.5		30	8	20	7	2	1.1	
							31	7		7	2		

Daily discharge, in second-feet, of Middle Ranch Creek at lower station near Arrow, Colo., for 1907–1909—Continued.

Day.	July.	Aug.	Sept.	Oct.	Nov.	Day.	July.	Aug.	Sept.	Oct.	Nov.
1909.						1909.					
1......	31.5	9	8	9	9	16......	17.5	8	8	8	....
2......	31.5	9	8	9	9	17......	16	8	8	8	....
3......	30	9	8	9	9	18......	14.5	8	8	9	....
4......	30	9	8	9	9	19......	13	8	9	9	....
5......	30	8	8	9	10	20......	13	8	9	9	....
6......	28.5	8	8	9	10	21......	13	8	9	9	....
7......	28.5	8	8	9	12	22......	11	8	9	9	....
8......	28.5	8	8	9	12	23......	11	8	10	9	....
9......	26.5	8	8	8	12	24......	12	8	10	9	....
10......	25	8	8	8	12	25......	11	8	10	9	....
11......	23.5	8	8	8	12	26......	9	8	10	9	....
12......	21.5	8	8	8	10	27......	9	8	10	9	....
13......	20	8	8	8	10	28......	9	8	10	9	....
14......	20	8	8	8	9	29......	9	8	9	9	....
15......	18.5	8	8	8	....	30......	9	8	9	9	....
						31......	9	8	....	9	....

NOTE.—Discharges estimated for days on which gage was not read.

Monthly discharge of Middle Ranch Creek at lower station near Arrow, Colo., for 1907–1909.

Month.	Discharge in second-feet.			Run-off (total in acre-feet).
	Maximum.	Minimum.	Mean.	
1907.				
June.....................	64	19	45	2,690
July.....................	48	7	21	1,290
August...................	7	2	5	270
1908.				
May.....................	12	7	11.3	695
June.....................	36	20	24.4	1,450
July.....................	18	7	9.1	437
August...................	7	2	4.5	277
September................	2.5	1.1	2.20	130
Oct., 1–14................	1.7	1.7	1.70	47
1909.				
July.....................	31.5	9	18.7	1,150
August...................	9	8	8.1	498
September................	10	8	8.6	512
October..................	9	8	8.7	535
Nov., 1–14...............	12	9	10.4	289

SOUTH RANCH CREEK AT UPPER STATION NEAR ARROW, COLO.

This station, which was established May 1, 1908, and discontinued November 8, 1909, is located 2 miles from Arrow, on the Colonial wagon road, and at an elevation of about 10,000 feet above sea level. The drainage area is about 3 squares miles. The datum and location of the Lallie automatic gage remained constant while the station was being maintained.

Discharge measurements of South Ranch Creek at upper station near Arrow, Colo., in 1908–1909.

Date.	Hydrographer.	Gage height.	Discharge.
1908.		*Feet.*	*Sec.-ft.*
May 6	Stanley Krajicek	1.00	4.2
19	....do...	1.75	33.2
26	....do...	1.37	19.5
June 6	....do	1.32	16.3
20	....do...	1.30	14.4
July 7	....do	1.07	7.5
19	....do...	1.00	3.9
27	....do...	.98	3.7
Aug. 31	....do...	.85	1.8
Sept. 29	....do	.88	2.3
29	....do...	.85	2.6
1909.			
July 19	N. O'Daniels...	6.40	9
Aug. 2	....do...	.30	6
6	....do...	.25	5
13	....do...	.20	4
18	....do...	.20	4

Daily gage height, in feet, of South Ranch Creek at upper station near Arrow, Colo., for 1908–1909.

Day.	May.	June.	July.	Aug.	Sept.	Oct.	Day.	May.	June.	July.	Aug.	Sept.	Oct.
1908.							**1908.**						
1..........	1.00	1.25	1.15	0.97	0.81	0.84	16..........	1.30	1.45	1.04	0.92	0.85	......
2..........	1.00	1.25	1.15	.97	.82	.84	17..........	1.40	1.41	1.04	.92	.85	......
3..........	1.00	1.29	1.12	.95	.85	.84	18..........	1.40	1.38	1.00	.90	.85	......
4..........	1.00	1.30	1.11	.95	.85	.84	19..........	1.60	1.32	1.00	.90	.85	......
5..........	1.00	1.31	1.08	.95	.85	.84	20..........	1.33	1.30	1.00	.90	.85	......
6..........	1.00	1.33	1.08	.94	.85	.84	21..........	1.33	1.30	1.00	.90	.85	......
7..........	1.00	1.48	1.06	.94	.85	.84	22..........	1.33	1.30	1.00	.90	.85	......
8..........	1.00	1.42	1.08	.94	.85	.84	23..........	1.33	1.30	.98	.90	.85	......
9..........	1.00	1.42	1.08	.94	.85	.82	24..........	1.33	1.28	.98	.89	.82	......
10..........	1.00	1.50	1.08	.94	.85	.82	25..........	1.33	1.27	.98	.87	.83	......
11..........	1.00	1.51	1.08	.92	.85	.82	26..........	1.33	1.25	.98	.85	.84	......
12..........	1.45	1.52	1.08	.92	.85	.82	27..........	1.28	1.24	.98	.85	.83	......
13..........	1.40	1.51	1.08	.92	.85	.82	28..........	1.28	1.23	.97	.85	.82	......
14..........	1.40	1.49	1.08	.92	.85	.82	29..........	1.25	1.17	.97	.85	.84	......
15..........	1.30	1.47	1.05	.92	.85	......	30..........	1.24	1.17	.97	.85	.85	......
							31..........	1.24	......	.97	.84	......	......

Day.	July.	Aug.	Sept.	Oct.	Nov.	Day.	July.	Aug.	Sept.	Oct.	Nov.
1909.						**1909.**					
1......	....	....	0.20	....	....	16......	....	....	....	0.25	....
2......	....	0.30	.20	....	....	17......	....	....	....	.25	....
3......	....	.25	.20	0.25	....	18......	....	0.20	....	.25	....
4......	....	.25	.20	.25	0.20	19......	....	....	....	.25	....
5......	....	.25	.20	.25	....	20......	....	....	....	.25	....
6......	....	.25	.20	.25	....	21......	....	.20	....	.25	....
7......	....	.25	.25	.25	....	22......	....	.20	....	..25	....
8......	0.70	....	.20	.25	.20	23......	....	.20	....	.25	....
9......	....	....	.20	.25	....	24......	....	.20	....	....	....
10......	....	....	.20	.25	....	25......	....	.20	..	....	....
11......	....	....	.20	.25	....	26......	....	.20	....	....	....
12......	....	....	.20	.25	....	27......	....	.20	0.25	.20	....
13......	.55	.20	.20	.25	....	28......	....	.20	....	....	....
14......	....	....	....	.25	....	29......	....	.20	....	....	....
15......	....	....	....	.25	....	30......	....	.20	.25	....	....
						31......	0.30	.20	....	....	....

Daily discharge, in second-feet, of South Ranch Creek at upper station near Arrow, Colo., for 1908-9.

Day.	May.	June.	July.	Aug.	Sept.	Oct.	Day.	May.	June.	July.	Aug.	Sept.	Oct.
1908.							**1908.**						
1	5	12	8	3.9	1.3	1.7	16	13	20	5	2.9	1.8	......
2	5	12	8	3.9	1.4	1.7	17	17	18	5	2.9	1.8	......
3	5	13	8	3.8	1.8	1.7	18	17	17	3.9	2.5	1.8	......
4	5	13	7	3.8	1.8	1.7	19	27	14	3.9	2.5	1.8	......
5	5	14	6	3.8	1.8	1.7	20	14	13	3.9	2.5	1.8	......
6	5	14	6	3.0	1.8	1.7	21	14	13	3.9	2.5	1.8	......
7	5	20	6	3.0	1.8	1.7	22	14	13	3.9	2.5	1.8	......
8	5	18	6	3.0	1.8	1.7	23	14	13	3.8	2.5	1.8	......
9	5	18	6	3.0	1.8	1.4	24	14	13	3.8	2.4	1.4	......
10	5	22	6	3.0	1.8	1.4	25	14	12	3.8	2.1	1.5	......
11	5	22	6	2.9	1.8	1.4	26	14	12	3.8	1.8	1.7	......
12	20	23	6	2.9	1.8	1.4	27	13	11	3.8	1.8	1.5	......
13	17	22	6	2.9	1.8	1.4	28	13	11	3.7	1.8	1.4	......
14	17	22	6	2.9	1.8	1.4	29	12	9	3.7	1.8	1.4	......
15	13	21	6	2.9	1.8	......	30	11	9	3.7	1.8	1.8	......
							31	11	......	3.7	1.8	......	......

Day.	July.	Aug.	Sept.	Oct.	Nov.	Day.	July.	Aug.	Sept.	Oct.	Nov.
1909.						**1909.**					
1	27	6	4	5	4	16	13	4	4	5	....
2	27	6	4	5	4	17	11	4	4	5	....
3	27	5	4	5	4	18	11	4	4	5	....
4	24.5	5	4	5	4	19	9	4	5	5	....
5	24.5	5	4	5	4	20	9	4	5	5	....
6	22.5	5	4	5	4	21	9	4	5	5	....
7	22.5	5	5	5	4	22	9	4	5	5	....
8	22.5	5	4	5	4	23	7.5	4	5	5	....
9	20	5	4	5	4	24	7.5	4	5	4	....
10	20	5	4	5	4	25	7.5	4	5	4	....
11	17.5	4	4	5	4	26	7.5	4	5	4	....
12	17.5	4	4	5	4	27	7.5	4	5	4	....
13	15	4	4	5	4	28	6	4	5	4	....
14	15	4	4	5	4	29	6	4	5	4	....
15	13	4	4	5	....	30	6	4	5	4	....
						31	6	4	....	4	....

NOTE.—Discharges estimated for days on which gage was not read.

Monthly discharge of South Ranch Creek at upper station near Arrow, Colo., for 1908-9.

Month.	Discharge in second-feet.			Run-off (total in acre-feet).
	Maximum.	Minimum.	Mean.	
1908.				
May	27	5	11.4	701
June	23	9	15.5	922
July	8	3.7	5.17	318
August	3.9	1.8	2.74	168
September	1.8	1.3	1.71	102
Oct. 1-14	1.7	1.4	1.57	44
1909.				
July	27	6	14.5	892
August	6	4	4.39	270
September	5	4	4.43	264
October	5	4	4.74	291
Nov. 1-14	4	4	4.0	111

SOUTH RANCH CREEK AT LOWER STATION NEAR ARROW, COLO.

This station, which was established June 3, 1907, and discontinued November 8, 1909, is located about 2 miles below the upper station and just above the junction of the South Fork with the Middle Fork. The drainage area is about 5 square miles.

The datum and location of the rod gage remained constant while the station was being maintained.

Discharge measurements of South Ranch Creek at lower station near Arrow, Colo., in 1907–1909.

Date.	Hydrographer.	Gage height.	Discharge.
		Feet.	*Sec.-ft.*
1907.			
June 4	G. M. Bull..	1.75	28
9	.....do..	2.00	36
26	.....do..	1.95	31
July 5	.....do..	1.55	17.1
12	.....do..	1.25	10.3
Aug. 5	.....do..	.65	4.1
13	.....do..	.55	2.7
20	.....do..	.60	2.5
1908.			
June 12	Stanley Krajicek......................................	1.60	18.2
July 4	.....do..	.90	8.8
11	.....do..	.70	4.9
27	.....do..	.50	3.2
Aug. 31	.....do..	.49	3.3
Sept. 8	.....do..	.49	3.2
29	.....do..	.50	1.3
1909.			
July 13	N. O'Daniels...	.60	13
Aug. 2	.....do..	.20	5
6	.....do..	.15	4
13	.....do..	.15	4

Daily gage height, in feet, of South Ranch Creek at lower station near Arrow, Colo., for 1907–1909.

Day.	June.	July.	Aug.	Day.	June.	July.	Aug.	Day.	June.	July.	Aug.
1907.				1907.				1907.			
1............			0.7	11............	2.05			21............	2.0		
2............				12............		1.25	0.6	22............			
3............	1.75	1.6	.65	13............			.55	23............	2.0		
4............	1.75			14............	2.1	1.35		24............			
5............		1.55	.65	15............	2.2			25............			
6............				16............				26............		1.95	
7............			.6	17............	2.15			27............			
8............	2.05			18............		.9		28............		1.8	
9............	2.0	1.5		19............	2.15			29............			
10............	2.0	1.35	.6	20............		.85	.6	30............			
								31............			

Daily gage height, in feet, of South Ranch Creek at lower station near Arrow, Colo., for 1907–1909—Continued.

Day.	May.	June.	July.	Aug.	Sept.	Oct.	Day.	May.	June.	July.	Aug.	Sept.	Oct.
1908.							**1908.**						
1	1.50	1.15	1.00	0.50	0.49	0.49	16	1.50	1.48	0.60	0.50	0.48	.....
2	1.50	1.22	.99	.50	.49	.49	17	1.52	1.45	.60	.50	.49	.....
3	1.50	1.28	.96	.50	.49	.49	18	1.54	1.42	.60	.50	.49	.....
4	1.50	1.34	.93	.50	.49	.49	19	1.56	1.39	.60	.50	.49	.....
5	1.50	1.40	.90	.50	.49	.49	20	1.58	1.37	.60	.49	.49	.....
6	1.50	1.47	.86	.50	.49	.49	21	1.59	1.33	.60	.49	.49	.....
7	1.50	1.53	.82	.50	.49	.49	23	1.60	1.30	.60	.49	.49	.....
8	1.50	1.60	.79	.50	.49	.49	23	1.50	1.27	.55	.49	.49	.....
9	1.50	1.60	.76	.50	.49	.49	24	1.40	1.25	.55	.49	.49	.....
10	1.50	1.60	.73	.50	.48	.49	25	1.30	1.23	.55	.49	.49	.....
11	1.50	1.58	.70	.50	.48	.49	26	1.20	1.20	.55	.49	.49	.....
12	1.50	1.58	.66	.50	48	.49	27	1.19	1.19	.55	.49	.49	.....
13	1.50	1.55	.63	.50	.48	.49	28	1.18	1.18	.55	.49	.49	.....
14	1.50	1.53	.60	.50	.48	.49	29	1.18	1.17	.55	.49	.49	.....
15	1.50	1.50	.60	.50	.48	.....	30	1.16	1.16	.55	.49	.49	.....
							31	1.15	.....	.55	.49	.....	.....

Day.	July.	Aug.	Sept.	Oct.	Nov.	Day.	July.	Aug.	Sept.	Oct.	Nov.
1909.						**1909.**					
1	......	......	......	......	......	16	......	......	......	......	......
2	......	0.20	0.10	......	......	17	......	......	......	......	......
3	......	......	......	0.25	......	18	......	0.20	......	......	......
4	......	......	......	......	0.25	19	......	......	......	......	......
5	......	......	......	......	......	20	......	.15	......	......	......
6	......	.15	......	......	......	21	......	......	......	......	......
7	......	......	.20	......	......	22	......	......	......	......	......
8	0.80	......	......	......	.15	23	......	.15	0.30	......	......
9	......	......	.15	......	......	24	......	......	......	......	......
10	......	......	.15	......	......	25	......	......	......	0.25	......
11	......	......	......	......	......	26	......	.10	......	......	......
12	......	......	......	.25	......	27	......	......	.25	......	......
13	......	.15	......	......	......	28	......	......	......	......	......
14	.60	......	......	......	......	29	......	......	......	......	......
15	......	......	......	......	......	30	......	.10	.25	......	......
						31	0.20	......	......	......	......

Daily discharge, in second-feet, of South Ranch Creek at lower station near Arrow, Colo., for 1907–1909.

Day.	June.	July.	Aug.	Day.	June.	July.	Aug.	Day.	June.	July.	Aug.
1907.				**1907.**				**1907.**			
1	27	24	4	11	39	12	3	21	36	5	2
2	27	23	4	12	40	10	3	22	36	5	2
3	28	22	4	13	41	12	2.7	23	36	5	2
4	28	20	4	14	42	14	3	24	34	5	2
5	31	17	4.1	15	46	12	3	25	32	5	2
6	34	17	3	16	45	10	3	26	31	5	2
7	37	17	3	17	44	8	3	27	29	5	2
8	39	17	3	18	44	6	2	28	27	4	2
9	36	17	3	19	44	6	2	29	26	4	2
10	36	14	3	20	40	5	2.5	30	25	4	1
								31	......	4	1

Daily discharge, in second-feet, of South Ranch Creek at lower station near Arrow, Colo., for 1907–1909—Continued.

Day.	May.	June.	July.	Aug.	Sept.	Oct.	Day.	May.	June.	July.	Aug.	Sept.	Oct.
1908.							**1908.**						
1	15	12	8	3.2	3	3	16	15	16	4.2	3.2	3	....
2	15	13	8	3.2	3	3	17	16	15	4.2	3.2	3	....
3	15	14	8	3.2	3	3	18	17	15	4.2	3.2	3	....
4	15	14	7	3.2	3	3	19	17	15	4.2	3.2	3	....
5	15	15	7	3.2	3	3	20	18	16	4.2	3.1	3	....
6	15	16	7.6	3.2	3	3	21	18	14	4.2	3.1	3	....
7	15	18	6.5	3.2	3	3	22	18	13	4.2	3.1	3	....
8	15	19	6.5	3.2	3	3	23	16	12	4	3.1	3	....
9	15	19	6.5	3.2	3	3	24	14	12	4	3.1	3	....
10	15	19	5.4	3.2	3	3	25	13	12	4	3.1	3	....
11	15	18	5.4	3.2	3	3	26	11	12	4	3.1	3	....
12	15	18	5.4	3.2	3	3	27	11	11	4	3.1	3	....
13	15	17	4.5	3.2	3	3	28	11	11	4	3.1	3	....
14	15	17	4.2	3.2	3	3	29	11	11	4	3.1	3	....
15	15	16	4.2	3.2	3	....	30	10	11	4	3.1	....	....
							31	10		4	3.1		

Day.	July.	Aug.	Sept.	Oct.	Nov.	Day.	July.	Aug.	Sept.	Oct.	Nov.
1909.						**1909.**					
1	26.5	5	3	6	6	16	13.5	4	4	6	....
2	26.5	5	3	6	6	17	11.5	5	5	6	....
3	25	5	3	6	6	18	11.5	5	5	6	....
4	25	5	3	6	6	19	11.5	5	5	6	....
5	23.5	4	3	6	6	20	10.5	4	5	6	....
6	23.5	4	4	6	5	21	10.5	4	6	6	....
7	21.5	4	5	6	5	22	9	4	6	6	....
8	21.5	4	5	6	4	23	9	4	6.5	6	....
9	21.5	4	4	6	4	24	7.5	4	6.5	6	....
10	20	4	4	6	4	25	7.5	3	6.5	6	....
11	18	4	4	6	4	26	6.5	3	6	6	....
12	16.5	4	4	6	4	27	6.5	3	6	6	....
13	15	4	4	6	4	28	6	3	6	6	....
14	15	4	4	6	4	29	5	3	6	6	....
15	13.5	4	4	6	....	30	5	3	6	6	....
						31	5	3	....	6	....

NOTE.—Discharges estimated for days on which gage was not read.

Monthly discharge of South Ranch Creek at lower station near Arrow, Colo., for 1907–1909.

Month.	Discharge in second-feet.			Run-off (total in acre-feet).
	Maximum.	Minimum.	Mean.	
1907.				
June	46	25	35	2,120
July	24	4	11	668
August	4	1	3	164
1908.				
May	18	10	14.5	892
June	19	11	14.7	875
July	8	4	5.15	317
August	3.2	3.1	3.16	194
September	3	3	3.0	179
Oct. 1–14	3	3	3.0	84
1909.				
July	26.5	5	14.5	892
August	5	3	4.0	245
September	6.5	3	4.75	283
October	6	6	6.0	369
Nov. 1–14	6	4	4.86	135

WILLIAMS FORK BASIN.

WILLIAMS FORK NEAR SULPHUR SPRINGS, COLO.

Williams Fork rises in the Williams River Mountains and flows northwestward, joining Grand River in the central part of Middle Park.

The gaging station was established July 25, 1904, to obtain data for use in connection with the development of the stream for power, storage, and irrigation. It is located near the mouth of the stream, at the wagon bridge on the ranch of F. A. Field, about 9 miles west of Hot Sulphur Springs, Colo. The nearest railroad point is Parshall, a station on the Denver, Northwestern & Pacific Railroad.

The drainage area is about 200 square miles.

The station is below all tributaries. A number of irrigation ditches divert water below the station, and it is possible that in the future a tunnel will divert water from the headwaters of Williams Fork to the headwaters of Clear Creek in the Platte drainage basin. Some work has been done toward the construction of a reservoir and power plant a couple of miles downstream from the station. (See Pl. IV, A.)

Springs keep the ice from getting very thick at this station, but slush ice occurs frequently throughout the winter. The morning readings are usually distorted as the result of ice at the gage, but the afternoon readings indicate the open-water stage closely.

No change has been made in the location or in the datum of the staff gage at the bridge during the maintenance of the station.

Results are satisfactory. During low stages in the winter the flow is constant, being nearly all from springs.

Discharge measurements of Williams Fork near Sulphur Springs, Colo., in 1909.

Date.	Hydrographers.	Width.	Area of section.	Gage height.	Dis- charge.
		Feet.	*Sq. ft.*	*Feet.*	*Sec.-ft.*
Mar. 11	C. L. Chatfield..	26	42	3.20	44
Apr. 30	.....do..	47	60	3.40	64
May 9	.....do..	49	82	3.76	152
June 18	Chatfield and Snelson, jr.................................	54	139	5.00	1,110
Aug. 8	Freeman and Howe..	48	91	3.62	145
Oct. 14	Woolsey and Urquhart....................................	44	72.4	3.47	105

Daily gage height, in feet, of Williams Fork near Sulphur Springs, Colo., for 1909.

[F. A. Field, observer.]

Day.	Jan.	Feb.	Mar.	Apr.	May.	June.	July.	Aug.	Sept.	Oct.	Nov.	Dec.
1	3.20	3.20	3.13	3.22	3.43	4.11	4.88	3.85	3.76	3.48	3.33	3.26
2	3.20	3.20	3.15	3.28	3.41	4.18	4.85	3.81	3.68	3.46	3.35	3.20
3	3.25	3.18	3.17	3.30	3.43	4.22	4.77	3.75	3.61	3.44	3.32	3.15
4	3.23	3.16	3.20	3.25	3.60	4.40	4.94	3.69	3.60	3.44	3.32	3.12
5	3.22	3.16	3.15	3.27	3.75	4.56	5.00	3.64	3.61	3.43	3.26	3.10
6	3.22	3.12	3.18	3.30	3.77	4.66	4.76	3.66	3.67	3.46	3.22	3.12
7	3.16	3.14	3.15	3.32	3.75	4.75	4.68	3.69	3.82	3.50	3.24	3.12
8	3.16	3.14	3.15	3.28	3.79	4.85	4.66	3.65	3.71	3.50	3.18	3.10
9	3.14	3.12	3.16	3.32	3.85	4.90	4.62	3.65	3.72	3.39	3.12	3.10
10	3.10	3.10	3.20	3.40	3.90	4.78	4.58	3.64	3.78	3.49	3.14	3.10
11	3.10	3.12	3.21	3.32	3.96	4.77	4.46	3.66	3.72	3.48	3.18	3.10
12	3.12	3.12	3.22	3.30	3.98	4.74	4.36	3.65	3.62	3.44	3.35	3.12
13	3.12	3.10	3.22	3.27	3.88	4.80	4.32	3.71	3.66	3.46	3.18	3.12
14	3.14	3.12	3.18	3.27	3.88	4.76	4.23	3.62	3.66	3.46	3.18	3.10
15	3.14	3.12	3.22	3.34	3.82	4.75	4.20	3.59	3.74	3.43	3.20	3.10
16	3.14	3.12	3.20	3.35	3.92	4.80	4.14	3.56	3.70	3.42	3.12	3.10
17	3.18	3.12	3.22	3.38	3.97	4.98	4.10	......	3.68	3.41	3.10	3.10
18	3.20	3.12	3.22	3.42	4.05	5.15	4.10	......	3.63	3.40	3.10	3.10
19	3.20	3.12	3.22	3.46	4.04	5.28	4.10	3.93	3.61	3.40	3.12	3.10
20	3.18	3.10	3.22	3.39	4.09	5.28	4.10	3.76	3.59	3.39	3.40	3.12
21	3.15	3.10	3.23	3.36	4.17	5.12	4.06	3.80	3.58	3.40	3.32	3.12
22	3.15	3.11	3.20	3.36	4.22	5.12	3.98	3.72	3.60	3.39	3.28	3.10
23	3.16	3.12	3.18	3.43	4.25	5.15	3.90	3.68	3.60	3.41	3.30	3.10
24	3.12	3.12	3.22	3.41	4.18	5.16	4.02	3.69	3.58	3.38	3.36	3.12
25	3.12	3.12	3.25	3.42	4.09	5.20	4.10	3.66	3.57	3.39	3.36	3.12
26	3.14	3.13	3.22	3.44	4.02	5.14	4.07	3.61	3.54	3.34	3.32	3.10
27	3.12	3.12	3.25	3.54	4.01	4.96	4.02	3.57	3.52	3.36	3.30	3.15
28	3.12	3.13	3.28	3.56	4.13	4.96	3.96	3.60	3.51	3.35	3.10	3.15
29	3.15	......	3.28	3.50	4.18	4.92	3.87	3.57	3.50	3.34	3.38	3.18
30	3.15	......	3.28	3.45	4.04	4.96	3.85	3.58	3.49	3.33	3.30	3.16
31	3.20	......	3.27	......	4.12	......	3.70	3.70	......	3.32	......	3.18

NOTE.—Ice conditions from Jan. 1 to Apr. 15 and from Nov. 11 to Dec. 31. The afternoon readings are given as the mean for the day during these ice periods, as they are less affected by ice.

Daily discharge, in second-feet, of Williams Fork near Sulphur Springs, Colo., for 1909.

Day.	Jan.	Feb.	Mar.	Apr.	May.	June.	July.	Aug.	Sept.	Oct.	Nov.	Dec.
1	44	44	36	47	69	306	985	230	195	111	84	75
2	44	44	38	55	66	348	955	214	166	107	88	68
3	51	42	40	58	69	374	875	191	143	103	83	63
4	48	39	44	51	104	510	1,050	169	140	103	83	60
5	47	39	38	54	148	656	1,110	153	143	101	75	58
6	47	34	42	58	155	755	865	159	162	107	70	60
7	39	37	38	61	148	848	786	169	218	115	73	60
8	39	34	38	55	162	952	767	156	176	115	66	58
9	37	34	39	61	184	1,000	729	156	180	94	60	58
10	32	32	44	75	204	879	692	153	202	113	62	58
11	32	34	45	61	230	868	588	159	180	111	66	58
12	34	34	47	58	239	837	510	156	146	103	88	60
13	34	32	47	54	196	900	480	176	159	107	66	60
14	37	34	42	54	196	858	420	146	159	107	66	58
15	37	34	47	65	173	848	400	138	187	101	68	58
16	37	34	44	56	213	905	367	130	172	99	60	58
17	42	34	47	61	235	1,090	345	175	166	97	58	58
18	44	34	47	68	274	1,270	345	219	150	95	58	58
19	44	34	47	75	269	1,410	345	264	143	95	60	58
20	42	32	47	62	295	1,410	345	195	138	94	95	60
21	38	32	48	58	342	1,240	325	210	135	95	83	60
22	38	33	44	58	374	1,240	286	180	140	94	78	58
23	39	34	42	69	395	1,270	250	166	140	97	80	58
24	34	34	47	66	348	1,280	305	169	135	92	89	60
25	34	34	51	68	295	1,320	345	159	132	94	89	60
26	37	36	47	71	258	1,260	330	143	125	86	83	60
27	34	34	51	91	253	1,070	305	132	120	89	80	63
28	34	36	55	95	318	1,070	277	140	118	88	58	63
29	38	......	55	82	348	1,030	238	132	115	86	92	66
30	38	......	55	73	269	1,070	230	135	113	84	80	64
31	44	......	54	......	312	......	172	172	......	83	......	66

NOTE.—These discharges are based on rating curves applicable as follows: Jan. 1 to Apr. 15, not well defined; Apr. 16 to June 15 fairly well defined; June 16 to Dec. 31, well defined.

Monthly discharge of Williams Fork near Sulphur Springs, Colo., for 1909.

Month.	Discharge in second-feet.			Run-off (total in acre-feet).	Accuracy.
	Maximum.	Minimum.	Mean.		
January...	51	32	39.3	2,420	C.
February..	44	32	35.3	1,960	C.
March..	55	36	45.4	2,790	C.
April...	95	47	64.0	3,810	B.
May..	395	66	230	14,100	B.
June..	1,410	306	962	57,200	A.
July..	1,110	172	517	31,800	A.
August...	264	130	169	10,400	A.
September......................................	218	113	153	9,100	A.
October..	113	83	98.9	6,080	A.
November.......................................	95	58	74.7	4,440	B.
December.......................................	75	58	60.7	3,730	B.
The year......................................	1,410	32	204	148,000	

EAGLE RIVER BASIN.

EAGLE RIVER AT GYPSUM, COLO.

This station, which was established February 7, 1907, to replace the station at Eagle, a few miles above, is located at the highway bridge one-fourth mile north of the Denver & Rio Grande Railroad station at Gypsum. It was discontinued December 31, 1909.

Gypsum Creek, the only tributary of consequence, enters about one-eighth mile below the station. The drainage area above mouth of Gypsum Creek is about 800 square miles.

A number of ditches divert water for irrigation, but such ditches are small, as in most places the valley is narrow.

The stream is frozen along the edges during the winter, and some slush and anchor ice also forms, but the river usually remains open in midstream.

The location and datum of the chain gage on the highway bridge have remained the same during the maintenance of the station. Discharge measurements are made from the bridge.

The channel is permanent, but measurements are only fairly satisfactory because of the rough and rocky bottom. Winter records are affected by ice and are only fairly accurate.

Discharge measurements of Eagle River at Gypsum, Colo., in 1909.

Date.	Hydrographer.	Width.	Area of section.	Gage height.	Discharge.
		Feet.	*Sq. ft.*	*Feet.*	*Sec.-ft.*
Jan. 12a	C. L. Chatfield.................................	92	....	2.80	107
Mar. 14	.....do.......................................	80	138	2.35	120
Apr. 19	.....do.......................................	94	213	3.20	408
June 20b	W. H. Snelson, jr.............................	96	632	7.72	5,720
July 14	.....do.......................................	96	342	4.52	1,300
July 24	W. B. Freeman................................	96	330	4.43	1,190
Aug. 8	W. H. Snelson, jr.............................	96	255	3.60	564
Oct. 9	G. H. Russell.................................	85	206	3.2	341

a Ice; made by wading.
b Subsurface velocity method used. Coefficient of 80 per cent. Stay line used.

*Daily gage height, in feet, of Eagle River at **Gypsum**, Colo., for 1909.* —

[J. F. Greenland, observer.]

Day.	Jan.	Feb.	Mar.	Apr.	May.	June.	July.	Aug.	Sept.	Oct.	Nov.	Dec.
1	2.9	2.45	2.5	2.5	3.2	4.4	6.3	3.7	3.8	3.0	2.8	2.8
2	2.75	2.5	2.5	2.5	3.1	4.45	6.3	3.6	3.75	3.0	2.8	2.8
3	2.7	2.6	2.55	2.5	3.0	4.5	6.5	3.6	3.7	3.0	2.8	2.8
4	2.7	2.6	2.7	2.5	3.15	5.7	6.55	3.5	3.7	3.0	2.8	2.8
5	2.65	2.55	2.65	2.5	3.55	6.2	6.4	3.5	3.7	3.0	2.8	2.8
6	2.6	2.55	2.5	2.5	3.85	6.6	6.05	3.45	3.8	2.95	2.8	2.8
7	2.65	2.55	2.5	2.5	3.95	6.55	5.7	3.75	3.95	2.95	2.8	2.8
8	2.65	2.55	2.5	2.5	4.00	6.8	5.4	3.55	4.0	2.95	2.75	2.8
9	2.6	2.5	2.5	2.5	3.95	6.65	5.4	3.7	3.85	3.0	2.75	2.8
10	2.6	2.5	2.5	2.5	3.95	6.6	5.1	3.75	3.7	3.05	2.8	2.8
11	2.55	2.45	2.45	2.55	4.2	6.4	5.0	3.8	3.6	3.05	2.8	2.8
12	2.8	2.45	2.45	2.6	4.3	6.35	4.8	3.8	3.5	3.0	2.8	2.8
13	2.5	2.4	2.5	2.6	4.0	6.4	4.5	3.8	5.5	3.0	2.8	2.8
14	2.5	2.4	2.45	2.6	4.05	6.4	4.5	3.7	3.5	3.0	2.75	2.8
15	2.45	2.4	2.45	2.6	4.0	6.3	4.5	3.55	3.5	2.95	2.8	2.8
16	2.4	2.4	2.45	2.65	4.2	6.35	4.5	3.45	3.5	2.95	2.8	2.8
17	2.4	2.4	2.6	3.15	4.05	6.25	4.4	3.45	3.45	2.95	2.8	2.8
18	2.4	2.4	2.5	3.2	4.4	6.2	4.4	4.1	3.4	2.95	2.8	2.8
19	2.4	2.4	2.5	3.2	4.65	6.55	4.45	4.4	3.4	2.95	2.8	2.8
20	2.4	2.4	2.5	3.15	4.85	7.40	4.5	4.15	3.4	2.95	2.8	2.8
21	2.4	2.4	2.55	3.05	4.95	6.65	4.3	3.7	3.4	2.95	2.8	2.8
22	2.45	2.4	2.55	2.9	5.4	6.35	4.2	3.6	3.35	2.9	2.8	2.8
23	2.45	2.45	2.55	2.85	5.45	6.75	4.2	3.65	3.3	2.9	2.8	2.75
24	2.4	2.45	2.5	2.9	5.35	6.95	4.35	3.65	3.3	2.9	2.8	2.75
25	2.35	2.45	2.5	3.05	4.7	6.9	4.25	3.65	3.25	2.9	2.8	2.75
26	2.4	2.5	2.5	3.2	4.6	6.75	4.1	3.5	3.2	2.9	2.8	2.75
27	2.4	2.5	2.5	3.25	4.75	6.75	4.05	3.55	3.1	2.85	2.8	2.75
28	2.4	2.5	2.5	3.45	5.3	6.5	3.95	3.55	3.1	2.85	2.8	2.75
29	2.4	....	2.5	3.4	5.5	6.35	4.0	3.6	3.1	2.85	2.8	2.75
30	2.4	....	2.5	3.3	4.95	6.4	4.0	3.65	3.1	2.8	2.8	2.75
31	2.35	....	2.5	....	4.7	....	4.0	3.85	....	2.8	....	2.75

NOTE.—Ice conditions from Jan. 1 to 14

Daily discharge, in second-feet, of Eagle River at Gypsum, Colo., for 1909.

Day.	Jan.	Feb.	Mar.	Apr.	May.	June.	July.	Aug.	Sept.	Oct.	Nov.	Dec.
1		144	155	155	389	1,240	3,480	610	675	265	200	200
2		155	155	155	346	1,290	3,480	550	642	265	200	200
3		181	168	155	307	1,340	3,780	550	610	265	200	200
4		181	209	155	368	2,790	3,860	490	610	265	200	200
5		168	195	155	578	3,330	3,630	490	610	265	200	200
6		168	155	155	779	3,930	3,110	462	675	248	200	200
7		168	155	155	852	3,860	2,630	642	782	248	200	200,
8		168	155	155	890	4,230	2,250	520	820	248	185	200
9		155	155	155	852	4,000	2,250	610	710	265	185	200
10		155	155	155	852	3,930	1,900	642	610	282	200	200
11		144	144	168	1,060	3,630	1,790	675	550	282	200	200
12		144	144	181	1,140	3,560	1,570	675	490	265	200	200
13		132	155	181	890	3,630	1,270	675	490	265	200	200
14		132	144	181	930	3,630	1,270	610	490	265	185	200
15	144	132	144	181	890	3,480	1,270	520	490	248	200	200
16	132	132	144	195	1,060	3,560	1,270	462	490	248	200	200
17	132	132	181	368	930	3,400	1,170	462	462	248	200	200
18	132	132	155	389	1,240	3,330	1,170	900	435	248	200	200
19	132	132	155	389	1,500	3,860	1,220	1,170	435	248	200	200
20	132	132	155	368	1,720	5,190	1,270	942	435	248	200	200
21	132	132	168	326	1,840	4,000	1,080	610	435	248	200	200
22	144	132	168	272	2,400	3,560	985	550	410	230	200	200
23	144	144	168	256	2,460	4,160	985	580	385	230	200	185
24	132	144	155	272	2,330	4,460	1,120	580	385	230	200	185
25	122	144	155	326	1,560	4,380	1,030	580	362	230	200	185
26	132	155	155	389	1,440	4,160	900	490	340	230	200	185
27	132	155	155	414	1,610	4,160	860	520	300	215	200	185
28	132	155	155	520	2,270	3,780	782	520	300	215	200	185
29	132	......	155	491	2,520	3,560	820	550	300	215	200	185
30	132	......	155	438	1,840	3,630	820	580	300	200	200	185
31	122	......	155	......	1,560	......	820	710	......	200	......	185

NOTE.—These discharges are based on rating curves applicable as follows: Jan. 15 to June 4, well defined above a discharge of 100 second-feet. June 5 to Dec. 31, well defined between discharges of 300 and 1,800 second-feet. Jan. 1 to 14, ice conditions; daily discharge estimated at 110 second-feet.

Monthly discharge of Eagle River at Gypsum, Colo., for 1909.

nth.	Discharge in second-feet.			Run-off (total in acre-feet).	Accuracy.
	Maximum.	Minimum.	Mean.		
January..	144	....	123	7,560	B.
February...	181	132	148	8,220	A.
March..	209	144	159	9,780	A.
April..	520	155	262	15,600	A.
May...	2,520	307	1,270	78,100	A.
June..	5,190	1,240	3,570	212,000	A.
July..	3,860	782	1,740	107,000	A.
August..	1,170	462	611	37,600	A.
September...	820	300	501	29,800	A.
October...	282	200	246	15,100	A.
November..	200	185	198	11,800	B.
December..	200	185	196	12,100	B.
The year.....................................	5,190	....	752	545,000	

ROARING FORK BASIN.

DESCRIPTION.

Roaring Fork, which enters the Grand at Glenwood Springs, drains a large area lying chiefly in Pitkin County and reaching to the Continental Divide. It is one of the largest tributaries of the Grand. Frying Pan and Crystal rivers are its most important branches.

ROARING FORK NEAR EMMA, COLO.

This station, which is located on a steel highway bridge about 1½ miles below Emma, a station on the Aspen branch of the Denver & Rio Grande Railroad, was established July 19, 1908. It was discontinued September 30, 1909.

Frying Pan River joins Roaring Fork about 3 miles above the station, and Sopris Creek comes in above the station near Emma. The drainage area at the station is more than 500 square miles.

A few small ditches divert water above the station, principally for meadow irrigation. The only important power plant above the station is that at Aspen, Colo., which generates 1,000 horsepower.

The stream at this point is not so much affected by ice as at other stations in that vicinity, and the channel is frequently open during the winter. Slush ice is common.

A staff gage was used until August 27, 1908, at which time a chain gage was established on the bridge with a datum 0.10 foot above that of the staff gage. Gage heights are all referred to the datum of the chain gage.

Results at this station are satisfactory except when affected by ice during very severe winters.

Discharge measurements of Roaring Fork near Emma, Colo., in 1909.

Date.	Hydrographer.	Width.	Area of section.	Gage height.	Discharge.
		Feet.	*Sq.ft.*	*Feet.*	*Sec.ft.*
Jan. 16a	C. L. Chatfield..	80	125	3.40	206
Mar. 19	.....do...	100	133	3.50	225
Apr. 23	.....do...	100	184	3.96	375
June 23	Wm. H. Snelson, jr..	117	957	9.85	7,080
July 12	.....do...	117	657	7.15	2,870
Aug. 7	.....do...	104	402	5.30	939
Oct. 12	Russell and Weaver..	64	249	4.50	586

a Open channel.

Daily gage height, in feet, of Roaring Fork near Emma, Colo., for 1909.

[W. R. Hood, observer.]

Day.	Jan.	Feb.	Mar.	Apr.	May.	June.	July.	Aug.	Sept.
1......................	3.7	3.5	3.45	3.5	4.2	6.1	8.75	5.4	5.65
2......................	3.6	3.55	3.4	3.5	4.2	6.0	9.15	5.5	5.45
3......................	3.6	3.5	3.4	3.65	4.15	6.5	9.1	5.4	5.4
4......................	3.6	3.6	3.4	3.75	4.5	7.2	8.8	5.35	5.4
5......................	3.5	3.5	3.45	3.7	4.95	8.05	9.3	5.15	5.75
6......................	3.6	3.5	3.45	3.5	5.2	8.5	8.65	5.3	6.1
7......................	3.6	3.5	3.5	3.5	5.4	8.6	8.2	5.3	6.4
8......................	3.55	3.5	3.45	3.4	5.45	9.0	7.8	5.3	6.0
9......................	3.55	3.5	3.35	3.45	5.45	9.25	7.65	5.5	5.85
10......................	3.5	3.5	3.35	3.45	5.6	9.15	7.5	5.45	5.7
11......................	3.4	3.5	3.4	3.6	5.75	9.05	7.2	5.4	=.55
12......................	3.3	3.45	3.35	3.5	5.75	9.10	7.05	5.35	5.5
13......................	3.5	3.45	3.35	3.5	5.6	9.4	6.9	5.5	5.6
14......................	3.65	3.45	3.35	3.5	5.55	9.15	6.8	5.3	5.55
15......................	3.55	3.35	3.4	3.6	5.4	9.2	6.7	5.1	5.6
16......................	3.5	3.4	3.4	3.85	5.7	9.0	6.6	5.1	5.65
17......................	3.5	3.45	3.5	4.05	5.65	9.6	6.4	5.45	5.55
18......................	3.45	3.4	3.5	4.3	6.0	9.7	6.3	6.1	5.45
19......................	3.5	3.4	3.5	4.5	6.2	10.1	6.4	6.1	5.4
20......................	3.45	3.3	3.5	4.5	6.45	10.1	6.3	5.65	5.3
21......................	3.5	3.4	3.6	4.2	6.15	9.85	6.2	5.7	5.2
22......................	3.5	3.45	3.6	4.05	6.75	9.75	6.2	5.45	5.15
23......................	3.5	3.4	3.6	3.9	6.85	9.6	6.3	5.4	5.1
24......................	3.4	3.4	3.55	3.95	6.7	9.7	6.3	5.5	5.1
25......................	3.4	3.45	3.55	4.0	6.25	9.7	6.2	5.4	5.0
26......................	3.4	3.4	3.55	4.3	6.25	9.4	6.05	5.2	4.9
27......................	3.5	3.3	3.6	4.55	6.6	9.35	6.1	5.2	4.8
28......................	3.5	3.35	3.6	4.65	7.0	9.15	5.9	5.1	4.85
29......................	3.45	....	3.55	4.55	7.05	8.95	5.7	5.35	4.75
30......................	3.35	....	3.55	4.3	6.2	8.95	5.6	5.5	4.7
31......................	3.5	....	3.6	....	6.15	....	5.5	5.85	....

NOTE.—Mush ice running during portions of January, February, and first few days of March. Effect on monthly means probably slight.

Daily discharge, in second-feet, of Roaring Fork near Emma, Colo., for 1909.

Day.	Jan.	Feb.	Mar.	Apr.	May.	June.	July.	Aug.	Sept.
1	285	225	212	225	465	1,740	5,160	1,140	1,340
2	255	240	200	225	465	1,650	5,840	1,220	1,180
3	255	225	200	270	445	2,140	5,760	1,140	1,140
4	255	255	200	302	595	2,930	5,250	1,100	1,140
5	225	225	212	285	840	4,060	6,100	965	1,420
6	255	225	212	225	1,000	4,750	5,000	1,070	1,740
7	255	225	225	225	1,140	4,910	4,280	1,070	2,040
8	240	225	212	200	1,340	5,590	3,700	1,070	1,650
9	240	225	188	212	1,180	6,020	3,500	1,220	1,520
10	225	225	188	212	1,300	5,840	3,300	1,180	1,380
11	200	225	200	255	1,420	5,680	2,930	1,140	1,260
12	175	212	188	225	1,420	5,760	2,750	1,100	1,220
13	225	212	188	225	1,300	6,270	2,580	1,220	1,300
14	270	212	188	225	1,260	5,840	2,470	1,070	1,260
15	240	188	200	255	1,140	5,930	2,360	930	1,300
16	225	200	200	338	1,380	5,590	2,250	930	1,340
17	225	212	225	408	1,340	6,630	2,040	1,180	1,260
18	212	200	225	505	1,650	6,810	1,940	1,740	1,180
19	225	200	225	595	1,840	7,530	2,040	1,740	1,140
20	212	175	225	595	2,090	7,530	1,940	1,340	1,070
21	225	200	255	465	2,090	7,080	1,840	1,380	1,000
22	225	212	255	408	2,420	6,900	1,840	1,180	965
23	225	200	255	355	2,520	6,630	1,940	1,140	930
24	200	200	240	372	2,360	6,810	1,890	1,220	930
25	200	212	240	390	1,890	6,810	1,840	1,140	870
26	200	200	240	505	1,890	6,270	1,700	1,000	810
27	225	175	255	620	2,250	6,180	1,740	1,000	750
28	225	188	255	670	2,690	5,840	1,560	930	780
29	212	....	240	620	2,750	5,500	1,380	1,100	722
30	188	....	240	505	1,840	5,500	1,300	1,220	695
31	225	....	255	....	1,790	....	1,220	1,520	....

NOTE.—These discharges are based on a rating curve well defined between 150 and 7,500 second-feet.

Monthly discharge of Roaring Fork near Emma, Colo., for 1909.

Month.	Discharge in second-feet.			Run-off (total in acre-feet).	Accuracy.
	Maximum.	Minimum.	Mean.		
January	285	175	227	14,000	B.
February	255	175	211	11,700	B.
March	255	188	221	13,600	A.
April	670	200	364	21,700	A.
May	2,750	445	1,490	91,600	A.
June	7,530	1,650	5,560	331,000	A.
July	6,100	1,220	2,890	178,000	A.
August	1,740	930	1,170	71,900	A.
September	2,040	695	1,180	70,200	A.
The period	....	....	....	804,000	

ROARING FORK AT GLENWOOD SPRINGS, COLO.

This station, which was established April 6, 1906, is located at the mouth of the stream on a single-span wooden road bridge, about 500 feet above the junction of Grand River and Roaring Fork, and about four blocks west of Grand Avenue, Glenwood Springs. It was discontinued September 30, 1909.

A number of small irrigation ditches divert water from the main stream and tributaries. Three important power plants, located on Crystal River, Yule Creek, and Maroon and Castle creeks, develop about 2,100 horsepower. A number of smaller plants are also in operation in this drainage area.

Surface ice rarely forms solid across the river at this station, although slush and anchor ice are common. Extremely high stages of Grand River may affect the flow at this station to a small degree.

Neither the location nor the datum of the chain gage on the bridge has been changed during the maintenance of the station.

As the stream bed is very rough conditions are unfavorable for accurate measurements. The channel is, however, fairly permanent, and the results are satisfactory.

Discharge measurements of Roaring Fork at Glenwood Springs, Colo., in 1909.

Date.	Hydrographer.	Width.	Area of section.	Gage height.	Dis-charge.
		Feet.	*Sq. ft.*	*Feet.*	*Sec. ft.*
Jan. 13a	C. L. Chatfield...	140	142	1.05	389
Mar. 16a	.....do..	45	148	1.20	362
Apr. 21	.....do..	55	291	2.00	1,010
June 21	Wm. H. Snelson, jr..	179	1,200	7.46	10,500
30	W. B. Freeman..	172	1,030	6.68	8,410
July 13	Wm. H. Snelson, jr..	166	601	4.25	3,370
Aug. 5	.....do..	157	359	2.70	1,310
Oct. 10	G. H. Russell...	153	292	2.25	908

a Slush ice in sections. Results rough.

Daily gage height, in feet, of Roaring Fork at Glenwood Springs, Colo., for 1909.

[Mrs. J. W. Johnson, observer.]

Day.	Jan.	Feb.	Mar.	Apr.	May.	June.	July.	Aug.	Sept.
1...................	1.15	1.15	1.15	1.3	2.0	3.7	6.35	3.0	3.0
2...................	1.2	1.2	1.2	1.35	2.0	3.55	6.25	3.05	2.8
3...................	1.2	1.2	1.15	1.4	2.05	4.2	6.35	3.0	2.75
4...................	1.3	1.15	1.2	1.45	2.3	4.85	6.4	2.75	2.8
5...................	1.35	1.15	1.2	1.5	2.8	5.6	6.8	2.65	2.9
6...................	1.4	1.2	1.25	1.45	3.05	6.35	6.2	2.6	3.9
7...................	1.35	1.2	1.25	1.4	3.1	6.6	5.75	2.7	3.9
8...................	1.25	1.2	1.2	1.35	3.35	6.8	5.3	2.7	3.7
9...................	1.3	1.15	1.1	1.35	3.4	6.75	5.15	2.8	3.4
10...................	1.25	1.2	1.15	1.4	3.5	6.65	4.85	2.75	3.45
11...................	1.2	1.2	1.15	1.45	3.7	6.65	4.8	2.75	3.2
12...................	1.2	1.2	1.2	1.4	3.7	6.35	4.55	2.8	3.0
13...................	1.15	1.2	1.2	1.3	3.4	6.6	4.4	2.9	3.05
14...................	1.3	1.2	1.3	1.4	3.2	6.35	4.3	2.8	2.9
15...................	1.3	1.1	1.25	1.4	3.1	6.3	4.2	2.65	3.0
16...................	1.25	1.1	1.25	1.6	3.3	6.2	4.15	2.7	2.9
17...................	1.2	1.1	1.3	1.7	3.4	6.75	4.1	3.15	2.85
18...................	1.15	1.1	1.3	2.2	3.5	7.15	4.0	3.3	2.75
19...................	1.2	1.15	1.25	2.25	3.85	7.8	4.05	3.2	2.7
20...................	1.2	1.1	1.3	2.3	4.2	7.85	3.85	3.25	2.6
21...................	1.15	1.05	1.3	2.0	4.15	7.65	3.65	2.9	2.6
22...................	1.25	1.0	1.3	1.8	4.45	7.3	3.5	2.8	2.6
23...................	1.3	1.0	1.25	1.8	4.45	7.15	3.5	2.7	2.5
24...................	1.25	1.0	1.3	1.8	4.3	7.3	3.8	2.85	2.5
25...................	1.15	1.05	1.25	1.85	3.95	7.35	3.65	2.75	2.4
26...................	1.1	1.1	1.3	2.05	3.85	7.2	3.55	2.75	2.4
27...................	1.2	1.0	1.3	2.35	4.45	6.85	3.55	2.7	2.4
28...................	1.2	1.0	1.3	2.5	4.65	6.85	3.4	2.6	2.3
29...................	1.15	....	1.25	2.4	4.7	6.65	3.35	2.65	2.3
30...................	1.1	....	1.3	2.2	4.1	6.45	3.25	2.8	2.3
31...................	1.0		1.3	....	3.8	....	3.05	3.15	....

NOTE.—Probable ice conditions during January, February, and March.

Daily discharge, in second-feet, of Roaring Fork at Glenwood Springs, Colo., for 1909.

Day.	Jan.	Feb.	Mar.	Apr.	May.	June.	July.	Aug.	Sept.
1	375	375	375	450	995	3,040	7,660	1,640	1,640
2	400	400	400	480	995	2,820	7,440	1,700	1,420
3	400	400	375	510	1,040	3,850	7,660	1,640	1,370
4	450	375	400	545	1,290	5,020	7,780	1,370	1,420
5	480	375	400	580	1,860	6,530	8,740	1,270	1,530
6	510	400	425	545	2,160	8,160	7,320	1,220	2,840
7	480	400	425	510	2,220	8,730	6,310	1,320	2,840
8	425	400	400	480	2,550	9,190	5,350	1,320	2,540
9	450	375	350	480	2,620	9,080	5,040	1,420	2,130
10	425	400	375	510	2,750	8,840	4,450	1,370	2,200
11	400	400	375	545	3,040	8,840	4,350	1,370	1,880
12	400	400	400	510	3,040	8,160	3,890	1,420	1,640
13	375	400	400	450	2,620	8,730	3,630	1,530	1,700
14	450	400	450	510	2,350	8,160	3,460	1,420	1,530
15	450	350	425	510	2,220	8,040	3,300	1,270	1,640
16	425	350	425	650	2,480	7,820	3,220	1,320	1,530
17	400	350	450	730	2,620	9,080	3,140	1,820	1,480
18	375	350	450	1,190	2,750	9,800	2,990	2,000	1,370
19	400	375	425	1,240	3,270	11,200	3,060	1,880	1,320
20	400	350	450	1,290	3,850	11,300	2,760	1,940	1,220
21	375	325	450	995	3,760	10,800	2,470	1,530	1,220
22	425	300	450	810	4,280	9,940	2,200	1,420	1,220
23	450	300	425	810	4,280	9,580	2,200	1,320	1,120
24	425	300	450	810	4,020	9,940	2,690	1,480	1,120
25	375	325	425	855	3,430	10,100	2,470	1,370	1,030
26	350	350	450	1,040	3,270	9,700	2,330	1,370	1,030
27	400	300	450	1,340	4,280	8,860	2,330	1,320	1,030
28	400	300	450	1,500	4,640	8,860	2,130	1,320	935
29	375	....	425	1,400	4,730	8,380	2,060	1,270	935
30	350	....	450	1,190	3,680	7,900	1,940	1,420	935
31	300	....	450	....	3,190	....	1,700	1,820	....

NOTE.—These discharges are based on rating curves applicable as follows: Jan. 1 to June 18, well defined. June 19 to Sept. 30, well defined.

Monthly discharge of Roaring Fork at Glenwood Springs, Colo., for 1909.

Month.	Discharge in second-feet.			Run-off (total in acre-feet).	Accuracy.
	Maximum.	Minimum.	Mean.		
January	510	300	410	25,200	C.
February	400	300	362	20,100	C.
March	450	350	421	25,900	C.
April	1,500	450	782	46,500	A.
May	4,730	995	2,910	179,000	A.
June	11,300	2,820	8,350	497,000	B.
July	8.740	1,700	4,070	250,000	A.
August	2,000	1,220	1,480	91,000	A.
September	2,840	935	1,530	91,000	A.
The period	....	....	....	1,230,000	

FRYING PAN RIVER AT BASALT, COLO.

This station, which was established July 19, 1908, is located at the wooden highway bridge about 100 yards from the Colorado Midland Railroad depot and about 75 yards downstream from a concrete arch bridge in the town of Basalt. It was discontinued September 30, 1909.

This station is near the mouth of the stream and the records show the total run-off from the drainage basin.

Diversions above are limited to ditches for meadow irrigation. The stream affords good opportunities for power and storage development.

Heavy ice prevails during the winter months, and gage readings are at times distorted by the backwater caused by the freezing of the river below. Slush ice also interferes with the determination of discharge.

The location and datum of the staff gage at the bridge have remained unchanged during the maintenance of the station.

The accuracy of the measurements is affected by the roughness of the stream bed and the high current velocities during flood periods.

Discharge measurements of Frying Pan River at Basalt, Colo., in 1909.

Date.	Hydrographer.	Width.	Area of section.	Gage height.	Discharge.
		Feet.	Sq. ft.	Feet.	Sec.-ft.
Jan. 16a	C. L. Chatfield............................	46.5	68	1.32	67
Mar. 19a	.....do.................................	46.5	75	1.40	76
Apr. 23	.....do.................................	54.7	92	1.80	147
June 23	Chatfield and Snelson, jr.................	62.5	252	4.50	2,080
July 12	W. H. Snelson, jr........................	60.8	178	3.25	760
Aug 6	.....do.................................	46.5	105	2.00	204
Oct. 12	Russell and Weaver......................	45.5	97.5	1.8	159

a Ice along edges.

Daily gage height, in feet, of Frying Pan River at Basalt, Colo., for 1909.

[J. G. Ould, observer.]

Day.	Jan.	Feb.	Mar.	Apr.	May.	June.	July.	Aug.	Sept.
1..............................	3.4	1.95	1.4	1.45	2.0	3.1	4.0	2.05	2.55
2..............................	3.4	2.05	1.4	1.5	1.95	3.15	3.95	2.15	2.5
3..............................	3.45	2.15	1.4	1.5	2.0	3.45	3.95	2.1	2.4
4..............................	3.25	2.0	1.4	1.5	2.25	3.85	4.0	2.15	2.25
5..............................	3.15	2.05	1.4	1.6	2.65	4.35	4.35	2.05	2.4
6..............................	1.55	2.1	1.4	1.55	2.7	4.65	3.9	2.1	2.45
7..............................	1.65	2.1	1.4	1.5	2.8	4.8	3.6	2.15	2.45
8..............................	1.55	2.1	1.4	1.35	2.9	4.9	3.4	2.1	2.4
9..............................	1.5	2.0	1.4	1.4	2.85	4.85	3.3	2.2	2.3
10.............................	1.45	2.0	1.45	1.45	2.95	4.6	3.2	2.15	2.25
11.............................	1.4	1.85	1.55	1.55	3.05	4.6	3.15	2.15	2.2
12.............................	2.45	1.65	1.6	1.5	3.1	4.5	3.0	2.15	2.2
13.............................	2.4	1.55	1.6	1.5	2.95	4.6	2.95	2.25	2.2
14.............................	2.3	1.5	1.55	1.5	2.8	4.6	2.9	2.15	2.2
15.............................	1.5	1.4	1.6	1.6	2.8	4.6	2.9	2.0	2.2
16.............................	1.45	1.4	1.55	1.7	3.05	4.55	2.8	2.0	2.2
17.............................	1.3	1.4	1.45	1.95	2.95	4.9	2.75	2.2	2.2
18.............................	1.3	1.45	1.4	2.1	3.1	5.1	2.65	2.5	2.2
19.............................	1.35	1.45	1.4	2.25	3.25	5.35	2.6	2.5	2.15
20.............................	1.3	1.45	1.5	2.35	3.45	5.2	2.55	2.3	2.1
21.............................	1.3	1.5	1.5	2.15	3.5	5.15	2.6	2.2	2.0
22.............................	1.3	1.55	1.45	1.85	3.65	4.75	2.55	2.1	2.0
23.............................	1.3	1.6	1.4	1.85	3.7	4.75	2.55	2.1	2.0
24.............................	1.35	1.5	1.45	1.85	3.5	4.75	2.55	2.1	1.95
25.............................	1.4	1.4	1.45	1.85	3.4	4.75	2.5	2.1	1.9
26.............................	1.5	1.4	1.5	2.05	3.4	4.45	2.4	2.0	1.9
27.............................	1.5	1.45	1.45	2.25	3.55	4.45	2.4	2.0	1.8
28.............................	1.5	1.4	1.4	2.35	3.8	4.35	2.25	2.1	1.8
29.............................	1.6	....	1.4	2.25	3.8	4.15	2.25	2.1	1.8
30.............................	1.65	....	1.45	2.1	3.45	4.15	2.2	2.2	1.7
31.............................	1.75	....	1.45	....	3.2	....	2.1	2.55	....

Note.—Ice conditions from Jan. 1 to 15 and from Jan. 26 to Feb. 14.

Daily discharge, in second-feet, of Frying Pan River at Basalt, Colo., for 1909.

Day.	Jan.	Feb.	Mar.	Apr.	May.	June.	July.	Aug.	Sept.
1			65	72	190	660	1,460	204	375
2			65	80	177	692	1,400	233	355
3			65	80	190	910	1,400	218	315
4			65	80	264	1,300	1,460	233	264
5			65	98	417	1,880	1,880	204	315
6			65	89	440	2,280	1,350	218	335
7			65	80	490	2,490	1,040	233	335
8			65	60	540	2,630	870	218	315
9			65	65	515	2,560	795	248	280
10			72	72	570	2,210	725	233	264
11			89	89	630	2,210	692	233	248
12			98	80	660	2,080	600	233	248
13			98	80	570	2,210	570	264	248
14			89	80	490	2,210	540	233	248
15		65	98	98	490	2,210	540	190	248
16	72	65	89	118	630	2,140	490	190	248
17	55	65	72	177	570	2,630	465	248	248
18	55	72	65	218	660	2,930	418	355	248
19	60	72	65	264	760	3 300	395	355	233
20	55	72	80	298	910	3,080	375	280	218
21	55	80	80	233	950	3,000	395	248	190
22	55	89	72	152	1,090	2,420	375	218	190
23	55	98	65	152	1,140	2,420	375	218	190
24	60	80	72	152	950	2,420	375	218	177
25	65	65	72	152	870	2,420	355	218	164
26		65	80	204	870	2,020	315	190	164
27		72	72	264	995	2,020	315	190	140
28		65	65	298	1,240	1,880	264	218	140
29			65	264	1,240	1,640	264	218	140
30			72	218	910	1,640	248	248	118
31			72		725		218	375	

NOTE.—These discharges are based on a rating curve, well defined between 50 and 2,800 second-feet. Daily discharge for ice periods Jan. 1 to 15 and Jan. 26 to Feb. 14, estimated as 75 second-feet.

Monthly discharge of Frying Pan River at Basalt, Colo., for 1909.

Month.	Discharge in second-feet.			Run-off (total in acre-feet).	Accuracy.
	Maximum.	Minimum.	Mean.		
January			69.8	4,290	C.
February			74.1	4,120	C.
March	98	65	73.8	4,540	B.
April	298	60	146	8,690	B.
May	1,240	177	682	41,900	A.
June	3,300	660	2,150	128,000	B.
July	1,880	218	676	41,600	A.
August	375	190	238	14,600	A.
September	375	118	240	14,300	A.
The period				262,000	

CRYSTAL RIVER NEAR CARBONDALE, COLO.

This station, which was established July 18, 1908, is located on a single-span highway bridge 150 feet above a section house, at a railroad point known as Sewell, on the Redstone branch of the Denver & Rio Grande Railroad. It is about 5 miles above Carbondale. Station was discontinued September 30, 1909.

No important tributaries enter below the station, but Thompson Creek comes in a short distance above. The drainage area of the river below the mouth of Thompson Creek is about 300 square miles.

Several irrigation ditches, with a combined maximum capacity of probably 100 second-feet, divert water above. The fall, run-off, and storage sites, however, make this essentially a power stream, especially on the upper reaches. Present power plants above the station generate 1,150 horsepower. (See Pl. V, *A*.)

Ice during the winter months prevents the use of open-channel methods of calculating the discharge.

No change has occurred in either location or datum of the staff gage at the bridge during the maintenance of the station.

Except as affected by ice, the results at this station are satisfactory.

Discharge measurements of Crystal River near Carbondale, Colo., in 1909.

Date.	Hydrographer.	Width.	Area of section.	Gage height.	Discharge.
		Feet.	*Sq. ft.*	*Feet.*	*Sec. ft.*
Jan. 15	C. L. Chatfield	52	126	1.85	111
Mar. 18	.....do	52	128	1.80	97
Apr. 22	.....do	54	147	2.50	286
June 22	Chatfield and Snelson, jr	59.5	386	5.48	3,160
July 12	.....do	57	275	3.95	1,170
Aug. 7	.....do	55	210	2.80	445
Oct. 11	Russell and Weaver	48.5	134	2.20	185

Daily gage height, in feet, of Crystal River near Carbondale, Colo., for 1909.

[Alfred Blotham, observer.]

Day.	Jan.	Feb.	Mar.	Apr.	May.	June.	July.	Aug.	Sept.
1	1.60	2.20	1.75	1.85	2.75	3.80	5.45	3.05	2.65
2	1.55	2.15	1.75	1.90	2.70	3.80	5.45	2.95	2.75
3	1.60	2.00	1.80	2.10	2.75	4.40	5.35	2.85	2.80
4	1.60	2.00	1.80	2.10	2.95	5.50	5.45	2.80	2.95
5	1.65	1.90	1.80	2.05	3.45	5.90	5.45	2.85	3.10
6	1.70	1.90	1.80	2.05	3.70	5.90	5.25	2.70	3.40
7	1.70	1.90	1.75	2.10	3.70	6.40	5.05	2.65	3.60
8	1.60	1.90	1.75	2.00	3.80	5.80	4.90	2.55	3.45
9	1.60	1.90	1.75	2.00	3.85	5.55	4.75	2.65	3.50
10	1.70	1.90	1.75	1.90	4.05	5.45	4.45	2.65	3.55
11	1.70	1.90	1.75	1.90	3.95	5.60	4.20	2.95	3.40
12	1.70	1.90	1.80	1.90	3.85	5.50	4.05	2.90	3.35
13	1.80	1.80	1.80	1.90	3.65	5.40	4.15	3.05	3 25
14	1.90	1.70	1.80	1.95	3.50	5.20	3.95	2.85	2.90
15	1.90	1.65	1.70	2.10	3.45	5.30	3.90	2.65	2.70
16	1.90	1.70	1.75	2.30	3.65	5.45	3.80	2.80	2.45
17	1.75	1.70	1.85	2.75	5.65	5.65	3.65	3.05	2.45
18	1.65	1.65	1.85	2.90	3.90	5.60	3.50	3.05	2.15
19	1.60	1.60	1.85	3.00	4.25	5.95	3.00	3.00	2.05
20	1.60	1.60	1.85	2.85	4.40	6.15	3.35	2.85	2.00
21	1.75	1.60	1.90	2.80	4.55	6.05	3.40	2.65	1.95
22	1.85	1.65	1.90	2.70	4.50	5.90	3.35	2.50	1.90
23	1.80	1.70	1.90	2.60	4.45	5.75	3.45	2.50	1.85
24	1.80	1.65	1.85	2.60	4.50	5.70	3.50	2.50	1.85
25	1.75	1.80	1.85	2.65	4.55	5.90	3.40	2.70	1.80
26	1.75	1.75	1.90	2.75	4.45	5.90	3.30	2.85	1.80
27	1.75	1.80	1.90	2.95	4.60	5.80	3.35	2.80	1.75
28	1.80	1.70	1.85	3.10	4.60	5.60	3.30	2.55	1.70
29	1.80	....	1.85	2.95	4.30	5.35	3.20	2.55	1.65
30	1.80	....	1.85	2.75	4.20	5.25	3.10	2.65	1.60
31	2.20	....	1.85	....	4.05	....	3.00	2.70	....

Note.—Open-water conditions probably prevailed during the winter months except the period Jan. 31 to Feb. 4.

B. MIDDLE BOX CANYON OF GILA RIVER, NEAR REDROCK, N. MEX.

A. POWER PLANT OF COLORADO YULE MARBLE CO., ON CRYSTAL RIVER NEAR MARBLE, COLO.

Daily discharge, in second-feet, of Crystal River near Carbondale, Colo., for 1909.

Day.	Jan.	Feb.	Mar.	Apr.	May.	June.	July.	Aug.	Sept.
1	73	105	95	112	380	1,040	3,100	531	336
2	67	105	95	121	357	1,040	3,100	477	380
3	73	105	103	164	380	1,610	2,940	427	403
4	73	105	103	164	477	3,190	3,100	403	477
5	80	121	103	152	780	3,950	3,100	427	559
6	87	121	103	152	965	3,950	2,760	357	745
7	87	121	95	164	965	4,970	2,440	336	890
8	73	121	95	141	1,040	3,750	2,230	296	780
9	73	121	95	141	1,080	3,280	2,020	336	815
10	87	121	95	121	1,260	3,100	1,660	336	852
11	87	121	95	121	1,170	3,370	1,400	477	745
12	87	121	103	121	1,080	3,190	1,260	451	712
13	103	103	103	121	928	3,020	1,350	531	648
14	121	87	103	131	815	2,680	1,170	427	451
15	121	80	87	164	780	2,850	1,120	336	357
16	121	87	95	216	928	3,100	1,040	403	262
17	95	87	112	380	1,000	3,460	928	531	262
18	80	80	112	451	1,120	3,370	815	531	176
19	73	73	112	503	1,450	4,050	745	503	152
20	73	73	112	427	1,610	4,450	712	427	141
21	95	73	121	403	1,780	4,250	745	336	131
22	112	80	121	357	1,720	3,950	712	278	121
23	103	87	121	315	1,660	3,660	780	278	112
24	103	80	112	315	1,720	3,560	815	278	112
25	95	103	112	336	1,780	3,950	745	357	103
26	95	95	121	380	1,660	3,950	679	427	103
27	95	103	121	477	1,840	3,750	712	403	95
28	103	87	112	559	1,840	3,370	679	296	87
29	103	....	112	477	1,500	2,940	617	296	80
30	103	....	112	380	1,400	2,760	559	336	73
31	105	....	112	....	1,260	....	503	357	....

NOTE.—These discharges are based on a rating curve that is well defined below a discharge of 1,000 second feet. Discharges estimated Jan. 31 to Feb. 4.

Monthly discharge of Crystal River near Carbondale, Colo., for 1909.

Month.	Discharge in second-feet.			Run-off (total in acre-feet).	Accuracy.
	Maximum.	Minimum.	Mean.		
January	121	67	91.8	5,640	B.
February	121	73	98.8	5,490	B.
March	121	87	106	6,520	A.
April	559	112	269	16,000	A.
May	1,840	357	1,180	72,600	A.
June	4,970	1,040	3,320	198,000	A.
July	3,100	503	1,440	88,500	A.
August	531	278	393	24,200	A.
September	890	73	372	22,100	A.
The period	....	....	....	439,000	

DIVIDE CREEK BASIN.

DESCRIPTION.

Divide Creek enters Grand River from the south side, about 6 miles below Newcastle, Colo. It is formed by the junction of East and West Divide creeks a few miles above the mouth of the stream. The run-off is derived chiefly from the melting snows in the spring and from rain.

WEST DIVIDE CREEK AT HOSTUTLER'S RANCH, NEAR RAVEN, COLO.

This station, which was established July 27, 1909, is located 5 miles above Raven post office, Colo., one-fourth of a mile below the head-gates of the High Line ditch and 2 miles above the head of the Porter ditch.

Gage readings were discontinued for the season on September 20, 1909.

The vertical staff gage is fastened to a large overhanging cotton-wood tree on the right bank about 300 feet west of J. K. Hostutler's house.

Discharge measurements are made by wading in the vicinity of the gage.

The following discharge measurement was made by W. B. Freeman and R. E. Vickery:

July 27, 1909; width, 10 feet; area, 89 square feet; gage height, 1.00 feet; discharge, 15.1 second-feet.

Daily gage height, in feet, of West Divide Creek at Hostutler's ranch near Raven, Colo., for 1909.

[J. K. Hostutler, observer.]

Day.	July.	Aug.	Sept.	Day.	July.	Aug.	Sept.	Day.	July.	Aug.	Sept.
1		0.65	0.85	11		0.8	0.85	21		0.85	
2		.7	.75	12		.85	1.1	22		.85	
3		.7	.7	13		.7	1.1	23		.9	
4		.6	.8	14		.6	1.05	24		1.05	
5		.6	.85	15		.55	.95	25		.8	
6		.6	1.1	16		.6	.9	26		.7	
7		.85	1.2	17		.85	.8	27	1.0	.7	
8		.7	1.15	18		1.0	.75	28	.95	.6	
9		.6	1.0	19		1.1	.7	29	.85	.6	
10		.7	.9	20		.95	.7	30	.8	.65	
								31	.7	.8	

WEST DIVIDE CREEK AT RAVEN, COLO.

This station, which was established July 27, 1909, is located at Raven post office, Colo., 14 miles south of Newcastle.

The vertical staff gage is about 150 feet downstream from a highway bridge, at which high-water discharge measurements will be taken. Numerous ditches, of which the High Line and the Porter are the most important, divert water above this station.

Fair results should be obtained at this station. Ice affects the gage heights for a few months during the winter season.

The following discharge measurement was made by W. B. Freeman:

July 27, 1909; width, 4.6 feet; area, 3.7 square feet; gage height, 1.18 feet; discharge, 4.2 second-feet. Made by wading below gage.

Daily gage height, in feet, of West Divide Creek at Raven, Colo., for 1909.

[John F. Collins, observer.]

Day.	July.	Aug.	Sept.	Oct.	Nov.	Dec.	Day.	July.	Aug.	Sept.	Oct.	Nov.	Dec.
1		1.1	1.3	......	1.2	1.2	16		1.1	1.4	1.2	1.8	1.2
2		1.1	1.3	......	1.2	1.2	17		1.2	1.4	1.2	1.	1.2
3		1.1	1.3	1.2	1.2	1.2	18		1.25	1.3	.2	.	1.2
4		1.1	1.3	1.2	1.2	1.2	19		2	1.3	.2	.	1.2
5		1.1	1.3	1.2	1.2	1.2	20		1.5	1.3	.2	.	1.2
6		1.1	1.3	1.2	1.2	1.2	21		1.5	1.3	1.2	1.2	1.2
7		1.1	1.65	1.2	1.2	1.2	22		1.2	1.3	1.2	1.2	1.2
8		1	1.5	1.2	1.2	1.2	23		1.25	1.3	1.2	1.2	1.2
9		1	1.5	1.2	1.2	1.2	24		1.35	1.2	1.2	1.2	1.2
10		1.1	1.4	1.2	1.2	1.2	25		1.35	1.2	1.2	1.2	1.2
11		1.2	1.4	1.2	1.2	1.2	26		1.15	1.2	1.2	1.2	1.2
12		1.2	1.65	1.2	1.2	1.2	27	1.15	1.2	1.2	1.2	1.2	1.2
13		1.1	1.55	1.2	1.2	1.2	28	1.1	1.1	1.2	1.2	1.2	1.2
14		1	1.55	1.2	1.2	1.2	29	1.0	1.2	1.2	1.2	1.2	1.2
15		1.05	1.5	1.2	1.2	1.2	30	1.2	1.2	1.2	1.2	1.2	1.2
							31	1.2	1.2	......	1.2	......	1.2

Daily discharge, in second-feet, of West Divide Creek at Raven, Colo., for 1909.

Day.	July.	Aug.	Sept.	Oct.	Nov.	Dec.	Day.	July.	Aug.	Sept.	Oct.	Nov.	Dec.
1		2.5	8.5	5.0	5.0	5.0	16		2.5	13	5.0	5.0	5.0
2		2.5	8.5	5.0	5.0	5.0	17		5.0	13	5.0	5.0	5.0
3		2.5	8.5	5.0	5.0	5.0	18		6.8	8.5	5.0	5.0	5.0
4		2.5	8.5	5.0	5.0	5.0	19		61	8.5	5.0	5.0	5.0
5		2.5	8.5	5.0	5.0	5.0	20		18	8.5	5.0	5.0	5.0
6		2.5	8.5	5.0	5.0	5.0	21		18	8.5	5.0	5.0	5.0
7		2.5	28	5.0	5.0	5.0	22		5.0	8.5	5.0	5.0	5.0
8		1.0	18	5.0	5.0	5.0	23		6.8	8.5	5.0	5.0	5.0
9		1.0	18	5.0	5.0	5.0	24		10.8	5.0	5.0	5.0	5.0
10		2.5	13	5.0	5.0	5.0	25		10.8	5.0	5.0	5.0	5.0
11		5.0	13	5.0	5.0	5.0	26		3.8	5.0	5.0	5.0	5.0
12		5.0	28	5.0	5.0	5.0	27	3.8	5.0	5.0	5.0	5.0	5.0
13		2.5	22	5.0	5.0	5.0	28	2.5	2.5	5.0	5.0	5.0	5.0
14		1.0	22	5.0	5.0	5.0	29	1.0	5.0	5.0	5.0	·5.0	5.0
15		1.8	18	5.0	5.0	5.0	30	5.0	5.0	5.0	5.0	5.0	5.0
							31	5.0	5.0	......	5.0	......	5.0

NOTE.—These discharges are based on a curve that is well defined by measurements made in 1910.

Monthly discharge of West Divide Creek at Raven, Colo., for 1909.

Month.	Discharge in second-feet.			Run-off (total in acre-feet).	Accuracy.
	Maximum.	Minimum.	Mean.		
July 27–31	5.0	1.0	3.46	34	B.
August	61	1.0	6.71	412	B.
September	28	5.0	11.4	678	B.
October	5.0	5.0	5.0	307	B.
November	5.0	5.0	5.0	298	B.
December	5.0	5.0	5.0	307	B.
The period	......	......	....	2,040	

MAMM CREEK BASIN.

WEST MAMM CREEK NEAR RIFLE, COLO.

Mamm Creek, which enters Grand River from the south about 4 miles upstream from Rifle, Colo., is formed by the junction of West Mamm, East Mamm, and Middle Mamm, all of which rise on the north side of Battlement Mesa.

The flow of this stream is supplied by melting snow in the spring and by rains. Its channel is usually dry or nearly so during the late summer and fall.

The gaging station, which was established July 26, 1909, is located just south of J. T. Selby's ranch house, 9 miles south of Rifle, Colo., about half a mile above the mouth of Quakenasp Gulch Creek, and three-fourths of a mile above the dam site of a proposed irrigation reservoir, which has a capacity of a few thousand acre-feet. One ditch, of less than 10 second-feet capacity, diverts water above the station. The waste water from this ditch is returned to the creek a few feet above the staff gage.

Beginning November 16, 1909, a 24-inch trapezoidal sharp-edge weir, with end contractions, was used to measure the flow of the stream. This weir is located about 50 feet downstream from the gage. Current-meter measurements are made at various sections in the vicinity of gage.

Ice exists to some extent for quite an extended period during the winter season. Conditions at this station are not conducive to the most accurate results, especially during the higher stages, when gage heights show great fluctuations and measurements are difficult to make.

Daily gage height, in feet, of West Mamm Creek near Rifle, Colo., for 1909.

[J. T. Selby, observer.]

Day.	July.	Aug.	Nov.	Dec.	Day.	July.	Aug.	Nov.	Dec.	Day.	July.	Aug.	Nov.	Dec.
1.....	....	0.5	....	0.21	11.....	....	1.2	....	0.21	21.....	....	1.15	0.50	0.21
2.....	....	1.0	....	.21	12.....	....	1.75	....	.21	22.....	....	.6	.75	.21
3.....	....	.95	....	.21	13.....	....	.9	....	.21	23.....	....	.5	.83	.21
4.....	....	.65	....	.21	14.....	....	.65	....	.21	24.....	....	.5	.67	.21
5.....	....	.5	....	.21	15.....	....	.5	....	.21	25.....	....	.5	.67	.21
6.....	....	.5	....	.21	16.....	....	1.2	0.21	.21	26.....	1.0	.5	.58	.21
7.....	....	.5	....	.21	17.....	....	2.0	.21	.21	27.....	.8	.5	.50	.21
8.....	....	.5	....	.21	18.....	....	.55	.21	.21	28.....	.5	.5	.33	.21
9.....	....	.5	....	.21	19.....	....	.8	.25	.21	29.....	.5	....	.25	.33
10.....	....	.5	....	.21	20.....	....	.8	.25	.21	30.....	.5	....	.25	.42
										31.....	.5	....	....	.42

NOTE.—Gage heights July 26 to Aug. 28 are from staff gage. Readings Nov. 16 to Dec. 31 are heads on the weir. Observer recorded in inches, and his readings have been converted into decimals of a foot.

Daily discharge, in second-feet, of West Mamm Creek near Rifle, Colo., for 1909.

Day.	Nov.	Dec.	Day.	Nov.	Dec.	Day.	Nov.	Dec.
1		0.6	11		0.6	21	2.4	0.6
2		.6	12		.6	22	4.3	.6
3		.6	13		.6	23	5.0	.6
4		.6	14		.6	24	3.6	.6
5		.6	15		.6	25	3.6	.6
6		.6	16	0.6	.6	26	2.9	.6
7		.6	17	.6	.6	27	2.4	.6
8		.6	18	.6	.6	28	1.3	.6
9		.6	19	.8	.6	29	.8	.8
10		.6	20	.8	.6	30	.8	1.8
						31		1·8

NOTE.—These discharges were obtained from the weir formula $Q=3.33LH^{\frac{3}{2}}$. See Water-Supply Paper 200, p. 162.

Monthly discharge of West Mamm Creek near Rifle, Colo., for 1909.

Month.	Discharge in second-feet.			Run-off (total in acre-feet).
	Maximum.	Minimum.	Mean.	
Nov. 16–30	2.9	0.6	2.03	60
December	1.8	.6	.68	42

NOTE.—These estimates are approximate.

GUNNISON RIVER BASIN.

DESCRIPTION.

Gunnison River is formed in Gunnison County, Colo., by the union of East and Taylor rivers, two streams that originate among the snow-covered peaks and on the slopes of the Continental Divide in the northeastern part of the county, descend through narrow mountain valleys, and unite about 12 miles above Gunnison. From the junction of these rivers the Gunnison flows west and southwest to the point where it enters Grand River at Grand Junction, in the central part of Mesa County, Colo.

The upper course of the river lies through a broad, mountainous valley, but near the mouth of Lake Fork the valley narrows and the river enters Black Canyon of the Gunnison, through which it winds in a tortuous course for 56 miles between granite walls that rise precipitously 3,000 feet above the water's edge. A short distance below the mouth of North Fork, the largest tributary of the river, the canyon walls break abruptly, and the valley is broad and fertile. Below Delta the river enters another narrow canyon, with walls averaging 800 feet in height, and this continues irregularly to Grand Junction, only a few tracts of narrow bottom land lying between the channel and the canyon walls.

The soil of the lower valleys is chiefly adobe, and the higher mesas contain much gravel and sand. Groves of quaking aspen interspersed with large, open grazing plots cover broad areas of this plateau region. On the top of the Grand Mesa are forests of pine and aspen, and piñon and cedar grow along the foothills.   In the valleys chico and sagebrush form the principal vegetation, except along the streams, which are bordered in places by cottonwood, willow, and undergrowth.

The chief tributaries of the Gunnison are Ohio, Tomichi, Lake Fork, and Cimarron creeks, and Smith, North Fork, and Uncompahgre rivers, North Fork being the largest.

North Fork rises in the Huntsman Hills, 20 miles south of Glenwood Springs, flows in a general southerly and southwesterly course, and unites with the Gunnison about 8 miles west of Hotchkiss.   The drainage area is mountainous, except for a small portion which lies below Paonia, extreme points reaching an altitude of 13,000 feet. The mesa lands at the lower end of the valley stand 5,500 feet above sea level.   The higher peaks are formed of granite rocks, but lower down, sedimentary formations occupy at least 80 per cent of the area of the basin.   The mountains are forested and the mesa lands are covered with sagebrush.   All the tillable lands of the North Fork and its tributaries have been brought under cultivation, and irrigation is practiced to such an extent that the entire flow is needed for existing systems.

Uncompahgre River, the principal tributary of the Gunnison from the south, rises among the snowy peaks of the highly serrated Uncompahgre Mountains and flows a little west of north to its junction with the Gunnison at Delta.   The basin embraces a mountainous plateau and valley area of 1,130 square miles, oblong in shape, the width increasing slightly at the lower end.   The mountain area occupies but a small part of the basin, but contributes the perennial waters of the stream.   The plateau area is greatest in extent and borders the valley on both sides, the larger Uncompahgre Plateau lying to the southwest.   Escarpments are conspicuous features of this plateau. The relief features are terraced mesas flanked by buttes and ridges and trenched by deep, narrow canyons.   Uncompahgre Valley proper begins at a point near Eldredge siding, on the Denver & Rio Grande Railroad.

The other tributaries of the Gunnison need not here be described. Ohio, Tomichi, Lake Fork, and Cimarron creeks are perennial streams, but almost their entire volume is diverted for irrigation during the growing season, so that very little water reaches the Gunnison except at times of heavy storms or during spring floods.

Precipitation records for the Gunnison are meager.   Those which exist show a range from 9 inches in the plateau region to about 25 inches in the mountains.

The run-off of the Gunnison drainage basin is conserved to a large extent by four forest reserves, which have a total area of about 5,700 square miles, of which approximately 3,800 square miles are located within the basin. About 65 per cent of this area is in standing timber, the remainder being classified as sagebrush, barren, and burned. Investigation of the headwaters of East River and other tributaries in Gunnison County several years ago discovered that many of the hills had been almost entirely denuded of their timber, a discovery to which may be attributed the setting aside of the areas as forest reserves.

Along Gunnison River proper above the mouth of Lake Fork a number of ditches divert water for meadow irrigation, and irrigation is extensively practiced in the vicinity of Delta. The largest irrigated area in the Gunnison drainage basin is the Uncompahgre Valley. In addition to the lands being irrigated by large private ditches, this valley contains about 150,000 acres, which are being reclaimed under the Uncompahgre project of the United States Reclamation Service. The greater part, of the water for this land is diverted from Gunnison River by means of the Gunnison tunnel, which has a capacity of 1,300 second-feet. The formal opening of Gunnison tunnel was celebrated at the west portal of the tunnel September 23, 1909, with the President of the United States, the Secretary of the Interior, and officials of the Reclamation Service in attendance. The construction of the tunnel was begun in January, 1905, and the actual opening through the tunnel was completed July 6, 1909. The present water rights consume the normal flow of Uncompahgre River, and the Uncompahgre Valley project will divert all the available water from Gunnison River during normal stages.

The country is not adapted for large reservoirs, the meadows having so much fall and the valleys being so narrow that construction would be expensive in proportion to reservoir capacity. However, a large number of small reservoirs exist on the Gunnison and its tributaries, which can be advantageously utilized for power.

Power plants at present in operation in this basin develop about 2,200 horsepower, and there are many sites unutilized. The fall along some of the streams is heavy, ranging from 50 to 150 feet to the mile. Along the Uncompahgre, from its source to the 8,000-foot contour, the fall is almost 300 feet to the mile. At the present time the waters in this basin are being used for domestic purposes, irrigation, and power. By utilizing all the available storage it would theoretically be possible to develop about 200,000 horsepower. Along the South canal of the United States Reclamation Service, which receives the water from the Gunnison tunnel and carries it into Uncompaghre River, a series of drops will make possible the development of from 5,000 to 10,000 horsepower.

But two gaging stations were maintained previous to 1900, but since that year a number have been established and discontinued.

The records, as a rule, cover about three-year periods, and show that 1904 was the driest and 1907 the wettest year.   By comparison with other drainage basins adjacent to the Gunnison, however, it is evident that 1902 was a drier year than 1904.

GUNNISON RIVER AT RIVER PORTAL, COLO.[1]

This station, which was established April 7, 1905, replaced the station located at Cimarron, about 12 miles above.   It is about 300 feet above the portal of the Gunnison tunnel and is about 21 miles northeast of Montrose.

The station is about 8 miles below the mouth of Crystal Creek, and is above North Fork and Uncompahgre River, the two most important tributaries.

A number of small ditches divert water for meadow irrigation above the station.   The largest diversion along the river, and also in Colorado, is the recently completed Gunnison tunnel, with a capacity of about 1,300 second-feet, which diverts the water from the Gunnison into the Uncompahgre Valley, where it will be used for irrigation.[2]

Ice covers the river for about four months each year and attains a thickness of 1 to 2 feet.   No winter records of discharge have been obtained.

The original staff gage, which was bolted to the cliff on the right bank of the river, was dislodged by driftwood on June 4, 1909.   Prior to this date no change occurred in the location or datum of this gage.

From June 5 to 19 an old high-water gage, about 100 feet upstream on the left bank, was read.   The datum of this gage is 10.08 feet lower than that of the original gage.   The readings were reduced to the original datum.   This auxiliary gage could not be used after June 19, as the water surface had fallen too low.   On August 9, 1909, a new staff gage was installed at the same location and datum as the original gage.   This gage was broken off by ice November 20, 1909.

Discharge measurements were made from a cable located a few feet downstream from the original gage site.

The accuracy of the estimates of daily and monthly discharge is impaired by lack of sufficient discharge measurements to cover slight shifting of channel, and changes resulting from the dumping of débris from the Gunnison tunnel into the river below the station.

This station is maintained under the supervision of the United States Reclamation Service.   Computations of discharge have been made by engineers of the United States Geological Survey.

---

[1] This station was referred to in previous reports as at east portal of Gunnison tunnel.
[2] The Gunnison tunnel and Uncompahgre project are described in the reports of the United States Reclamation Service.

Discharge measurements of Gunnison River at River Portal, Colo., in 1909.

Date.	Hydrographer.	Width.	Area of section.	Gage height.	Discharge.
		Feet.	*Sq. ft.*	*Feet.*	*Sec.-ft.*
June 8	A. L. B. Moser	210	2,750	15.78	13,000
9	...do	210	2,700	15.48	12,900
10	...do	208	2,570	14.98	11,900
11	...do	206	2,510	14.68	11,600
12	...do	205	2,460	14.48	11,400

NOTE.—The gage readings for above measurements are reduced approximately to original gage datum. The measurements were made by the subsurface method, the reduction coefficient used being 0.92.

Daily gage height, in feet, of Gunnison River at River Portal, Colo., for 1909.

[John Dill and W. T. Blight, observers.]

Day.	Apr.	May.	June.	Aug.	Sept.	Oct.	Nov.	Day.	Apr.	May.	June.	Aug.	Sept.	Oct.	Nov.
1	6.3	7.8	11	......	7.9	7	5.8	16	7.5	11.3	13.7	7.35	8.3	6.7	5.4
2	6.45	7.8	10.3		7.6	7	5.8	17	8	11.2	14.2	7.2	8.2	6.6	5
3	6.8	7.9	11.7		7.6	7	5.9	18	8.8	12	14.5	7.45	8.2	6.6	5.15
4	6.9	8.5	13.3		7.7	6.9	5.95	19	9.15	12.2	14.7	7.55	8	6.5	5.2
5	6.9	9.9	14.8		7.9	6.9	5.9	20	9	12.3	......	7.6	7.8	6.5	......
6	6.4	10.3	15.1		10.15	6.95	5.8	21	8.1	12		7.6	7.7	6.5	
7	6.35	11	15.4		9.7	7.15	5.7	22	7.85	12.7		7.4	7.6	6.4	
8	6	11.8	15.7		9.35	7.1	5.6	23	7.6	12.6		7.2	7.5	6.3	
9	6.2	10.9	15.5	7.2	8.95	7	5.5	24	7.4	12.1		7.4	7.4	6.2	
10	6.2	11.4	15	7.1	8.5	6.9	5.7	25	7.5	11.6		7.45	7.3	6.15	
11	6.1	11.8	14.6	7.3	8.4	7	5.9	26	7.9	12		7.2	7.2	6.1	
12	6.2	11.6	14.3	7.35	8.2	6.9	5.7	27	8.5	12.7		7.2	7.1	6	
13	6.3	11.1	14.5	7.4	8.5	6.9	5.75	28	8.9	13		7.1	7.1	6	
14	6.4	10.8	14	7.4	8.45	6.9	5.7	29	8.7	12.7		7	7	6	
15	6.85	10.7	13.9	7.4	8.4	6.7	5.7	30	8.2	12		7.45	7	5.9	
								31		11.5		7.7	......	5.85	

Daily discharge, in second-feet, of Gunnison River at River Portal, Colo., for 1909.

Day.	Apr.	May.	June.	Aug.	Sept.	Oct.	Nov.
1	1,050	2,300	6,660	....	2,410	1,560	760
2	1,150	2,300	5,500		2,100	1,560	760
3	1,400	2,410	7,880		2,100	1,560	810
4	1,480	3,070	10,700		2,200	1,480	838
5	1,480	4,890	10,800		2,410	1,480	810
6	1,120	5,500	12,200		5,270	1,520	760
7	1,080	6,660	12,500		4,600	1,690	710
8	865	8,060	13,000		4,120	1,650	665
9	985	6,490	12,700	1,740	3,600	1,560	620
10	985	7,360	12,100	1,650	3,070	1,480	710
11	925	8,060	11,500	1,820	2,960	1,560	810
12	985	7,710	11,100	1,870	2,730	1,480	710
13	1,050	6,840	11,400	1,920	3,070	1,480	735
14	1,120	6,320	10,700	1,920	3,010	1,480	710
15	1,440	6,150	10,500	1,920	2,960	1,330	710
16	2,010	7,180	10,200	1,870	2,840	1,330	580
17	2,520	7,010	11,000	1,740	2,730	1,260	440
18	3,480	8,410	11,400	1,960	2,730	1,260	488
19	3,790	8,760	11,600	2,060	2,520	1,180	505
20	3,660	8,940	....	2,100	2,300	1,180	....
21	2,620	8,410		2,100	2,200	1,180	
22	2,360	9,640		1,920	2,100	1,120	
23	2,100	9,460		1,740	2,010	1,050	
24	1,920	8,580		1,920	1,920	985	
25	2,010	7,710		1,960	1,820	955	
26	2,410	8,410		1,740	1,740	925	
27	3,070	9,640		1,740	1,650	865	
28	3,540	10,200		1,650	1,650	865	
29	3,300	9,640		1,560	1,560	865	
30	2,730	8,410		1,960	1,560	810	
31		7,540		2,200	....	785	

NOTE.—These discharges were obtained from rating curves applicable as follows: Apr. 1 to June 4, and Aug. 9 to Nov. 19, well defined below 8,000 second-feet; June 5 to June 19, well defined.

Monthly discharge of Gunnison River at River Portal, Colo., for 1909.

Month.	Discharge in second-feet.			Run-off (total in acre-feet).	Accuracy.
	Maximum.	Minimum.	Mean.		
April..........................	3,790	865	1,950	116,000	B.
May............................	10,200	2,300	7,160	440,000	B.
June 1–19......................	13,000	5,500	10,800	407,000	B.
Aug. 9–31......................	2,200	1,560	1,870	85,300	B.
September......................	5,270	1,560	2,600	155,000	B.
October........................	1,690	785	1,270	78,100	B.
Nov. 1–19......................	838	440	691	26,000	C.
The period....................	....	....	....	1,310,000	

UNCOMPAHGRE RIVER AT FORT CRAWFORD, COLO.

This station, which was established October 2, 1907, to obtain information concerning the water supply above the principal diversions in the Uncompahgre Valley, replaces the station near Colona which was established August 10, 1903. Its present location is at a highway bridge across Uncompahgre River about half a mile west of Fort Crawford, in sec. 36, T. 48 N., R. 9 W.

The station is located just below the mouth of Horsefly Creek. A number of large private irrigation ditches divert water above this station. Existing power plants generate about 1,800 horsepower. Opportunity for extensive power development is found on the headwaters.

Thick ice forms along the edges of the river during the winter months. The channel remains open at the station, but slush ice affects the accuracy of the results at times. The channel scours during high stages and silts during periods of low water.

On June 21, 1908, the rod gage, which was established October 2, 1907, was washed out. On July 7, 1908, a temporary chain gage was installed. The zero of this gage was placed 1.95 feet below the zero of the rod gage. On July 23, 1908, a permanent rod gage was installed, the zero of which corresponds to 0.70 foot on the first rod gage, and to 2.65 feet on the chain gage.

During 1909 the chain gage, established July 7, 1908, was used. In the early part of 1910 a new rod gage was established with a datum 3.20 feet lower than that of the chain gage. The readings for 1909 have been reduced to this datum.

This station was maintained under the supervision of the United States Reclamation Service.

Discharge measurements of Uncompahgre River at Fort Crawford, Colo., in 1909.

Date.	Hydrographer.	Width.	Area of section.	Gage height.	Dis-charge.
		Feet.	*Sq. ft.*	*Feet.*	*Sec.-ft.*
Apr. 16	R. M. Adams.........................	56.5	70	1.60	251
July 19	....do............................	54	97	2.00	361
27	....do............................	58	118	2.60	502
Aug. 19	....do............................	61	98	2.45	457
Oct. 6	....do............................	57.5	71	1.30	266

Daily gage height, in feet, of Uncompahgre River at Fort Crawford, Colo., for 1909.

Day.	Aug.	Sept.	Oct.	Day.	Aug.	Sept.	Oct.	Day.	Aug.	Sept.	Oct.
1........		2.1	1.2	11........	1.9	1.85	1.2	21........	2.25	1.7	
2........		2.25	1.2	12........	2.0	2.1		22........	2.2	1.6	
3........		2.4	1.2	13........	2.05	2.15		23........		1.6	
4........		2.05	1.2	14........	1.95	2.0		24........	2.3	1.5	
5........		2.9	1.4	15........	2.3	1.95		25........	2.25	1.5	
6........	2.0	2.6	1.35	16........	2.3	2.0		26........	2.25	1.4	
7........	2.15	2.5	1.25	17........	2.2	2.05		27........	2.25	1.4	0.9
8........	2.15	2.4	1.25	18........	2.3	1.9		28........	2.15	1.45	1.0
9........	2.15	2.15	1.25	19........	2.4	1.9		29........	2.2	1.35	
10........	2.0	2.0	1.2	20........	2.35	1.8		30........	2.3	1.25	
								31........	2.25		

Daily discharge, in second-feet, of Uncompahgre River at Fort Crawford, Colo., for 1909.

Day.	Aug.	Sept.	Oct.	Day.	Aug.	Sept.	Oct.	Day.	Aug.	Sept.	Oct.
1........	240	380	230	11........	210	315	230	21........	345	340	140
2........	240	480	230	12........	245	435	230	22........	320	300	140
3........	240	535	230	13........	260	515	230	23........	345	300	140
4........	240	355	230	14........	230	435	230	24........	365	260	140
5........	240	885	305	15........	365	410	230	25........	345	305	140
6........	245	660	285	16........	365	435	230	26........	400	265	140
7........	300	670	250	17........	320	460	230	27........	400	265	140
8........	300	605	250	18........	365	385	230	28........	350	285	165
9........	300	460	250	19........	410	440	140	29........	375	250	150
10........	245	380	230	20........	390	390	140	30........	420	210	150
								31........	400		150

NOTE.—These discharges were obtained by the indirect method for shifting channels. Discharges estimated for days when gage was not read.

Monthly discharge of Uncompahgre River at Fort Crawford, Colo., for 1909.

Month.	Discharge in second-feet.			Run-off (total in acre-feet).	Ac-cu-racy.
	Maximum.	Minimum.	Mean.		
August.................................	....	210	317	19,500	C.
September.............................	885	210	414	24,600	C.
October................................	....	....	200	12,300	D.

UNCOMPAHGRE RIVER AT MONTROSE, COLO.

This station, which was reestablished April 22, 1903, to obtain for the United States Reclamation Service definite information concerning the amount of water carried by the Uncompahgre, is located at the iron highway bridge just west of Montrose and one-fourth mile west of the Denver & Rio Grande Railroad.

The station is about 2 miles above Happy Canyon Creek and is also above Cedar and Spring creeks. Large irrigation ditches divert water between this station and that at Fort Crawford. Existing water rights control the normal flow of this river for irrigation. Above these diversions, however, opportunities exist for storage and power development. Established plants generate about 1,800 horsepower. Open-channel conditions obtain at this station, although thick ice usually forms along the edges. Slush and anchor ice sometimes influence the accuracy of the results. The flow is controlled during the irrigation season by the large diversions above. The flow at this point will also be affected by the inflow from the south canal of the United States Reclamation Service when that canal is completed.

Neither the location nor the datum of the staff gage, which is 20 feet upstream from the bridge, has been changed during the maintenance of the station.

Results obtained are good except during winter periods and extreme low water.

This station is maintained under the supervision of the United States Reclamation Service.

Discharge measurements of Uncompahgre River at Montrose, Colo., in 1909.

Date.	Hydrographer.	Width.	Area of section.	Gage height.	Discharge.
		Feet.	*Sq. ft.*	*Feet.*	*Sec.-ft.*
June 7	R. M. Adams	84	206	5.10	1,070
July 19	.....do.	28	53	2.70	109
27	.....do.	31.5	72	3.10	224
Aug. 5	.....do.	28	53	2.50	86
12	.....do.	28	59	2.70	124
20	.....do.	33	83	3.50	320
Oct. 7	.....do.	28	54	2.65	110

Daily gage height, in feet, of Uncompahgre River at Montrose, Colo., for 1909.

[Thomas Reeves, observer.]

Day.	Apr.	May.	June.	July.	Aug.	Sept.	Oct.	Nov.
1	2.3	2.15	2.1	4.75	2.8	3.5	2.45	2.5'
2	2.3	2.25	2.2	4.55	3.0	3.05	2.4	2.6
3	2.5	2.1	2.5	4.7	2.6	3.1	2.4	2.5
4	2.5	2.6	3.6	4.55	2.55	3.75	2.4	2.55
5	2.55	3.5	4.8	4.5	2.7	4.05	2.3	2.5
6	2.5	3.55	5.35	4.5	2.5	4.7	2.3	2.35
7	2.5	3.6	5.8	4.0	2.5	4.8	2.7	2.25
8	2.3	3.6	5.05	3.7	2.5	4.45	2.6	2.2
9	2.35	3.2	4.9	3.7	2.45	4.3	2.55	2.1
10	2.5	3.1	4.8	3.6	2.6	4.15	2.6	2.1
11	2.5	3.5	5.0	3.45	2.7	4.2	2.55	2.05
12	2.4	3.3	5.1	2.95	2.65	3.9	2.5	2.05
13	2.45	3.4	5.35	2.65	3.0	4.65	2.45	1.95
14	2.5	2.75	4.8	2.45	2.7	4.3	2.4	2.0
15	2.65	2.45	4.95	2.3	3.1	4.1	2.45	2.05
16	3.05	2.5	5.35	2.45	3.05	4.2	2.5	2.0
17	4.05	2.5	5.4	2.7	2.8	4.2	2.5	2.05
18	4.15	2.4	5.45	2.5	3.55	4.1	2.5	2.05
19	3.90	2.7	5.5	2.55	3.2	3.9	2.4	2:1
20	3.55	3.1	5.45	2.5	3.4	3.65	2.45	2:1
21	3.15	2.7	5.5	2.35	2.95	3.6	2.4	2.05
22	2.7	2.45	5.6	2.55	3.4	3.4	2.5	2.15
23	2.8	2.55	5.65	3.2	3.2	3.35	2.3	2.1
24	2.85	2.45	5.6	3.75	3.5	3.2	2.2	2.1
25	2.8	2.55	5.8	3.3	3.2	3.2	2.1	2.05
26	2.7	2.5	5.3	3.1	3.25	3.0	2.0	2.05
27	2.6	2.55	5.1	3.05	3.05	2.9	2.0	2.0
28	3.15	2.6	4.9	2.8	2.8	2.75	2.0	2.05
29	2.85	2.65	5.0	2.7	2.95	2.65	2.0	2.1
30	2.5	2.5	5.0	2.9	3.1	2.55	2.0	2.1
31	....	2.15	....	2.8	3.6	....	2.0	....

Daily discharge, in second-feet, of Uncompahgre River at Montrose, Colo., for 1909.

Day.	Apr.	May.	June.	July.	Aug.	Sept.	Oct.	Nov.
1	75	52	45	880	139	333	75	83
2	75	68	60	777	183	196	67	100
3	107	45	107	854	100	209	67	83
4	107	127	402	777	92	425	67	92
5	117	370	912	752	119	545	53	83
6	107	386	1,200	752	83	854	53	60
7	107	402	1,450	524	83	906	119	47
8	75	402	1,040	406	83	728	100	41
9	82	277	964	406	75	656	92	30
10	107	249	912	369	100	588	100	30
11	107	370	1,020	316	119	610	92	25
12	90	306	1,070	172	110	484	83	25
13	98	338	1,200	110	183	828	75	16
14	107	159	912	75	119	656	67	21
15	138	98	990	53	209	566	75	25
16	236	107	1,200	75	196	610	83	21
17	568	107	1,230	119	139	610	83	25
18	610	90	1,250	83	351	566	83	25
19	510	148	1,280	92	237	484	67	30
20	386	249	1,250	83	299	388	75	30
21	263	148	1,280	60	172	369	67	25
22	148	98	1,340	92	299	299	83	36
23	170	117	1,360	237	237	283	53	30
24	183	98	1,340	425	333	237	41	30
25	170	117	1,450	267	237	237	30	25
26	148	107	1,170	209	252	183	21	25
27	127	117	1,060	196	196	160	21	21
28	263	127	958	139	139	129	21	25
29	183	138	1,010	119	172	110	21	30
30	107	107	1,010	160	209	92	21	30
31	....	52	....	139	369	....	21	....

NOTE.—These discharges are based on rating curves applicable as follows: April 1 to June 25, well defined between 30 and 600 second-feet; June 26 to Nov. 30, not well defined.

Monthly discharge of Uncompahgre River at Montrose, Colo., for 1909.

Month.	Discharge in second-feet.			Run-off (total in acre-feet).	Accuracy.
	Maximum.	Minimum.	Mean.		
April..	610	75	186	11,100	B.
May...	402	45	180	11,100	B.
June..	1,450	45	1,020	60,700	A.
July..	880	53	313	19,200	A.
August...	369	75	182	11,200	A.
September......................................	906	92	445	26,500	A.
October..	119	21	63.7	3,920	B.
November.......................................	100	16	38.9	2,310	B.
The period.................................	....	....	....	146,000	

UNCOMPAHGRE RIVER NEAR DELTA, COLO.

This station was originally established April 29, 1903, at a highway bridge one-fourth mile above the Denver & Rio Grande Railroad bridge. On November 17, 1903, it was removed to the railroad bridge, one-fourth mile northwest of the depot at Delta, Colo. The vertical gage at this bridge was read until April 21, 1904, when an inclined gage was installed on the right bank near the bridge. This gage was read until November, 1906, when a staff gage was installed at the present location, on the second highway bridge 2 miles south of Delta. Observations were not begun at this gage until April 21, 1907. It was washed out September 6, 1909, and not repaired.

The station is located near the junction of the Uncompahgre with the Gunnison and is below all tributaries and diversions. At ordinary stages the flow of the river at this point is nearly all seepage water from the irrigation of ditches above. During the irrigation season the ditches consume all the normal flow.

Results are probably not materially affected by ice, as ice does not form very thick except along the edges of the stream. Slush ice frequently occurs.

There is no determined relation between the datum of the last established gage and the several earlier gages, and the datum of the gage used from April 22, 1904, to November, 1906, is different from that of the previous gage.

Records obtained at this station are good except during extremely low stages.

This station is maintained under the supervision of the United States Reclamation Service.

The daily and monthly estimates for this station are withheld until more high-water measurements have been obtained.

Discharge measurements of Uncompahgre River near Delta, Colo., in 1909.

Date.	Hydrographer.	Width.	Area of section.	Gage height.	Discharge.
		Feet.	*Sq. ft.*	*Feet.*	*Sec.-ft.*
Apr. 28	R. M. Adams............................	42	101	2.54	242
June 5	.....do.................................	48	98	2.45	294
July 20	.....do.................................	40	42.4	1.57	21.1
29	.....do.................................	48	61.4	1.70	43.3
Aug. 20	.....do.................................	48	108	2.80	356
Oct. 22	.....do.................................	50	69	....	109

Daily gage height, in feet, of Uncompahgre River near Delta, Colo., for 1909.

Day.	Apr.	May.	June.	July.	Aug.	Sept.	Day.	Apr.	May.	June.	July.	Aug.	Sept.
1..........	1.8	1.9	1.95	3.65	1.7	2.5	16..........	2.05	2.05	3 85	1.7	2.15	......
2..........	1.8	1.75	1.9	3.45	1.7	2.5	17..........	2.4	2.0	3.85	1.7	2.3	......
3..........	1.8	1.95	1.9	3.35	1.65	3.1	18..........	3.2	2.15	4.3	1.7	2.65	......
4..........	1.8	2.4	2.0	3.4	1.6	3.5	19..........	3.4	2.55	4.6	1.7	3.0	......
5..........	1.9	2.75	3.0	3.85	1.6	3.65	20..........	3.3	2.55	4.4	1.75	2.65	......
6..........	2.0	3.4	3.5	3.8	1.6	4.5	21..........	3.1	2.55	4.35	1.6	2.6	......
7..........	2.1	3.45	3.95	3.55	1.7	......	22..........	2.55	2.25	4.1	2.25	2.55	......
8..........	1.95	3.65	4.15	3.15	1.7	......	23..........	2.2	2.45	4.2	2.05	2.35	......
9..........	1.9	3.4	4.25	2.75	1.65	......	24..........	1.9	2.3	4.25	2.3	2.45	......
10..........	1.9	3.25	4.1	2.5	2.3	......	25..........	1.75	2.2	4.1	2.0	2.5	......
11..........	1.9	3.25	3.95	2.45	2.5	......	26..........	1.85	2.1	4.05	2.0	2.45	......
12..........	1.9	3.25	3.9	2.0	2.0	......	27..........	2.25	2.3	3.95	1.95	2.6	......
13..........	1.9	3.05	3.9	2.0	1.85	......	28..........	2.6	2.5	3.85	1.8	2.25	......
14..........	1.8	2.65	3.8	2.0	2.35	......	29..........	2.5	2.45	3.6	1.7	2.0	......
15..........	1.8	2.25	3.75	1.8	2.15	......	30..........	2.1	2.3	3.65	1.7	2.0	......
							31..........	......	2.15	......	1.65	2.2	......

NOTE.—Gage washed out on Sept. 6.

FREMONT RIVER BASIN.

DESCRIPTION.

Fremont River heads in the eastern slopes of the Wasatch Mountains in Sevier County, Utah, one of its sources being Fish Lake. It flows in a general southerly direction to Thurber, from which point it traverses the central portion of Wayne County in a general easterly direction to Hawksville, where it turns southward; it joins Colorado River about 8 miles above Hite, Utah. It receives one important tributary, Curtis Creek,[1] from the north and a number of smaller streams, including Tantalus and Lewis creeks, from the south. The lower half of its course is through two deep canyons separated by a valley. On the upper water of the main river is what is known as Rabbit Valley. Both Fremont River and Curtis Creek are considerably augmented in volume by springs in their canyons, but they derive the greater part of their waters from melting snows on the plateau.

FREMONT RIVER NEAR THURBER, UTAH.

This station, which was established May 13, 1909, is located about 2 miles (by road) south of Thurber, Utah. The records show the total amount of water available for storage in a reservoir proposed at this point.

---

[1] Called Muddy River on General Land Office maps.

Pine Creek enters about 2 miles above the station. This creek and springs in the valley just above the station furnish much of the low-water flow. Most of the normal low-water flow is diverted above and below the station for irrigation.

The staff gage is on the left bank about 2,000 feet above a gristmill. The gage height records are probably not much affected by ice. Discharge measurements are made from a cable at the gage.

As the bed of the stream is of a shifting character, frequent measurements are necessary for reliable estimates of daily and monthly discharge.

Discharge measurements of Fremont River near Thurber, Utah, in 1909.

Date.	Hydrographer.	Width.	Area of section.	Gage height.	Discharge.
		Feet.	*Sq. ft.*	*Feet.*	*Sec.-ft.*
May 13	E. A. Porter..........................	29	61	4.25	116
June 30	.....do...............................	28	64.4	4.35	67.3
Aug. 11	.....do...............................	28.5	58	4.97	111
30	.....do...............................	28.7	37	5.02	132

Daily gage height, in feet, of Fremont River near Thurber, Utah, for 1909.

[John Smith, observer.]

Day.	May.	June.	July.	Aug.	Sept.	Oct.	Nov.	Dec.
1.................................		4.7	4.4	4.65	5.55	5.8	5.35	5.55
2.................................		4.75	4.5	4.7	6.0	5.85	5.3	5.6
3.................................		4.8	4.55	4.7	6.1	5.9	5.25	5.65
4.................................		4.8	4.6	4.65	5.9	5.85	5.25	5.6
5.................................		4.85	4.6	4.7	5.9	5.8	5.2	5.5
6.................................		4.85	4.6	4.8	5.85	5.8	5.0	5.6
7.................................		4.9	4.6	4.9	5.9	5.75	5.1	5.5
8.................................		5.0	4.55	5.25	5.8	5.8	5.15	5.45
9.................................		5.0	4.6	5.2	5.75	5.75	5.2	5.5
10................................		5.05	4.6	5.15	5.6	5.65	5.2	5.55
11................................		4.95	4.65	5.1	5.7	5.7	5.25	5.5
12................................		4.8	4.7	6.0	5.8	5.6	5.3	5.55
13................................	4.25	4.6	4.65	5.5	5.85	5.55	5.35	5.55
14................................	4.6	4.6	4.75	5.4	5.9	5.5	5.4	5.6
15................................	4.65	4.55	4.6	6.2	5.8	5.5	5.4	5.7
16................................	4.5	4.65	4.6	6.25	5.75	5.45	5.45	5.65
17................................	4.5	4.4	4.65	6.3	5.65	5.4	5.4	5.7
18................................	4.7	4.6	4.65	6.25	5.6	5.45	5.45	5.75
19................................	4.8	4.6	4.7	6.3	5.5	5.5	5.5	5.8
20................................	4.9	4.7	5.1	6.25	5.45	5.4	5.45	5.9
21................................	4.9	4.7	5.0	6.1	5.3	5.35	5.4	5.85
22................................	4.9	4.75	4.9	6.0	5.4	5.3	5.45	5.8
23................................	4.8	4.7	4.95	5.9	5.35	5.25	5.5	5.75
24................................	4.75	4.65	4.9	5.8	5.4	5.3	5.45	5.8
25................................	4.8	4.45	4.9	5.7	5.4	5.35	5.5	5.75
26................................	4.75	4.35	4.95	5.6	5.4	5.4	5.5	5.7
27................................	4.6	4.5	4.9	5.6	5.35	5.5	5.55	5.65
28................................	4.7	4.5	4.85	5.5	5.5	5.45	5.5	5.7
29................................	4.5	4.5	4.7	5.25	5.6	5.45	5.55	5.75
30................................	4.5	4.5	4.75	5.0	5.7	5.45	5.6	5.7
31................................	4.55	....	4.7	5.5	....	5.4	....	5.7

Daily discharge, in second-feet, of Fremont River near Thurber, Utah, for 1909.

Day.	May.	June.	July.	Aug.	Sept.	Oct.	Nov.	Dec.
1		136	71	83	185	207	153	171
2		140	78	88	235	213	148	176
3		144	82	88	240	218	143	181
4		143	86	83	223	213	143	176
5		145	86	88	223	209	138	166
6		143	86	96	218	209	143	176
7		145	86	106	223	202	128	166
8		154	81	139	212	209	133	161
9		152	85	134	207	201	138	166
10		154	85	128	191	190	138	171
11		142	88	119	202	196	142	166
12		127	92	226	213	185	149	171
13	116	107	88	165	218	180	152	171
14	150	105	97	156	223	175	157	176
15	145	100	84	228	213	175	157	174
16	137	108	83	244	207	169	162	179
17	136	85	87	251	194	163	157	174
18	150	100	87	246	189	169	162	189
19	163	100	91	253	179	175	168	195
20	172	108	128	248	174	163	162	205
21	171	107	119	234	159	158	157	200
22	170	111	110	226	169	153	162	195
23	168	105	113	216	164	148	167	189
24	153	99	109	207	169	153	162	195
25	156	81	109	198	169	158	167	189
26	150	70	112	189	169	163	167	184
27	134	83	107	190	164	164	172	179
28	141	82	102	179	179	169	167	184
29	121	81	89	154	189	163	172	189
30	120	67	93	132	189	163	176	184
31	120		89	184		163		184

NOTE.—These discharges were obtained by the indirect method for shifting channels.

Monthly discharge of Fremont River near Thurber, Utah, for 1909.

Month.	Discharge in second-feet.			Run-off (total in acre-feet).	Accuracy.
	Maximum.	Minimum.	Mean.		
May 13–31	172	116	146	5,500	D.
June	154	67	114	6,780	C.
July	128	71	93.6	5,760	B.
August	253	83	170	10,500	B.
September	240	159	196	11,700	C.
October	218	148	180	11,100	C.
November	176	128	155	9,220	C.
December	205	161	180	11,100	C.
The period				71,600	

MUDDY CREEK NEAR EMERY, UTAH.

Muddy Creek rises in the eastern slopes of the Wasatch Mountains in the extreme southern corner of Sanpete County and joins Curtis Creek about 8 miles below Emery, Utah. Curtis Creek flows southeasterly across Emery County and enters Fremont River near Hawksville, Utah.

The station was originally established April 29, 1909, at Jacobson's ranch, about 7 miles above and northwest of the town of Emery, Utah.

The station is below all tributaries and above all diversions.

Prior to August 25, 1909, records were obtained by measuring down to the water surface from a reference point on a flume in which discharge measurements were made. A staff gage was installed at the same location and datum on August 25. During August great variations in discharge were caused by heavy rains, which washed out the gage and so altered the section that the station had to be reestablished at a new location September 18 several hundred feet upstream. A staff gage was installed at a different datum and a cable erected, from which cable section discharge measurements will be made.

Discharge measurements of Muddy Creek near Emery, Utah, in 1909.

Date.	Hydrographer.	Width.	Area of section.	Gage height.	Discharge.
		Feet.	*Sq. ft.*	*Feet.*	*Sec.-ft.*
June 13	E. A. Porter...	24	41	3.40	247
Aug. 25	.....do..	17	15	1.90	48
Sept. 18	.....do..	17	18	a1.98	55
Dec. 3b	.....do..			a1.20	c3.8

a Gage height from new gage set Sept. 18 at a different location and datum from the one previously used.
b Creek frozen over along the entire course; 6 inches of ice at gage.
c Estimated.

Daily gage height, in feet, and discharge, in second-feet, of Muddy Creek near Emery, Utah, for 1909.

[Melvin Sorenson, observer.]

Day.	April.		May.		June.		July.		August.	
	Gage height.	Discharge.	Gage height.	Discharge.	Gage height.	Discharge.	Gage height.	Discharge.	Gage height.	Discharge.
1.............			1.85	44	3.25	224	2.75	153	2.35	....
2.............			1.95	53	3.75	300	2.7	146	2.3	....
3.............			2.5	119	3.7	292	2.55	126	2.2	....
4.............			2.7	146	3.75	300	2.5	119	2.3	....
5.............			2.6	132	3.7	292	2.35	100	2.3	....
6.............			1.4	12	3.7	292	2.35	100		....
7.............				15	3.6	277	2.3	94	2.1	....
8.............				18	3.65	284	2.3	94	2.05	....
9.............				21	3.5	262	2.25	88		....
10.............			1.6	24	3.5	262	2.25	88		....
11.............			1.95	53	3.5	262	2.45	112		....
12.............			2.45	112	3.45	254	2.5	119		....
13.............			2.5	119	3.5	262	2.5	119	2.1	....
14.............			2.5	119	3.45	254	2.45	112	2.0	....
15.............			2.4	106	3.45	254	2.45	112	2.0	....
16.............			2.95	181	3.45	254	2.4	106	2.0	....
17.............			3.45	254	3.45	254	2.4	106		....
18.............			3.55	270	3.35	240	2.45	112		....
19.............			3.25	224	3.35	240	2.45	112		....
20.............			3.00	188	3.35	240	2.45	112		....
21.............			3.3	232	3.3	232	2.45	112		....
22.............			3.05	195	3.3	232	2.4	106		....
23.............			3.0	188	3.2	217	2.5	119		....
24.............			2.9	174	3.2	217	2.45	112	2.1	....
25.............			3.15	210	3.15	210	2.45	112	1.9	....
26.............			3.7	292	3.1	202	2.45	112	1.9	....
27.............			3.25	224	3.1	202	2.35	100	1.9	....
28.............			3.2	217	3.0	188	2.35	100	1.9	....
29.............	1.9	48	3.1	202	2.95	181	2.35	100	1.9	....
30.............	1.8	40	2.9	174	2.8	160	2.3	94		....
31.............			3.1	202			2.3	94		....

NOTE.—The daily discharges are based on a rating curve that is fairly well defined between 40 and 300 second-feet.

Monthly discharge of Muddy Creek near Emery, Utah, for 1909.

Month.	Discharge in second-feet.			Run-off (total in acre-feet).	Accuracy.
	Maximum.	Minimum.	Mean.		
May...	292	12	146	8,980	B.
June...	300	160	245	14,600	A.
July...	153	88	109	6,700	A.

ESCALANTE RIVER BASIN.

ESCALANTE CREEK NEAR ESCALANTE, UTAH.

Escalante River rises in the southern part of Garfield County, Utah, under the walls forming the east face of the Table Cliff Plateau; flows first northeast, then east, and finally southeast, and enters the Colorado in Kane County about 12 miles above the mouth of the San Juan. It is 90 miles long, and the lower three-fourths of its course is through a narrow canyon whose vertical walls range in height from 900 to 1,200 feet. Through this gorge the river sweeps, in places filling the whole space from wall to wall, in places winding from side to side in a flood plain of sand and shifting its position more or less with every freshet.

In the upper part of its course it is joined by several tributaries, all of which flow through close canyons.

A gaging station, established August 5, 1909, is located on Escalante Creek, one of the headwaters of Escalante River, at the head of the canyon, about 2 miles below the town of Escalante, Utah. The records show the total amount of water available for storage in an excellent reservoir site at this point.

The principal tributaries above are Birch Creek, entering about 6 miles upstream, and Pine Creek, which enters just above the station. Practically all the normal low-water flow is diverted above the station for irrigation in and near Escalante, the run-off at the station representing only the surplus water.

Estimates of winter discharge are very unreliable. The shifting nature of the stream bed makes accurate interpretation of the results difficult.

The first gage used was located about 20 feet below the mouth of Pine Creek. It was washed out by a severe flood August 31, which scoured out the bed of the creek about 3 feet and changed the location of the channel. From September 1 to November 12 records were kept of the depth of water at a point near the gage site. On November 13 a new gage was set 35 feet above the old one and the observer's readings for the intervening period referred to the new datum. The records for this period are only approximate.

Discharge measurements of Escalante Creek near Escalante, Utah, in 1909.—

Date.	Hydrographer.	Width.	Area of section.	Gage height.	Discharge.
		Feet.	Sq.ft.	Feet.	Sec.-ft.
Aug. 5a	E. A. Porter			1.85	8.0
Nov. 13b	....do	11	7.9	2.06	13.5

a Made by floats, coefficient 0.8 used.
b Made by wading below gage installed this date at different datum.

Daily gage height, in feet, of Escalante Creek near Escalante, Utah, for 1909.

[D. C. Shurtz, observer.]

Day.	Aug.	Sept.	Oct.	Nov.	Dec.	Day.	Aug.	Sept.	Oct.	Nov.	Dec.
1		3.7	2.2		2.0	16	6.2	2.0	1.6	2.0	
2		4.7		1.8		17	2.0			2.0	2.6
3		3.2	2.15		2.0	18	5.2	2.2	1.55		2.0
4		4.1		1.8		19	3.5			2.0	2.4
5	1.8		2.2		2.0	20	2.7	2.1	1.8		
6	1.9	2.0		1.8		21	3.7			2.0	2.6
7	3.7	3.55	2.2		2.05	22	2.1	2.1	1.8		
8	1.6	2.0		1.8		23	2.5		1.8	2.0	2.8
9	3.1		1.75		2.1	24	3.3	2.2			
10	1.8	2.1		2.2		25			1.75	2.0	3.1
11	1.75	2.0	1.7		2.1	26	2.4	2.15			
12		2.0	1.75	1.9		27			1.75	2.0	3.1
13	1.9	2.0	1.65	2.0	2.2	28	2.8	2.2			
14	1.7		1.6			29	8.2		1.75	1.95	3.05
15		2.0		2.0	2.3	30	4.7	2.3			
						31	10.2		1.75		3.15

NOTE.—Gage heights Aug. 5 to 31 refer to gage installed Aug. 5. Gage heights Sept. 1 to Nov. 12, reduced to datum of new gage installed Nov. 13 and are approximate. See description.

Daily discharge, in second-feet, of Escalante Creek near Escalante, Utah, for 1909.

Day.	Aug.	Sept.	Oct.	Nov.	Dec.	Day.	Aug.	Sept.	Oct.	Nov.	Dec.
1		127	18	6	12	16	443	12	3	12	29
2		243	17	7	12	17	12	15	2	12	36
3		79	16	7	12	18	304	18	2	12	12
4		170	17	7	12	19	107	16	4	12	26
5	7	91	18	7	12	20	42	15	7	12	31
6	9.5	12	18	7	13	21	127	15	7	12	36
7	127	112	18	7	14	22	15	15	7	12	42
8	3	12	12	7	14	23	31	16	7	12	48
9	70	14	6	12	15	24	88	18	6	12	59
10	7	15	6	18	15	25	57	17	6	12	70
11	6	12	5	14	15	26	26	16	6	12	70
12	8	12	6	9.5	16	27	37	17	6	12	70
13	9.5	12	4	12	18	28	48	18	6	12	68
14	5	12	3	12	20	29	520	20	6	11	66
15	112	12	3	12	22	30	243	22	6	11	70
						31	670		6		74

NOTE.—These discharges are based on a rating curve that is not well defined.

Monthly discharge of Escalante Creek near Escalante, Utah, for 1909.

Month.	Mean discharge in second-feet.	Run-off (total in acre-feet).
August 5–31	116	6,210
September	39.5	2,350
October	8.19	504
November	10.8	643
December	33.2	2,040

NOTE.—These estimates are only approximate. See description.

SAN JUAN RIVER BASIN.

DESCRIPTION.

San Juan River rises among the snow masses that crown the high peaks of the San Juan Mountains in southwestern Colorado, flows southwestward into New Mexico, then swings to the west and north-west, passing from San Juan County, N. Mex., across the extreme southwestern corner of Colorado into San Juan County, Utah, in the southwestern part of which it unites with the Colorado.

For the first 75 miles of its course the San Juan is a typical mountain stream, but at Canyon Largo, N. Mex., where it turns westward, its character changes, and it occupies a broad, winding, sandy channel in an arid valley, bordered on each side by terraced mesas. Below the mouth of Mancos River the valley narrows and the river bottom is bounded by abrupt bluffs, broken and cut by dry water channels, and merging farther on into the walls of a deep, narrow, box canyon, in which the river flows to its end.

The drainage area includes portions of four States and Territories. Its topography ranges from the mountainous types at the headwaters in Colorado to the types exemplified in the valleys, plateaus, and eroded mesas of Utah, New Mexico, and Arizona. Large areas of eruptive rocks occur in the highest portions of the basin, but the predominating formations are of sedimentary origin. The headwater streams are protected by fine forests of spruce and yellow pine and at lower elevations large areas of aspen. The lower basin is practically barren except for an extensive growth of sagebrush, scattered cedars, piñons, and range grasses.

The principal tributaries of the San Juan are Navajo, Piedra, Pine, Florida, Animas, and La Plata rivers, the Animas being the most important.

Animas River has its source in the region above Silverton, draining portions of the Needle and La Plata Mountains, the former being the most rugged of the Rocky Mountain ranges. The river flows southward to the Colorado-New Mexico line and thence southwestward to the point where it joins the San Juan at Farmington, N. Mex. The upper portion of the basin, above Durango, is very mountainous and furnishes the principal part of the run-off. Much of this region is well timbered with pine, spruce, and aspen, but large areas consist of naked granite peaks. Immediately above and below Durango the valley broadens and is bordered by mesas and bluffs cut by narrow canyons and covered with sagebrush and scattered pines and piñons; along the stream channels cottonwoods predominate. The rocks of this region are chiefly of sedimentary origin. The soils of the lower valleys consist of sandy loam and are very fertile.

La Plata River rises in the granite masses known as La Plata Mountains, about 25 miles northwest of Durango, Colo., and flows southward to its point of junction with the San Juan. Its drainage basin is a narrow strip parallel to and adjoining the Animas basin. The upper portion of the basin is a well-watered and forest-clad mountain region which merges southward into an arid mesa, plateau, and canyon country. La Plata Valley proper is a narrow, shallow depression from Hesperus down, bounded on both sides by high, broken tablelands and deeply eroded mountains. The lower mountain slopes are covered with piñon, scrub oak, and cedar; the lower valleys support heavy growths of sagebrush and chico; the upper mountain slopes were at one time heavily timbered with spruce and yellow and white pine, but these forests have been largely removed by lumbermen.

The other tributaries of the San Juan need not here be described. Those mentioned are perennial streams, but much of their water is diverted for irrigation and never reaches the main river. In addition to the perennial streams are many intermittent creeks throughout New Mexico, which contribute large volumes of water during heavy storms.

The altitudes in this drainage basin range from over 13,000 feet in the highest mountains to between 6,000 and 7,000 feet at the Colorado-New Mexico line. The San Juan at the mouth of the Animas has an elevation of about 5,300 feet; at its junction with Colorado River the elevation is about 3,500 feet.

Most of the timbered land in the San Juan drainage basin is included in the San Juan National Forest, which contains nearly 2,000 square miles of merchantable timber, 100 square miles of woodland, 300 square miles of sagebrush, and 200 square miles of barren and burnt area.

In a small area in the high mountains the annual precipitation exceeds 25 inches, and over a considerable area the average exceeds 20 inches; but for the remainder of the area the average in Colorado seems to be about 15 inches, that in New Mexico about 10 inches, and in Utah about 15 inches.

Above an altitude of 7,500 feet the winters are severe and snowfalls are heavy; below an elevation of 6,000 feet the winters are comparatively open and mild. The upper mountain streams flow under a thick ice cover, but in the more open country, in the vicinity of Aztec, it is rather unusual for the rivers to freeze over entirely, though much ice forms along the edges, and slush ice is often seen.

Much land along the valleys of San Juan, Animas, Pine, Florida, and La Plata rivers and the smaller tributaries in Colorado is now under cultivation, and also a few thousand acres of valley land in New Mexico. Up to this time irrigation has largely been confined to the bottom land. The greatest opportunities for future development are in San Juan County, N. Mex., where exceptionally large areas, aggregating probably a million acres of fertile lands, are excellently adapted to irrigation. The rivers there are bordered by broad mesas and benches, sloping back for miles in many places and easily reached by irrigation canals, and the water supply is ample.

Numerous small lakes, high up in the mountains, tend to equalize the flow of some of the tributaries, and many large and small storage reservoir sites are available. Among others may be mentioned the Turley reservoir site, on San Juan River below the mouth of the Pine, which has a storage capacity of about one and a half million acre-feet.

Excellent opportunities for power development are presented. Theoretically, with proper storage, it will be possible to develop nearly 300,000 horsepower. Falls of 100 to 300 feet per mile are common on the upper reaches of the stream. The San Juan has an average fall of about 13 feet to the mile from the mouth of the Piedra to the mouth of the Mancos, a distance of about 115 miles, while the fall above the mouth of the Piedra is very much greater. The Animas has a fall of over 70 feet to the mile from Silverton to Durango, a distance of about 40 miles, and from Durango to its mouth the average fall is over 20 feet to the mile. Present developments are practically limited to two power plants on Animas River, of 6,000 and 1,000 horsepower. Several other plants are contemplated.

The largest deposits of lignite and bituminous and coking coal in the West are in this drainage area.

FLOOD IN SAN JUAN VALLEY, SEPTEMBER 5 AND 6, 1909.[1]

GENERAL FEATURES.

A very heavy flood occurred on the San Juan and its tributaries in southwestern Colorado and northwestern New Mexico on September 5 and 6, 1909. The oldest residents in the vicinity of Aztec and Farmington stated that both the Animas and San Juan reached higher stages during this flood than at any time since the country has been settled, and even the Indians are said to have had no memory of a time when the San Juan was so high.

The flood was caused by continuous heavy rains in the headwater region and all along the courses of these streams. The upper San Juan itself and its tributaries from the north, from the Piedra west to and beyond La Plata, were in flood simultaneously with Canyon Largo and other large intermittent tributaries from the south, all of which are capable of carrying enormous quantities of water. The main watercourse of Canyon Largo, for instance, is more than 70 miles long, and some of its tributaries are 40 or 50 miles long, so that it drains a very large area which at times is subject to an excessive run-off. The combined floods caused the usual damage.

---

[1] This report by Mr. Freeman was also published in the Monthly Weather Review for September, 1909.

In Colorado the Denver & Rio Grande Railroad sustained the greater part of the loss. Along Animas River between Silverton and Durango many stretches of the railroad tracks were washed out and others were covered with immense deposits of sediment. Traffic on this branch was suspended for about three weeks. The cost to the railroad company amounted to many thousand dollars. Animas River also did considerable damage in the town of Durango and surrounding country.

Pine River, in Colorado, was in heavy flood and overflowed its banks in many places. At La Boca, the river took out the approach embankment to railroad bridge on right bank. This embankment has since been replaced by a pile span. Spring Creek, which flows into Pine River near this point, washed out the railroad track in several places. A locomotive went into the ditch at a small railroad trestle on this creek. The Piedra and upper San Juan were also at very high stages.

In San Juan County, N. Mex., the principal loss was sustained by the Farmington branch of the Denver & Rio Grande Railroad; by the county, through the washing out of the suspension bridge across the San Juan at Blanco; and by the farmers, whose lands were overflowed by the flood waters of Animas and San Juan Rivers. The total loss in the county is believed to have been not less than $20,000; the Blanco bridge alone cost about $4,200. The Flora Vista glade, which runs into Animas River near Flora Vista, carried a great deal of water and sediment and did not a little damage to the railroad track and the surrounding country, both by washouts and by the deposition of sediment. Other glades and arroyos were correspondingly high, though most of them are not in a position to do much damage. The streams in this section began to rise rapidly on Sunday, September 5, and by Sunday night they were at high stages. Maximum heights were attained early Monday morning in the vicinity of Farmington.

PRECIPITATION.

The rainfall was general. (See Pl. VI.) As far west as Fruitland there was about 0.7 inch of precipitation during the first six days of September, most of which fell on the 6th; and as far east as Chama, the rainfall amounted to 1.6 inches from September 1 to 6. At Durango, Colo., the rainfall was 3.6 inches during the above period, of which 1.64 inches fell on September 5 and 0.82 inch on September 6. The combined precipitation on September 5 and 6 at several points was as follows:

Inches.

Silverton	2.50
Hesperus	2.68
Mancos	1.41
Ignacio	.87
Pagosa Springs	1.87

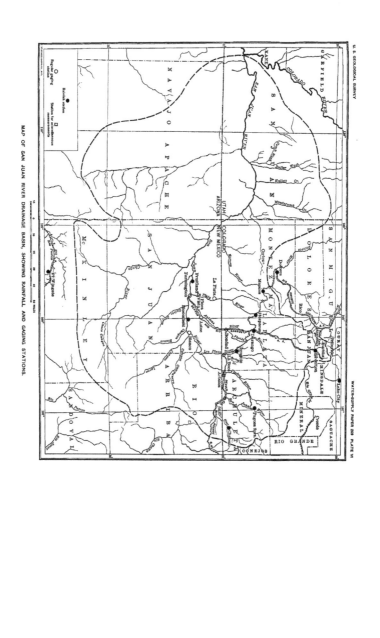

MAP OF SAN JUAN RIVER DRAINAGE BASIN, SHOWING RAINFALL AND GAGING STATIONS.

San Miguel River was in heavy flood at this same time, and the bursting of the Trout Lake reservoir dam caused the destruction of a great deal of property below. Several miles of railroad track were damaged or entirely washed away, causing the suspension of traffic for a long period, and no little damage was done to highways and bottom lands in San Miguel County, Colo.

The following table shows the daily precipitation at various points in San Juan County from September 1 to 7, inclusive:

Precipitation in inches in San Juan drainage basin, September 1–7, 1909.

Stations.	September—							Total.
	1	2	3	4	5	6	7	
Silverton, Colo.	0.23	0	0	0.31	2.00	0.50	0.30	3.34
Lake City, Colo.	0	T.	.10	.05	.60	.09	.02	.86
Dolores, Colo.	.11	.16	0	.10	.41	.48	0	1.26
Hesperus, Colo.	.40	.06	.15	.04	1.99	.69	.29	3.62
Durango, Colo.	.50	.19	.04	.37	1.64	.82	.01	3.57
Ignacio, Colo.	.06	T.	T.	.04	.50	.37	.12	1.09
Chromo, Colo.	.15	.03	.27	.18	.79	.37	.30	2.09
Mancos, Colo.	0	.13	.29	.04	.99	.42	T.	1.87
Pagosa Springs, Colo.	.01	0	.18	.30	1.40	.47	.35	2.71
Bloomfield, N. Mex.	0	0	0	0	1.00	.17	0	1.17
Fruitland, N. Mex.	T.	T.	0	0	.30	.65	0	.95
Fort Wingate, N. Mex.	.40	0	.32	1.20	.59	1.20	0	3.71

SAN JUAN RIVER PROPER.

At Arboles, Colo., above the most important tributaries, the San Juan was at a very high stage and probably reached its maximum on Sunday night. At Blanco, N. Mex., above Canyon Largo, the gage read 10 feet on Sunday evening, and at 6 a. m. Monday, September 6, when the county suspension bridge was washed out, the maximum gage height was about 11 feet. The maximum discharge there was not less than 15,000 second-feet. At Bloomfield, 12 miles below the mouth of Canyon Largo, the river had a flood cross section of 1,800 square feet and a maximum discharge estimated from 18,000 to 20,000 second-feet. The river overflowed the bottom lands all along its course, and in many places the course of the channel was entirely changed. At one place part of an orchard was washed into the river, and at another a house was carried away when the bank caved in. At the county bridge at Farmington, below the mouth of the Animas, the river reached a maximum height late Monday evening. The gage height was in the neighborhood of 12.3 feet on the old United States Geological Survey gage, the cross sectional area was 3,200 square feet, and the estimated discharge was not less than 35,000 second-feet. At this place the river overflowed the right bank for a distance of about 100 yards beyond the end of the bridge, but the bridge was not damaged in any way.

ANIMAS RIVER.

At Durango, Animas River reached a maximum height of 8.50 feet at the power company's gage between 4 and 6 o'clock Monday morning, September 6 (Pl. VII, *B*). The maximum discharge was estimated to be over 10,000 second-feet. Between 6 and 7 o'clock on Monday evening the maximum height of 11 feet was reached on the United States Geological Survey gage at Aztec, N. Mex., and between 10 and 11 p. m. the maximum gage height at the old United States Geological Survey gage at Farmington was 11.1 feet. At Aztec (Pl. VII, *A*) the river overflowed the right bank for a distance of 800 feet beyond the end of the suspension bridge, the water being from 0.5 foot to 2 feet in depth; and at Farmington the river was over its right bank for a distance of a quarter of a mile; the fair grounds and race track there were entirely submerged.

The flood cross section at Aztec was about 3,100 square feet, and the estimated discharge was 16,000 second-feet. At Farmington the maximum discharge was probably from 18,000 to 20,000 second-feet.

LA PLATA RIVER.

La Plata River attained a maximum gage height of 7.60 feet on the United States Geological Survey and Territorial engineer's gage at La Plata, N. Mex., on Sunday night, September 5, and it is estimated that the maximum discharge was from 6,000 to 7,000 second-feet. The left bank was overflowed for a distance of 1,000 feet, and the cross-sectional area of the overflow was 1,300 square feet, while the cross-sectional area of the main channel was only 224 square feet. It is estimated that the channel under the bridge at La Plata carried a maximum of 3,000 second-feet, while the overflow amounted to as much or more than that amount.

RÉSUMÉ.

During this flood on the San Juan River probably over 150,000 acre-feet of flood waters went to waste and were allowed to do a great deal of damage to the surrounding country. Had these waters been properly stored in any of the numerous reservoirs contemplated in San Juan County, such as the proposed San Juan reservoir, all of this damage could have been prevented and the waters used for the irrigation of a large area.

It is true that a flood of this magnitude is not liable to occur for a number of years, and it is also true that it is not necessary to store these excessive flood waters for irrigation. The San Juan is a valuable stream and its normal flow is ample for the irrigation of all the lands which it will be possible to put under cultivation for a

A. OVERFLOW OF LEFT BANK IN VICINITY OF GAGING STATION AT AZTEC, N. MEX.

B. VIEW AT FOOTBRIDGE NEAR DURANGO POWER PLANT, DURANGO, N. MEX.

FLOOD ON ANIMAS RIVER, SEPTEMBER 6, 1909.

number of years to come. Some storage, however, is necessary on the San Juan, as on any other river in the West. In the latter part of September, 1909, the flow of the San Juan above the mouth of the Animas was only 1,100 second-feet, and that of Animas River at Aztec 800 second-feet. These flows are considerably above the normal for that priod of the year.

SAN JUAN RIVER AT BLANCO, N. MEX.

This station was established December 9, 1908, to take the place of the station at Turley, which was discontinued November 30, 1908. It was located at a new suspension bridge, which crossed the San Juan at Blanco, about 4 miles below Turley post office, 16 miles southeast of the Denver & Rio Grande Railroad at Aztec, N. Mex., and half a mile above the mouth of Canyon Largo.

The suspension bridge and chain gage were washed out by a flood on September 6, 1909, and on September 29, 1909, a temporary staff gage was established about 30 feet upstream from location of the bridge and at a new datum. Discharge measurements after September 6, 1909, were taken at the suspension bridge at Bloomfield about 9 miles downstream, where a wire gage was installed on September 28, 1909. The flow of the river at Blanco and Bloomfield should be the same except for the inflow of Canyon Largo. This stream carried practically no water in the days when discharge measurements were taken between September 28, 1909, and December 31, 1909.

Discharge measurements of San Juan River at Blanco, N. Mex., in 1908–9.

Date.	Hydrographer.	Width.	Area of section.	Gage height.	Discharge.
1908.		*Feet.*	*Sq. ft.*	*Feet.*	*Sec.-ft.*
Dec. 10a	J. B. Stewart	194	194	2.60	157
1909.					
Jan. 11	C. D. Miller	210	298	2.80	526
Mar. 30	J. B. Stewart	238	507	4.40	2,010
May 14	....do	246	915	6.00	5,820
June 26	V. L. Sullivan	246	1,060	6.70	5,970
July 27	J. B. Stewart	243	794	5.80	4,020
Sept. 28b	W. B. Freeman			c4.70	1,100
Nov. 17b	J. B. Stewart			d4.20	435

a Some slush ice.
b Measurement made at Bloomfield.
c New rod gage installed Sept. 29, 1909, at different datum.
d Observer's gage reading at Blanco.

Daily gage height, in feet, of San Juan River at Blanco, N. Mex., for 1909.

[Cleofes Valdez, observer.]

Day.	Jan.	Feb.	Mar.	Apr.	May.	June.	July.	Aug.	Sept.	Oct.	Nov.	Dec.
1.............	2.8	2.5	3.3	4.1	5.1	5.5	5.7	4.5	6.8	4.6	4.3	4.2
2.............	2.8	2.6	3.7	4.85	5.1	5.4	5.6	4.7	6.4	4.6	4.3	4.2
3.............	2.8	2.9	3.9	5.6	5.5	5.9	5.6	4.6	6.0	4.6	4.3	4.2
4.............	2.8	2.9	4.3	5.6	6.0	6.6	5.5	4.7	7.2	4.5	4.2	4.3
5.............	2.8	2.9	4.4	4.8	6.4	7.2	5.7	5.3	8.55	5.6	4.2	4.2
6.............	2.8	2.8	4.6	4.55	6.6	7.6	5.7	4.7	a 9.85	5.6	4.2	4.2
7.............	2.9	2.8	4.4	4.1	6.7	7.7	5.6	4.6	....	6.1	4.2	4.2
8.............	2.8	2.8	4.7	4.1	6.8	7.6	5.2	4.6	....	5.2	4.2	4.2
9.............	2.9	2.9	4.7	3.85	6.65	7.8	5.2	4.5	....	5.1	4.2	4.2
10............	2.9	2.7	3.6	3.95	6.3	7.2	5.0	4.2	....	4.9	4.2	4.2
11............	2.7	2.7	3.3	4.7	6.1	7.0	4.8	4.5	....	4.8	4.2	4.2
12.....	2.6	2.7	3.3	4.7	6.0	6.9	4.7	4.3	....	4.8	4.2	4.2
13............	2.6	3.0	3.4	4.5	5.9	6.9	4.5	4.3	....	4.8	4.2	4.2
14............	2.6	2.9	3.4	4.8	6.0	6.8	4.6	4.1	....	4.7	4.2	4.1
15............	2.7	2.9	3.7	6.0	6.0	6.7	4.5	4.8	....	4.7	4.2	4.1
16............	2.8	2.8	4.2	6.45	6.05	6.8	4.4	4.8	....	4.7	4.2	4.1
17............	3.0	2.8	4.7	7.25	6.1	6.8	4.2	4.8	....	4.6	4.2	4.1
18............	2.9	2.9	4.8	7.3	6.1	7.1	4.2	4.5	....	4.6	4.1	4.1
19............	2.8	2.9	4.4	7.3	6.4	6.9	4.2	5.9	....	4.6	4.1	4.1
20............	2.8	3.3	4.5	6.85	6.2	7.0	4.0	5.9	....	4.6	4.3	4.1
21............	2.9	2.9	4.5	6.05	6.2	7.0	4.3	6.7	....	4.5	4.4	4.2
22............	3.3	2.6	4.2	5.6	6.0	6.5	4.4	6.0	....	4.4	4.4	4.2
23............	3.6	2.5	4.7	5.5	5.85	6.4	6.0	5.9	....	4.4	4.4	4.2
24............	3.6	2.3	4.2	5.2	5.8	6.4	5.8	5.8	....	4.4	4.3	4.2
25............	3.5	2.9	4.5	5.0	5.6	6.4	5.5	6.5	....	4.4	4.3	4.2
26............	3.3	2.8	4.7	4.9	5.6	6.5	4.9	6.3	....	4.4	4.3	4.2
27............	3.0	3.0	4.6	5.2	5.5	6.2	5.5	7.5	....	4.4	4.2	4.2
28............	2.6	3.3	4.4	5.8	5.55	5.9	5.3	5.8	4.7	4.4	4.2	4.3
29............	2.6	....	4.4	5.6	5.9	5.9	4.9	7.1	4.7	4.4	4.2	4.3
30............	2.6	....	4.2	5.4	5.7	5.9	4.6	7.4	4.6	4.4	4.2	4.3
31............	2.6	....	4.2	....	5.5	....	4.6	7.0	....	4.4	....	4.3

a Estimated maximum gage height of 11.0 feet.

Daily discharge, in second-feet, of San Juan River at Blanco, N. Mex., for 1908–9.

Day.	Dec.	Day.	Dec.	Day.	Dec.
1908.		1908.		1908.	
1.....................	....	11.....................	165	21.....................	270
2.....................	....	12.....................	90	22.....................	300
3.....................	....	13.....................	90	23.....................	450
4.....................	....	14.....................	90	24.....................	450
5.....................	....	15.....................	210	25.....................	350
6.....................	....	16.....................	400	26.....................	350
7.....................	....	17.....................	100	27.....................	300
8.....................	....	18.....................	400	28.....................	300
9.....................	165	19.....................	280	29.....................	250
10....................	165	20.....................	280	30.....................	320
				31.....................	320

Daily discharge, in second-feet, of San Juan River at Blanco, N. Mex., for 1908-9—Contd.

Day.	Jan.	Feb.	Mar.	Apr.	May.	June.	July.	Aug.	Sept.	Oct.	Nov.	Dec.
1909.												
1...............	410	230	650	1,530	3,350	4,390	3,250	1,480	7,050	950	570	470
2...............	420	280	1,060	2,820	3,350	4,110	3,100	1,800	5,770	950	570	470
3...............	425	450	1,280	4,670	4,390	5,560	3,100	1,540	4,560	950	570	470
4...............	435	450	1,840	4,670	5,860	7,820	2,880	1,800	8,400	830	480	550
5...............	440	450	2,000	2,720	7,140	10,000	3,330	2,850	....	2,400	480	470
6...............	450	380	2,350	2,260	7,820	11,500	3,330	1,800	....	2,400	480	470
7...............	550	380	2,000	1,530	8,170	11,900	3,200	1,540	....	3,400	480	470
8...............	475	380	2,530	1,530	8,520	11,500	2,400	1,540	....	1,700	480	470
9...............	570	450	2,530	1,190	8,000	12,300	2,400	1,480	....	1,550	460	470
10..............	575	300	900	1,320	6,810	10,000	2,050	1,080	....	1,250	460	470
11..............	450	300	610	2,530	6,170	9,100	1,720	1,480	....	1,150	460	470
12..............	370	300	610	2,530	5,860	8,570	1,640	1,210	....	1,150	450	470
13..............	370	500	700	2,170	5,560	8,450	1,350	1,210	....	1,150	450	470
14..............	370	420	700	2,720	5,860	7,950	1,480	970	....	1,000	450	420
15..............	425	420	1,010	5,860	5,860	7,500	1,350	1,940	....	1,000	450	420
16..............	500	350	1,680	7,310	6,020	7,670	1,240	1,940	....	1,000	450	420
17..............	650	350	2,530	10,200	6,170	7,560	1,010	1,940	....	880	440	420
18..............	560	400	2,720	10,400	6,170	8,400	1,010	1,480	....	880	380	420
19..............	490	400	2,000	10,400	7,140	7,600	1,010	4,290	....	880	380	430
20..............	480	700	2,170	8,700	6,490	7,800	800	4,290	....	880	530	430
21..............	550	380	2,170	6,020	6,490	7,650	1,140	6,720	....	760	620	500
22..............	925	200	1,680	4,670	5,860	5,850	1,330	4,560	....	670	620	500
23..............	1,280	150	2,530	4,390	5,410	5,420	4,530	4,290	....	670	620	500
24..............	1,280	90	1,680	3,590	5,260	5,250	3,980	4,000	....	670	550	500
25..............	1,120	360	2,170	3,130	4,670	5,160	3,240	6,100	....	670	550	500
26..............	890	300	2,530	2,920	4,670	5,350	2,100	5,470	....	670	550	500
27..............	600	410	2,350	3,590	4,390	4,550	3,280	9,500	....	670	460	500
28..............	290	650	2,000	5,260	4,530	3,750	2,850	4,000	1,100	670	460	580
29..............	290	......	2,000	4,670	5,560	3,750	2,120	8,050	1,100	670	460	580
30..............	290	......	1,680	4,110	4,960	3,750	1,540	9,110	950	670	460	580
31..............	290	......	1,680	....	4,390	....	1,540	7,700	......	670	.....	580

NOTE.—These discharges are based on rating curves applicable as follows: Dec. 9 to 15, 1908, and Mar. 4 to June 9, 1909, not well defined; Dec. 16, 1908, to Mar. 3, 1909, June 10 to Sept. 4, and Sept. 28 to Dec. 31, 1909, indirect method for shifting channels used.

Discharge Sept. 5 and 6 estimated as 9,650 and 13,000 second-feet, respectively. Discharge Sept. 7 to 27 estimated as equivalent to 6,190 second-feet per day by hydrograph comparison with the flow of the Animas River at Aztec, N. Mex.

Monthly discharge of San Juan River at Blanco, N. Mex., for 1908-9.

Month.	Discharge in second-feet.			Run-off (total in acre-feet).	Accuracy.
	Maximum.	Minimum.	Mean.		
1908.					
December 9-31...................................	450	90	265	12,100	D.
1909.					
January..	1,280	290	555	34,100	D.
February.......................................	700	90	372	20,700	D.
March..	2,720	610	1,750	108,000	B.
April...	10,400	1,190	4,310	256,000	B.
May..	8,520	3,350	5,840	359,000	B.
June...	12,300	3,750	7,340	437,000	C.
July...	4,530	800	2,240	138,000	C.
August...	9,500	970	3,460	213,000	D.
September......................................	a 13,000	950	6,050	360,000	D.
October..	3,400	670	1,090	67,000	D.
November.......................................	620	380	494	29,400	D.
December.......................................	580	420	483	29,700	D.
The year......................................	13,000	90	2,830	2,050,000	

a Estimated.

ANIMAS RIVER AT AZTEC, N. MEX.

This station was originally established June 21, 1904, at a wooden-truss highway bridge about three-eighths of a mile west of Aztec, N. Mex. It was discontinued December 14, 1904, and reestablished at the same location on June 8, 1907. On September 13, 1908, it was moved to a new suspension bridge about half a mile above the old bridge, which was torn down on completion of the new bridge. The station is about one-third of a mile west of Aztec, on the main wagon road to Farmington and La Plata.

No change in the staff gage or gage datum occurred during the maintenance of the station at the old location. Beginning September 13, 1908, an inclined staff gage, installed a few feet downstream from the suspension bridge at an arbitrary datum, was read.

The station, although 20 miles above the mouth of the river, is below all important tributaries. The drainage is about 1,300 square miles. Between Durango and Aztec many large ditches divert water for irrigation and the discharge at this station does not represent the total run-off of the stream. Notwithstanding numerous existing water rights, an ample supply of water is available for future development.

Ice forms to a considerable depth along the edges during the greater part of the winter, but the river seldom freezes across. Slush ice occurs frequently during the winter months.

Results obtained at this station are good except during and after high water, when the shifting of the channel interferes with the accuracy of the data. The flood of September 6, 1909, made it necessary to use the indirect method for shifting channels in estimating the daily and monthly flow after that date.

Discharge measurements of Animas River at Aztec, N. Mex., in 1909.

Date.	Hydrographer.	Width.	Area of section.	Gage height.	Discharge.
		Feet.	*Sq. ft.*	*Feet.*	*Sec.-ft.*
Jan. 12a	C. D. Miller	75	101	3.30	177
Mar. 29	J. B. Stewart	145	228	4.15	657
May 13	....do	157	539	6.20	3,070
June 27	V. L. Sullivan	151	731	6.90	4,470
July 28	J. B. Stewart	155	457	5.65	2,010
Sept. 27	W. B. Freeman	150	252	4.65	793
Nov. 16	J. B. Stewart	149	152	4.05	342

a Made by wading, not at regular section.

Daily gage height, in feet, of Animas River at Aztec, N. Mex., for 1909.

[H. S. Wattles, observer.]

Day.	Jan.	Feb.	Mar.	Apr.	May.	June.	July.	Aug.	Sept.	Oct.	Nov.	Dec.
1	3.4	3.5	3.6	4.4	5.1	5.4	6.55	4.65	5.8	4.5	4.8	4.1
2	3.4	3.5	3.8	4.5	5.0	5.8	6.4	4.7	5.3	4.4	4.5	4.1
3	3.4	3.5	4.0	4.6	5.0	6.5	6.3	4.5	5.3	4.4	4.3	4.15
4	3.4	3.5	4.0	4.7	5.4	7.45	6.1	4.5	5.7	4.5	4.2	4.2
5	3.4	3.5	4.1	4.8	6.4	8.3	6.3	4.8	7.85	4.6	4.2	4.2
6	3.4	3.5	4.3	4.6	6.6	8.7	6.35	4.4	10.0	4.65	4.2	4.15
7	3.4	3.5	4.2	4.4	6.7	8.75	6.2	4.35	9.9	4.5	4.1	4.15
8	3.4	3.5	4.2	4.3	6.95	8.7	6.0	4.3	8.25	4.6	4.1	4.1
9	3.4	3.5	4.2	4.3	6.75	8.7	5.7	4.45	7.65	4.5	4.1	4.1
10	3.4	3.5	4.1	4.5	6.3	8.5	5.5	4.35	6.4	4.5	4.1	4.1
11	3.3	3.5	4.0	4.7	6.1	8.25	5.3	4.5	6.0	4.5	4.1	4.0
12	3.3	3.5	3.9	4.7	6.15	7.8	5.2	4.4	6.0	4.4	4.1	4.0
13	3.3	3.5	3.8	4.5	6.1	7.95	5.1	4.5	5.7	4.4	4.1	4.0
14	3.2	3.5	3.8	4.6	6.1	7.8	5.0	4.45	5.6	4.4	4.1	4.0
15	3.3	3.5	3.9	5.1	6.0	7.65	4.9	4.4	5.5	4.4	4.1	4.0
16	3.4	3.5	4.0	5.8	6.0	7.8	4.7	5.5	5.7	4.4	4.1	4.0
17	3.5	3.4	4.1	6.3	6.1	7.9	4.6	5.2	5.7	5.4	4.1	4.0
18	3.4	3.4	4.2	6.6	6.25	8.15	4.5	4.8	5.7	5.4	4.0	4.0
19	3.4	3.4	4.2	6.8	6.65	8.1	4.4	4.8	5.65	5.4	4.0	3.9
20	3.4	3.4	4.4	6.5	6.8	7.95	4.4	4.8	5.6	5.4	4.0	3.9
21	3.5	3.45	4.3	6.0	6.85	7.85	4.5	5.0	5.6	5.4	4.0	3.9
22	3.6	3.45	4.2	5.7	6.2	7.6	4.9	4.8	5.6	5.4	4.0	3.9
23	3.75	3.45	4.5	5.3	6.1	7.55	5.7	4.7	5.6	5.4	4.05	3.9
24	3.5	3.45	4.3	5.0	6.1	7.5	5.85	5.0	5.6	5.35	4.05	3.9
25	3.5	3.45	4.4	4.8	5.8	7.4	5.8	5.3	5.6	5.35	4.05	4.0
26	3.4	3.5	4.5	4.8	5.7	7.2	5.1	5.8	5.0	5.35	4.05	4.0
27	3.3	3.5	4.5	5.25	6.0	7.1	6.0	5.0	4.8	5.3	4.05	4.0
28	3.3	3.5	4.4	5.1	6.8	7.0	5.65	5.2	4.6	5.3	4.1	4.0
29	3.3	....	4.3	5.6	6.8	6.75	5.3	6.2	4.5	5.3	4.1	3.9
30	3.3	....	4.3	5.4	6.1	6.65	5.0	6.5	4.5	5.3	4.1	3.9
31	3.6	....	4.3	....	5.6	....	4.8	6.0	....	5.1	....	4.0

Daily discharge, in second-feet, of Animas River at Aztec, N. Mex., for 1909.

Day.	Jan.	Feb.	Mar.	Apr.	May.	June.	July.	Aug.	Sept.	Oct.	Nov.	Dec.
1	235	292	350	858	1,510	1,860	3,690	1,050	2,390	700	830	380
2	235	292	471	930	1,400	2,390	3,410	1,100	1,740	620	640	380
3	235	292	594	1,010	1,400	3,590	3,230	930	1,740	620	500	410
4	235	292	594	1,100	1,860	5,610	2,880	930	2,240	700	450	440
5	235	292	657	1,190	3,410	7,700	3,230	1,190	6,570	750	450	440
6	235	292	788	1,010	3,780	8,780	3,320	858	12,500	800	450	410
7	235	292	721	858	3,980	8,920	3,050	823	12,100	700	390	410
8	235	292	721	788	4,490	8,780	2,710	788	7,450	750	390	380
9	235	292	721	788	4,080	8,780	2,240	894	5,950	680	390	380
10	235	292	657	930	3,230	8,240	1,980	823	3,300	680	380	380
11	180	292	594	1,100	2,880	7,570	1,740	930	2,640	680	380	320
12	180	292	532	1,100	2,960	6,450	1,630	858	2,620	620	380	320
13	180	292	471	930	2,880	6,810	1,510	930	2,190	620	380	320
14	130	292	471	1,010	2,880	6,450	1,400	894	1,900	620	370	320
15	180	292	532	1,510	2,710	6,090	1,290	858	1,770	620	370	320
16	235	292	594	2,390	2,710	6,450	1,100	1,980	2,050	620	370	320
17	292	235	657	3,230	3,140	6,690	1,010	1,630	2,050	1,420	370	320
18	235	235	721	3,780	3,880	7,310	930	1,190	2,050	1,400	310	320
19	235	235	721	4,180	3,880	7,180	858	1,190	1,970	1,400	310	260
20	235	235	858	3,590	4,180	6,810	858	1,190	1,900	1,390	310	260
21	292	264	788	2,710	4,280	6,570	930	1,400	1,800	1,390	310	260
22	350	264	721	2,240	3,050	5,970	1,190	1,190	1,650	1,380	310	260
23	440	264	930	1,740	2,880	5,850	2,240	1,100	1,800	1,380	350	260
24	292	264	788	1,400	2,880	5,730	2,470	1,400	1,780	1,360	350	260
25	292	264	858	1,190	2,390	5,490	2,390	1,740	1,780	1,350	350	330
26	235	292	930	1,190	2,240	5,040	1,510	2,390	1,110	1,340	350	330
27	180	292	930	1,680	2,710	4,820	2,710	1,400	910	1,290	350	330
28	180	292	858	1,510	4,180	4,000	2,180	1,630	760	1,270	380	330
29	180	......	788	2,110	4,180	4,080	1,740	3,050	700	1,250	380	270
30	180	......	788	1,860	2,880	3,880	1,400	3,590	700	1,250	380	270
31	350	......	788	......	2,110	......	1,190	2,710	......	1,100	......	330

NOTE.—These discharges are based on rating curves applicable as follows: Jan. 1 to Sept. 6, fairly well defined between discharges of 130 and 4,600 second-feet. Sept. 7 to Dec. 31, indirect method for shifting channels used.

Monthly discharge of Animas River at Aztec, N. Mex., for 1909.

Month.	Discharge in second-feet.			Run-off (total in acre-feet).	Accuracy.
	Maximum.	Minimum.	Mean.		
January...	440	130	239	14,700	A.
February..	292	235	279	15,500	A.
March..	930	350	697	42,900	A.
April...	4,180	788	1,660	98,800	A.
May..	4,490	1,400	3,030	186,000	A.
June...	8,920	1,860	6,150	366,000	A.
July...	3,690	858	2,000	123,000	A.
August...	3,590	788	1,380	84,800	B.
September..	12,500	700	3,000	179,000	C.
October..	1,420	620	992	61,000	B.
November...	830	310	398	23,700	B.
December...	440	260	333	20,500	B.
The year......................................	12,500	130	1,680	1,220,000	

LA PLATA RIVER AT LA PLATA, N. MEX.

This station, which was established May 25, 1905, to obtain data for use by the United States Reclamation Service in connection with their proposed La Plata project, is located at a wooden, single-span highway bridge, about 16 miles northwest of Aztec, N. Mex., and 1 mile south of La Plata post office, in sec. 3, T. 31 N., R. 13 W. of New Mexico principal meridian. Being located below all the principal diversions, the station shows the amount of flood water available for storage and irrigation.

The station is below all tributaries and about 15 miles above the mouth of the La Plata. The drainage area is about 340 square miles.

Nearly all the normal flow of this stream is diverted for irrigation above the station, and there are a few small diversions below.

Thin ice frequently forms across the stream during the winter period, thick ice forms along the edges, and slush ice at times interferes with winter measurements.

On December 9, 1908, a chain gage was installed on the bridge and is read in place of the rod gage, as the latter does not record low stages. The datum remained unchanged.

Because of shifting conditions of channel and the uncertainty of certain gage heights, the results obtained at this station are not good. The discharge measurements and gage heights for 1905 to 1908 for this station have been published in Water-Supply Papers 174, 210, and 248.

The monthly estimates of the flow for 1905 to 1909 are included in this report. The data on which they are based are very meager, and owing to the torrential character of the stream and the extreme low stages during the greater part of each year the estimates given are very approximate and should be used with caution. They are, however, the best interpretation that can be made of the available data.

Discharge measurements of La Plata River at La Plata, N. Mex., in 1909.

Date.	Hydrographer.	Width.	Area of section.	Gage height.	Discharge.
		Feet.	*Sq. ft.*	*Feet.*	*Sec.-ft.*
Jan. 12	C. D. Miller...	7	1.8	1.50	3
Mar. 29	J. B. Stewart...	31	30	2.22	54
May 13	....do...	33	58	3.05	264
June 27	V. L. Sullivan..	29	15.1	1.80	12.4
July 26	J. B. Stewart...			1.45	a .8
Sept. 25	W. B. Freeman......................................	32	30	2.30	37
Nov. 16	J. B. Stewart...			1.90	a 1.2

a Estimated.

Daily gage height, in feet, of La Plata River at La Plata, N. Mex., for 1909.

[Frank Williams, observer.]

Day.	Jan.	Feb.	Mar.	Apr.	May.	June.	July.	Aug.	Sept.	Oct.	Nov.	Dec.
1...............	1.3	1.5	3.4	2.3	2.6	2.35	1.55	1.4	2.9	2.2	2.0	1.55
2...............	1.3	1.5	3.85	2.4	2.55	2.4	1.5	1.4	2.8	2.2	2.0	1.55
3...............	1.3	1.6	2.4	2.75	2.6	2.55	1.35	1.4	2.6	2.15	2.0	1.55
4...............	1.3	1.8	4.45	2.8	2.9	3.05	1.5	1.4	2.5	2.1	2.0	1.55
5...............	1.3	1.7	3.4	2.8	3.6	3.15	1.5	1.4	7.35	2.1	2.0	1.55
6...............	1.3	1.6	3.3	2.5	3.6	3.15	1.45	1.4	5.8	2.1	2.0	1.6
7...............	1.3	1.6	2.4	2.55	3.7	3.35	1.45	1.4	4.5	2.1	1.9	1.6
8...............	1.4	1.5	2.4	2.5	3.8	3.65	1.45	2.0	3.8	2.1	1.9	1.6
9...............	1.5	1.4	2.4	2.6	3.1	3.3	1.45	1.4	3.35	2.1	1.9	1.6
10...............	1.5	1.4	2.3	2.8	3.0	3.2	1.45	1.4	3.0	2.1	1.9	1.6
11...............	1.5	1.4	2.4	2.8	2.85	2.95	1.45	2.0	2.8	2.1	1.9	1.6
12...............	1.5	1.3	2.0	2.7	2.95	2.75	1.45	1.4	2.85	2.1	1.9	1.6
13...............	1.4	1.4	2.4	2.65	3.05	2.75	1.45	1.4	2.8	2.1	1.9	1.65
14...............	1.5	1.4	2.5	2.8	3.0	2.7	1.45	1.4	2.1	2.1	1.9	1.65
15...............	1.4	1.6	2.5	3.4	2.85	2.55	1.45	2.5	2.55	2.1	1.9	1.65
16...............	1.8	1.7	2.3	3.9	2.8	2.55	1.45	2.9	2.6	2.1	1.9	1.65
17...............	1.65	1.8	2.4	4.35	2.9	2.5	1.45	2.3	2.55	2.1	1.9	1.65
18...............	1.8	1.7	2.6	4.6	3.15	2.5	1.45	1.8	2.5	2.1	1.9	1.65
19...............	1.6	1.8	2.4	4.55	2.9	2.45	1.45	2.3	2.45	2.05	1.9	1.65
20...............	1.6	1.8	2.4	3.9	3.15	2.55	1.45	3.5	2.4	2.0	1.9	1.65
21...............	1.8	1.7	2.3	3.4	2.8	2.45	1.45	2.4	2.4	2.0	1.9	1.65
22...............	1.8	1.6	2.2	3.1	2.75	2.35	1.45	2.1	2.35	2.0	1.8	1.65
23...............	2.0	1.5	2.3	2.9	2.7	2.35	1.45	2.0	2.3	2.0	1.8	1.65
24...............	1.9	1.5	2.4	2.7	2.55	2.15	1.45	2.1	2.3	2.0	1.8	1.65
25...............	1.6	1.5	2.2	2.6	2.5	2.95	1.45	3.8	2.3	2.0	1.8	1.65
26...............	1.4	1.6	2.2	2.55	2.5	2.95	1.45	2.5	2.3	2.0	1.8	1.65
27...............	1.6	1.7	2.3	2.65	2.5	1.9	1.4	2.7	2.25	2.0	1.2	2.0
28...............	1.6	2.0	2.5	2.9	2.6	1.75	1.4	2.45	2.2	2.0	1.2	2.0
29...............	1.5	....	2.3	2.95	2.7	1.7	1.4	3.1	2.2	2.0	1.55	2.0
30...............	1.5	....	2.3	2.6	2.5	1.65	1.4	4.5	2.2	2.0	1.55	2.0
31...............	1.5	....	2.3	....	2.4	....	1.4	3.2	....	2.0	....	2.0

NOTE.—Maximum gage height on Sept. 5, 7.65 feet.

Daily discharge, in second-feet, of La Plata River at La Plata, N. Mex., for 1909.

Day.	Jan.	Feb.	Mar.	Apr.	May.	June.	July.	Aug.	Sept.	Oct.	Nov.	Dec.
1.	0.9	3	390	67	127	76	4	.5	156	23	3	0.1
2.	.9	3	595	85	116	85	3	.5	130	23	3	.1
3.	.9	5.5	85	166	127	116	6	.5	88	19	3	.1
4.	.9	12.4	a890	180	212	264	3	.5	70	16	3	.1
5.	.9	8.5	390	180	475	300	3	.5	a5,000	15	3	.1
6.	.9	5.5	352	105	475	300	2	.5	a1,550	12	3	.1
7.	.9	5.5	317	116	520	370	2	.5	920	12	1.5	.1
8.	1.5	3	85	105	570	498	2	14	480	12	1.5	.1
9.	3	1.5	85	127	282	352	1.5	.5	300	12	1.5	.1
10.	3	1.5	67	180	247	317	1.5	.5	184	11	1.5	.1
11.	3	1.5	85	180	196	230	1.5	14	130	11	1.5	.1
12.	3	.9	52	152	230	166	1.5	.5	142	11	1.5	.1
13.	1.5	1.5	85	140	264	166	1.5	.5	130	10	1.5	.1
14.	3.	1.5	105	180	247	152	1.5	.5	20	10	1.5	.1
15.	1.5	5.5	105	390	196	116	1	70	78	10	1.5	.1
16.	12.4	8.5	67	620	180	116	1	156	88	10	1.5	.1
17.	7	12.4	67	a850	212	105	1	40	80	9	1	.1
18.	12.4	8.5	127	a970	300	105	1	6	70	9	1	.1
19.	5.5	12.4	85	a930	212	95	1	40	62	7	1	.1
20.	5.5	12.4	85	620	300	116	1	360	54	6	1	.1
21.	12.4	8.5	67	390	180	95	.8	70	54	6	1	.1
22.	12.4	5.5	52	282	166	76	.8	20	46	6	.5	.1
23.	27	3	67	212	152	76	.8	13	40	5	.5	.1
24.	18.5	3	85	152	116	45	.8	20	40	5	.5	.1
25.	5.5	3	52	127	105	230	.8	500	40	5	.5	.1
26.	1.5	5.5	52	116	105	230	.8	70	32	4	.5	.1
27.	5.5	8.5	67	140	105	18.5	.5	108	28	4	0	2.5
28.	5.5	27	105	212	127	10.4	.5	13	23	4	0	2.5
29.	3	....	67	230	152	8.5	.5	216	23	4	.1	2.5
30.	3	....	67	127	105	7.0	.5	920	23	4	.1	2.5
31.	3	....	67	....	85	....	.5	250	....	3	....	2.5

a Estimated by extension of rating curve. Discharge, approximate.

NOTE.—These discharges are based on rating curves applicable as follows: Jan. 1 to June 30, not well defined; July 1 to Dec. 31, indirect method for shifting channels. Maximum flood discharge, Sept. 5, gauge height 7.65 feet, estimated by Kutter's formula as 7,000 second-feet.

Monthly discharge of La Plata River at La Plata, N. Mex., for 1905-1909.

Month.	Discharge in second-feet.			Run-off (total in acre-feet).	Accuracy.
	Maximum.	Minimum.	Mean.		
1905.					
May 25–31	565	165	337	4,680	C.
June	817	95	358	21,300	C.
July	107	....	7.8	480	D.
August	0	....	0	0	
September	970	....	64.0	3,810	D.
October	800	....	79.7	4,900	D.
November	8	....	1.2	71	D.
December	5	4	4.2	258	D.
The period	....	....	....	35,500	
1906.					
January			a 10	615	D.
February			a 5.0	278	D.
March	190	....	b 29.5	1,810	C.
April	360	44	160	9,520	B.
May	450	74	238	14,600	B.
June	320	2	154	9,160	C.
July	182	....	b 9.3	572	D.
August	....	....	a 2.0	123	D.
Sept. 1–24	....	....	a 2.0	95	D.
The period	....	....	....	36,800	
1907.					
June 7–30	260	.46	139	6,620	C.
July	750	.5	45.7	2,810	C.
August	1,280	.5	123	7,560	C.
September	246	1	38.7	2,300	C.
October			a .5	31	D.
November			a .5	30	D.
December	....	....	a .6	37	D.
The period	....	....	....	19,400	
1908.					
January			a 0.60	37	D.
February	600	.5	49.7	2,860	C.
March	106	.5	27.1	1,670	B.
April	85	.9	22.1	1,320	B.
May	12.4	....	3.17	195	C.
June	1.5	....	.14	8	D.
July	247	....	12.9	793	D.
August	2,300	....	154	9,470	D.
December	5.5	....	1.33	82	B.
The period	....	....	....	16,400	
1909.					
January	27	.9	5.35	329	B.
February	27	.9	6.38	354	A.
March	890	52	158	9,720	A.
April	970	67	278	16,500	B.
May	570	85	222	13,600	A.
June	498	7	161	9,580	A.
July	6	.5	1.53	94	C.
August	920	.5	93.7	5,760	C.
September	a 5,000	23	336	20,000	C.
October	23	3	9.6	590	C.
November	3	....	1.36	81	C.
December	2.5	.1	.49	30	D.
The year	a 5,000	....	....	76,600	

a Estimated.   b Partly estimated.

NOTE.—The monthly estimates for 1905 to 1908 were obtained from daily discharges found by using the 1909 rating curve as standard for shifting channels from May 25, 1905 to Apr. 10, 1908, and by using it direct from Apr. 11 to Dec. 31, 1908. The low-water discharges are liable to large percentages of error because of infrequent measurements and excessively low flow. It was necessary to make flat estimates for many of the low-water periods. The recorded maximum values are only roughly approximate as the high-water stages were not covered by discharge measurements.

LITTLE COLORADO RIVER BASIN.

DESCRIPTION.

The country drained by Little Colorado River consists of a high plateau with an elevation over 4,000 feet above sea leveı, extending from the Continental Divide in northwestern New Mexico westward to the San Francisco Mountains in Arizona and from the Grand Canyon of the Colorado southward to the Mogollon Mesa. The greater part of this plateau is composed of rolling plains with a few feet of soil at the surface underlain by rock. Through this plateau the river winds northwestward to its junction with the great Colorado.

The run-off from approximately 6,000 square miles of the drainage area finds its way into the Little Colorado above the mouth of Rio Puerco, the largest tributary, which joins the main stream 2 miles above the town of Holbrook, Ariz. Both the Little Colorado and Rio Puerco are flashy streams, seldom clear even during low stages. Their bottoms are shifting and sandy, and where not confined in canyons the stream beds are wide with abrupt earth banks. The discharge fluctuates greatly, being insignificant in dry seasons. The floods are short and violent and carry large quantities of silt in suspension.

LITTLE COLORADO RIVER AT ST. JOHNS, ARIZ.

This station, which was established April 18, 1906, to determine the amount of water available for irrigation, is located at the south end of the town of St. Johns, half a mile above the dam and county bridge. The bed of the stream is clean, sandy, and shifting. Frequent measurements are necessary to properly determine the daily flow at this station. The discharge measurements and daily gage heights were furnished by the United States Reclamation Service, who maintain this station. The daily and monthly estimates of discharge were made by engineers of the United States Geological Survey.

Discharge measurements of Little Colorado River at St. Johns, Ariz., in 1909.

[By W. D. Rencher.]

Date.	Area of section.	Gage height.	Discharge.	Date.	Area of section.	Gage height.	Discharge.
	Sq. ft.	*Feet.*	*Sec.-ft.*		*Sq. ft.*	*Feet.*	*Sec.-ft.*
Jan. 22...............	20	0.97	36	July 5...............	20	0.30	33
29...............	21	1.00	43	15...............	16	.20	22
Feb. 8...............	16	.90	26	22...............	114	3.00	601
15...............	16	.90	27	Aug. 2...............	50	1.30	150
23...............	16	.89	24	10...............	38	.90	74
Mar. 1...............	14	.90	21	16...............	197	5.80	1,640
10...............	26	1.00	41	26...............	38	.57	87
17...............	26	a 1.09	48	Sept. 6...............	88	1.90	525
26...............	26	a 1.03	45	13...............	48	.80	139
Apr. 8...............	25	1.02	43	20...............	25	.56	53
16...............	42	1.60	140	28...............	24	.50	44
24...............	36	1.20	124	Oct. 8...............	20	.60	35
May 1...............	68	.70	181	18...............	18	.65	29
9...............	46	.50	107	28...............	17	.64	27
17...............	32	.37	64	Nov. 10...............	18	.65	27
29...............	22	.23	32	24...............	18	.66	27
June 9...............	14	.17	14	Dec. 8...............	14	.85	27
19...............	8.0	.13	9.3	19...............	18	1.00	33
26...............	8.8	.10	8.2	31...............	17	1.15	36

a Gage height doubtful.

Daily gage height, in feet, of Little Colorado River at St. Johns, Ariz., for 1909.

Day.	Jan.	Feb.	Mar.	Apr.	May.	June.	July.	Aug.	Sept.	Oct.	Nov.	Dec.
1	0.85	0.98	0.90	1.01	0.65	0.20	0.09	0.95	0.90	0.54	0.65	0.78
2	.85	.96	.90	1.02	.60	.20	.10	1.18	1.12	.54	.65	.81
3	.85	.95	.86	1.03	.58	.18	.13	.97	.92	.55	.65	.82
4	.86	.94	.85	1.03	.58	.18	.28	.91	.75	.56	.65	.84
5	.86	.93	.88	1.03	.55	.18	.31	1.10	1.40	.57	.65	.84
6	.86	.90	.90	1.02	.55	.17	.22	1.00	1.85	.58	.65	.85
7	.88	.....	.94	1.02	.50	.17	.20	.90	2.12	.60	.65	.86
8	.87	.....	.98	1.02	.50	.17	.22	.82	1.95	.60	.66	.86
9	.88	.....	.99	1.01	.52	.17	.21	.98	1.08	.62	.66	.86
10	.90	.....	1.00	1.02	.50	.16	.20	.95	.92	.62	.65	.85
11	.90	.....	1.00	1.03	.50	.16	.20	.82	.90	.63	.65	.88
12	.91	.....	1.00	1.03	.47	.16	.20	.75	.86	.64	.64	.94
13	.91	.....	1.01	1.03	.45	.15	.20	.72	.80	.64	.64	.93
14	.92	.90	1.02	1.35	.44	.15	.24	.90	.75	.65	.65	.96
15	.94	.90	1.02	1.51	.40	.15	.20	1.75	.68	.65	.65	.95
16	.95	.88	1.03	1.72	.40	.14	.20	3.58	.66	.65	.65	.97
17	.95	.88	1.03	2.05	.38	.14	.28	2.15	.62	.65	.66	.98
18	.95	.90	1.03	1.68	.35	.14	.28	2.45	.60	.65	.66	1.00
19	.95	.90	1.03	1.65	.35	.14	.30	1.90	.58	.65	.66	1.00
20	.95	.90	1.01	1.58	.35	.13	.55	1.45	.56	.66	.66	1.01
21	.94	.89	1.01	1.45	.33	.13	.75	1.60	.55	.66	.66	1.06
22	.97	.88	1.02	1.30	.32	.12	2.60	1.20	.54	.66	.66	1.04
23	.96	.92	1.02	1.25	.31	.12	1.08	.95	.52	.65	.66	1.05
24	.95	.90	1.02	1.20	.30	.11	.80	.80	.51	.65	.66	1.06
25	.97	.90	1.02	1.04	.28	.10	.88	.70	.50	.65	.66	1.10
26	.98	.91	1.03	.85	.26	.10	.85	.56	.50	.65	.67	1.11
27	1.00	.92	1.03	.80	.25	.10	.80	.50	.50	.64	.68	1.12
28	1.00	.92	1.03	.77	.24	.10	.78	.52	.50	.64	.71	1.13
29	1.00	.....	1.03	.68	.23	.10	.55	.55	.50	.64	.74	1.14
30	1.00	.....	1.02	.69	.22	.10	.42	.54	.50	.64	.77	1.15
31	1.00	.....	1.01	.....	.20	.....	.55	.50	.....	.64	.....	1.15

Daily discharge, in second-feet, of Little Colorado River at St. Johns, Ariz., for 1909.

Day.	Jan.	Feb.	Mar.	Apr.	May.	June.	July.	Aug.	Sept.	Oct.	Nov.	Dec.
1	25	40	21	44	170	26	16	85	200	43	27	29
2	25	36	21	43	155	25	10	126	260	42	28	30
3	25	35	20	44	146	25	16	88	222	40	28	30
4	26	35	20	44	143	23	28	78	192	39	28	30
5	26	34	25	44	133	21	34	111	370	38	28	30
6	26	30	26	43	130	19	26	94	510	37	28	29
7	28	28	30	43	115	17	24	77	574	36	28	28
8	27	26	38	43	110	16	25	65	508	35	29	28
9	28	26	39	43	110	14	24	90	250	35	29	28
10	29	26	41	43	105	12	23	85	200	35	30	26
11	29	26	40	47	103	10	23	65	185	35	30	28
12	30	27	40	47	95	10	22	56	164	34	29	35
13	30	27	41	47	90	10	22	52	139	34	29	32
14	30	27	41	94	85	10	25	77	124	30	29	34
15	34	27	42	121	75	10	22	247	105	30	29	32
16	35	25	43	172	73	9	20	798	94	30	28	32
17	35	25	43	240	68	9	24	365	80	29	28	33
18	35	27	43	170	60	9	20	460	71	29	28	33
19	35	27	43	173	60	9	18	315	62	29	28	33
20	35	27	40	168	60	9	40	212	53	29	27	32
21	34	26	41	150	54	9	60	258	53	29	27	37
22	36	25	42	128	52	8	480	175	52	29	27	35
23	35	28	42	125	50	8	108	133	48	28	27	34
24	33	25	43	124	45	8	63	112	47	28	27	34
25	35	24	43	115	43	8	75	102	45	28	26	38
26	39	24	45	100	40	8	70	85	45	28	25	35
27	40	24	45	115	37	8	63	85	44	27	25	37
28	42	23	45	128	33	8	60	93	44	27	25	38
29	43	.....	45	135	32	9	33	105	42	27	28	36
30	42	.....	45	158	31	10	20	112	40	27	28	36
31	42	.....	44	.....	27	.....	33	112	.....	27	.....	36

NOTE.—These discharges were obtained by the indirect method for shifting channels.

Monthly discharge of Little Colorado River at St. Johns, Ariz., for 1909.

Month.	Discharge in second-feet.			Run-off (total in acre-feet).	Accuracy.
	Maximum.	Minimum.	Mean.		
January.........................	43	25	32.7	2,010	B.
February........................	40	23	27.9	1,550	B.
March..........................	45	20	38.0	2,340	B.
April...........................	240	43	99.7	5,930	B.
May............................	170	27	81.6	5,020	C.
June...........................	26	8	12.6	750	C.
July...........................	480	10	49.1	3,020	C.
August.........................	798	52	159	9,780	C.
September......................	574	40	161	9,580	B.
October........................	43	27	32.1	1,970	B.
November.......................	30	25	27.8	1,650	B.
December.......................	38	26	32.5	2,000	B.
The year......................	798	8	62.9	45,600	

LITTLE COLORADO RIVER AT WOODRUFF, ARIZ.

This station, which was established March 16, 1905, was located about 300 feet below the crossing of the Holbrook-Winslow wagon road and one-fourth mile below the Woodruff dam. It is maintained by the United States Reclamation Service.

The station equipment, which was carried away by the flood of November 26 and 27, 1905, was replaced March 24, 1906. The object of the station was to determine the amount of water available for irrigation. The bed of the stream is sandy and shifting, and frequent measurements are required to determine the daily flow.

No measurements were made in 1909 and only a few gage heights were recorded. The water surface was below the gage during the greater part of January.

LITTLE COLORADO RIVER AT HOLBROOK, ARIZ.

This station, which was established March 17, 1905, to determine the amount of water available for irrigation, is located at the county bridge across Little Colorado River at Holbrook, Ariz.

The bed of the stream is sandy and shifting, and frequent discharge measurements are required to properly determine the daily flow.

No discharge measurements were made in 1909.

This station is maintained by the United States Reclamation Service.

Daily gage height, in feet, of Little Colorado River at Holbrook, Ariz., for 1909.

[Anna Conner, observer.]

Day.	Jan.	Feb.	Mar.	Apr.	May.	June.	July.	Aug.	Sept.	Oct.	Nov.	Dec.
1	3.4	3.0	3.0	3.0	3.4	2.4	2.6	3.0	5.0	3.6	2.8	3.0
2	3.4	3.0	3.0	3.0	3.3	2.4	2.6	3.0	5.0	3.6	2.8	3.0
3	3.4	3.0	3.0	3.0	3.0	2.4	2.55	4.5	5.2	3.5	2.8	3.0
4	3.4	3.0	3.0	3.0	3.0	2.4	2.5	3.0	5.2	3.4	2.8	3.0
5	3.3	3.0	3.0	3.0	3.0	2.4	2.4	5.5	7.5	3.4	2.8	3.0
6	3.2	3.0	3.0	3.0	3.0	2.4	2.3	5.0	7.0	3.35	2.8	3.0
7	3.2	3.0	3.0	3.0	3.0	2.4	2.2	6.0	5.5	3.3	2.8	3.0
8	3.1	3.0	3.0	3.0	3.0	2.35	2.1	6.0	5.0	3.3	2.8	2.95
9	3.0	3.0	3.0	3.0	3.0	2.3	2.05	6.0	4.9	3.25	2.8	2.9
10	3.0	3.0	3.0	3.0	3.0	2.3	2.0	4.5	4.7	3.2	2.8	2.9
11	3.0	3.0	3.0	3.5	3.0	2.3	2.0	5.8	4.6	3.2	2.8	3.0
12	3.0	3.0	2.9	3.2	2.0	2.3	2.0	6.3	4.55	3.2	2.8	3.0
13	3.0	3.0	2.8	3.0	2.95	2.3	2.0	6.0	4.5	3.1	2.8	3.0
14	3.0	3.0	2.8	3.0	2.9	2.3	2.0	7.8	4.45	3.1	2.8	3.0
15	3.0	3.0	2.8	3.0	2.9	2.25	2.0	6.0	4.4	3.1	2.8	3.0
16	3.0	2.9	2.8	3.0	2.9	2.2	2.0	6.5	4.4	3.1	2.8	3.0
17	3.0	2.8	2.9	3.0	2.9	2.2	2.0	5.0	4.35	3.0	2.8	3.0
18	3.0	2.8	3.0	3.0	2.85	2.1	2.0	7.0	4.3	3.0	2.8	3.0
19	3.0	2.8	3.0	3.0	2.8	2.0	2.0	7.0	4.3	3.0	.2.8	4.0
20	3.55	2.8	3.0	3.0	2.8	2.0	2.0	7.5	4.2	3.0	2.8	3.5
21	3.5	2.8	3.0	3.3	2.8	2.0	3.5	7.5	4.1	3.0	2.8	3.0
22	3.4	2.8	3.0	3.5	2.8	2.0	3.5	7.0	4.1	3.0	3.25	3.0
23	3.2	2.8	3.0	3.2	2.8	2.0	3.0	7.5	4.0	2.95	3.15	3.0
24	3.0	2.8	3.0	3.0	2.8	2.0	3.0	7.45	4.0	2.9	3.0	3.0
25	3.0	3.0	3.0	3.0	2.7	2.0	5.5	7.5	4.0	2.9	3.0	2.9
26	3.0	3.0	3.0	3.0	2.6	2.0	4.5	7.8	4.0	2.9	3.0	2.8
27	3.0	3.0	3.0	3.0	2.6	2.8	3.5	7.8	3.9	2.85	3.0	2.9
28	3.0	3.0	3.0	3.0	2.6	2.8	3.0	7.8	3.9	2.8	3.0	3.0
29	3.0	....	3.0	3.0	2.6	2.6	3.0	7.5	3.7	2.8	3.0	3.0
30	3.0	....	3.0	3.4	2.6	2.6	3.0	6.5	3.8	2.8	3.0	3.0
31	3.0	....	3.0	....	2.6	....	3.0	5.5	....	2.8	....	3.0

EVAPORATION AT HOLBROOK, ARIZ.

Observations of evaporation at Holbrook, Ariz., were begun on August 2, 1905, and observations were made daily from that time until December 31, 1909, when they were discontinued.

As water in the river was insufficient to float an evaporation pan during the greater part of the summer months, a cement tank 8 feet square and 3 feet deep was built on the property of Mr. John Conner in the town of Holbrook. In this tank a regulation pan, 3 feet square and 18 inches deep, was floated, so secured as to keep it at all times practically in the very center of the tank. The tank was situated about 25 feet from the house and 50 feet from some large cotton-wood trees.

A regulation Weather Bureau rain gage was located in the immediate vicinity and at such distance as to receive the least possible influence from such obstructions as houses and fences.

The prevailing winds in this section are from the southwest to the northeast; the nearest trees on the southwest are about 150 feet away. The trees within 50 feet, above referred to, are practically due north of the site of the tank and exert very little influence on the observations.

The method used in making measurements was as follows: The pan was kept filled within a few inches of the top, the water in the tank outside being kept practically at the same level, the pan being filled to a height indicated by the pointed tip of a vertical rod in the center of the tank. Every morning and evening, after such change as might be brought about by evaporation or rainfall, water level was restored to the original height, the precise amount of water transferred being measured with a cup of such size that one cupful of water was equivalent to 0.01 inch in depth in the tank.

No interruptions occurred in the observations during the year 1905. During 1906 water in the pan froze on January 5 and remained so until January 30, no observations being taken during that period. During the remainder of the year very few interruptions occurred, so that for 1906 the records are fairly complete. The same is true for 1907, 1908, and 1909.

The records show considerable variation between a daily minimum of zero in cold weather and 0.5 inch in hot weather, the maximum figures being occasioned chiefly by days of excessively high wind, which prevails along the Little Colorado for a large part of the spring months.

The tables given below show the monthly evaporation, rainfall, average temperature of water in the pan and average temperature of air. The temperature of water outside the pan did not vary from the temperature of that inside by more than about 1 degree. The records of rainfall and air temperature were obtained from Weather Bureau reports for the station at Holbrook, and may not show precisely the conditions at the evaporation station. The other records were compiled from data furnished by the United States Reclamation Service.

The latitude of Holbrook is about 34.5° north; its longitude about 110° west; its elevation above sea level, 5,072 feet; the rainfall for the period from 1887 to 1908 shows a mean of 8.99 inches, the lowest being 4.58 inches in 1899 and the highest 17.63 inches in 1905.

Monthly evaporation, rainfall, and temperature at Holbrook, Ariz., for 1905–1909.

Month.	Evaporation.	Rainfall.	Temperature.		Month.	Evaporation.	Rainfall.	Temperature.	
			Water in pan.	Air.				Water in pan.	Air.
1905.	*Inches.*	*Inches.*	°F.	°F.	1906.	*Inches.*	*Inches.*	°F.	°F.
January............					January............	b 0.03	1.12	32.5	41.8
February...........					February...........	1.24	.23	41.9	38.6
March.............					March.............	3.28	.46	46.4	42.2
April..............					April..............	5.12	.50	52.2	50.4
May...............					May...............	6.93	.17	60.3	57.7
June..............					June..............	8.61	.00	65.6	69.6
July..............					July..............	7.42	1.11	70.5	76.4
August............	a 6.60	0.76	73.5	76.5	August............	6.24	1.25	70.4	74.8
September.........	4.57	1.37	66.4	68.2	September.........	5.12	.34	65.3	65.6
October...........	4.26	.05	56.9	53.6	October...........	3.44	T.	56.0	55.0
November..........	1.45	3.82	48.4	44.2	November..........	c 1.47	1.22	46.1	43.5
December..........	.22	1.11	37.5	25.9	December..........	d .94	2.32	42.4	31.0

a 30 days.   b 5 days.   c 26 days.   d 24 days.

Monthly evaporation, rainfall, and temperature at Holbrook, Ariz., for 1905–1909—Con.

Month.	Evapo-ration.	Rain-fall.	Temperature.		Month.	Evapo-ration.	Rain-fall.	Temperature.	
			Water in pan.	Air.				Water in pan.	Air.
1907.	*Inches.*	*Inches.*	° F.	° F.	1908.	*Inches.*	*Inches.*	°F.	°F.
January..........	a 0.72	1.64	41.5	38.8	August..........	5.18	2.71	52.4	74.4
February........	b 1.16	.44	42.4	44.6	September.......	4.07	.26	49.4	67.4
March..........	2.28	.71	44.6	48.8	October.........	3.53	1.31	46.5	52.0
April...........	3.82	.85	50.9	56.6	November........	3.23	.21	44.0	47.0
May............	3.98	.73	56.5	57.8	December........	3.29	1.51	41.1	39.0
June...........	3.92	.54	52.0	66.2					
July............	6.35	3.09	69.3	76.5	The year...	48.62	12.54	45.4	54.7
August..........	6.39	1.89	71.7	73.2					
September.......	4.62	.43	66.8	66.8	1909.				
October.........	3.95	3.44	65.6	56.6	January..........	4.13	.44	42.6	41.8
November........	c 2.51	1.40	51.2	42.8	February........	3.33	.17	42.2	38.6
December........	2.37	T.	44.5	37.9	March..........	1.99	.35	41.5	42.2
					April...........	2.78	.67	40.7	50.4
1908.					May............	4.26	.00	44.1	57.7
January..........	3.41	.67	44.3	37.0	June...........	6.79	.02	62.5	69.6
February........	3.40	1.49	43.1	39.2	July............	7.77	3.99	76.7	76.4
March..........	2.85	.52	41.4	47.2	August..........	4.10	2.70	71.9	74.8
April...........	2.85	.81	42.9	52.4	September.......	c 3.63	.31	71.4	65.6
May............	4.87	.66	46.3	55.8	October.........	3.11	.00	65.0	·55.0
June...........	5.18	.24	48.0	68.0	November........	2.12	.15	48.8	43.5
July............	6.76	2.15	53.4	77.0	December........	1.37	1.27	40.7	31.0

a 24 days. b 27 days. c 29 days.

CLEAR CREEK NEAR WINSLOW, ARIZ.

This station, which was established June 13, 1906, to determine the quantity of water available for irrigation, is located 6 miles from Winslow and 3 miles above the Clear Creek Irrigation Co.'s dam and the county bridge. It is one-half mile above the pump house. The bed of the stream is strewn with large bowlders and is permanent.

No discharge measurements were made in 1909.

The station is maintained, and the data have been furnished by the United States Reclamation Service.

Daily gage height, in feet, of Clear Creek near Winslow, Ariz., for 1909.

Day.	Jan.	Day.	Jan.	Day.	Jan.
1.....................	2.1	11.....................	2.1	21.....................	4.0
2.....................	2.1	12.....................	2.1	22.....................	4.0
3.....................	2.1	13.....................	2.1	23.....................	4.0
4.....................	2.1	14.....................	2.1	24.....................	4.0
5.....................	2.1	15.....................	2.1	25.....................	4.0
6.....................	2.1	16.....................	4.1	26.....................	6.5
7.....................	2.1	17.....................	5.0	27.....................	5.5
8.....................	2.1	18.....................	4.25	28.....................	5.0
9.....................	2.1	19.....................	4.0	29.....................	4.5
10.....................	2.1	20.....................	4.0	30.....................	4.0
				31.....................	4.0

NOTE.—Rise on Jan. 16 and 26 due to rain and melting snow in mountains. The stream was flowing more than usual at this season.

FLOOD ON ZUNI RIVER, ARIZ., SEPTEMBER 6, 1909.[1]

During the last 10 days of the month of July, all of August, and the first six days of September, 1909, the rains on the drainage basin of Zuni River in Arizona were heavy; but they were not general over the whole tract, being more in the nature of heavy showers, approaching at times what are termed "cloud bursts."

The Weather Bureau stated in its summary for the month of August:

For the district [No. 9, in which the Zuni drainage basin is situated], as a whole, the month was the wettest since the beginning of the record, and was even more remarkable for the abnormal distribution of the rainfall than for its intensity.

For the entire Little Colorado Basin the average was 1.782 above the normal.

These rains culminated in heavy discharges on both sides of the Zuni Mountains on September 5 and 6. On the afternoon of September 6 the Bluewater dam on the east slopes of this mountain range was destroyed, and within a few minutes of the same time the water undermined the mesa on the south end of the Zuni dam, on the west slopes, partly wrecking the reservoir. (See Pls. VIII and IX.)

Records were kept of the amount of water flowing into the Zuni reservoir by observing gage heights, the capacity of the reservoir at all depths being known. During the period from July 20 to September 6 there was impounded 9,540 acre-feet of water.

The drainage area of this stream above the reservoir is 650 square miles, at elevations ranging from 6,300 feet at the reservoir to 9,200 feet on the tops of the mountains, which form the Continental Divide.

No records of rainfall are available within this drainage area, the nearest one being at Zuni, below the dam site, at Fort Wingate some 35 miles northeast of the dam, and at Manuelito about the same distance northwest, and all at an elevation of about 6,300 feet; the two last mentioned are on other drainage basins. On the east side of the divide records are available from Bluewater station, Bluewater reservoir, and San Rafael. The rainfall on the mountains was undoubtedly heavier than at these lower elevations.

An effort has been made to determine the relation of the run-off to the rainfall by using the available records, but the nature of the country and the peculiar conditions of heavy rains in one section and no precipitation in others a few miles distant make it impossible.

The points where the records were kept on the west of the divide, form a triangle practically 30 miles on a side in an air line, though a

---

[1] By H. F. Robinson, superintendent of irrigation, United States Indian Irrigation Service; published also in Eng. News, vol. 64, Aug. 25, 1910, pp. 203-204.

A. SPILLWAY SURFACE, LOOKING DOWNSTREAM, SHOWING DEPRESSION CAVITY, WRECKED
SPILLWAY, AND BOTH WRECKED WALLS.

B. LOWER TOE OF DAM, SHOWING MOVEMENT AND BREAK IN MASONRY AT SPILLWAY.

FAILURE OF ZUNI DAM, N. MEX., SEPTEMBER 6, 1909.

third longer by any road, and there is nothing in the rainfall records to indicate a general storm of similar intensity at any time during the period. On days when rain was recorded at all three points the precipitation varied greatly—as on August 2: Manuelito 1.05, Fort Wingate trace, Blackrock 0.35; or as on the 29th: Manuelito 0.50, Fort Wingate 0.20, Blackrock 0.20; and on September 6: Manuelito

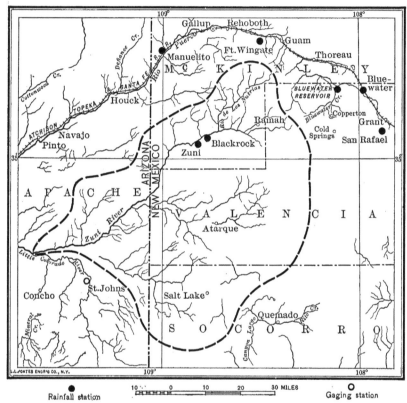

Rainfall station   Gaging station

FIGURE 1.—Map of Zuni River drainage basin, showing rainfall stations.

0.01, Fort Wingate 1.20, and Blackrock 0.46. On these same dates, on the east slope of the mountains the record shows:

August 2, Bluewater, 0.0; Bluewater reservoir, 0.0; San Rafael, 0.0.
August 29, Bluewater, 0.12; Bluewater reservoir, 0.0; San Rafael, 0.96.
September 6, Bluewater, record lost; Bluewater reservoir, Tr.; San Rafael, 0.70.

The accompanying tables show the rainfall at these six points and the inflow into the Zuni reservoir.

Precipitation, in inches, between July 20 and Sept. 6, 1909.

Date.	Blackrock.	San Rafael.	Bluewater reservoir.	Bluewater station.	Fort Wingate.	Manuelito.	Increase in reservoir in acre-feet.
July 20	0.02			0.24	0.40	0.33	
21	.03	0.14			.10	.20	
22	.04		0.35		Tr.	.10	
23	.05	1.38		.47	.40		1,160
24	.07	.06			.10	.70	
25	.20			.05	.40		
26	.20	.30					
27			Tr.				
28							
29							
30			Tr.			.01	
31	.20				.15		
Aug. 1		.51		.71	.30	.15	320
2	.35				Tr.	1.05	
3	.29	.29		Tr.	.18		20
4	.35	.82			.20	.10	280
5	.28	.17			.13	.20	440
6					.50	Tr.	380
7	.09			.20	.20	.50	
8					.10	.35	50
9	.13			.27	.20	.20	60
10	.10			.04		.70	
11	.04	1.52	Tr.	.05	Tr.		200
12		1.18	Tr.	.48	Tr.	.20	
13	Tr.						
14	.08	.50		.50			50
15	Tr.				Tr.		
16	Tr.	.74		.32	.50		360
17	.01	.19		.28	.10		80
18	.06	.60		.18	.20		180
19	.02	.76		.22	.38	Tr.	820
20	Tr.				.30		100
21		.90		.62	.28	.50	900
22	Tr.			.29			180
23	.63		Tr.	.37	.20	3.00	20
24	.07						780
25	Tr.			Tr.	.37	1.30	80
26	Tr.	.44		Tr.	.58	.50	
27	Tr.			.03	.20		20
28	Tr.	.45		Tr.	Tr.		
29	.13	.96		.12	.20	.50	
30	.20				.63	.30	
31				.66			
Sept. 1	.03	.18		Tr.	.40	.50	
2							260
3	.04	.38		.63	.32	.50	
4	.20			.91	1.20	.50	
5	.60	.88		.42	.59	.20	540
6	.46	.70		Tr.	1.20	.01	a 2,260
Total for period, July 20 to Sept. 6	4.97	14.05	6.04	5.00	10.51	10.10	

a To 4 p. m. when Zuni reservoir failed. Bluewater dam went out at practically the same time.

The rains began about July 20, and there had been no flow in Zuni River for some time previous to this date. A heavy rain was recorded at Fort Wingate on the 20th, and on the night of the 22d and the day of the 23d the amount of water impounded was 1,160 acre-feet.

The small value of existing rainfall records in this connection is shown by the reports for the 25th and 26th; these days 0.2 inch of rain was recorded at the fort, but there was no flow into the reservoir. No further rainfall was recorded until August 1, but on that day there was an increase of water in the reservoir of 320 acre-feet, showing a local rain in the drainage area.

A. VIEW OF SPILLWAY, LOOKING NORTHEAST.

B. VIEW OF INTAKE, LOOKING NORTHEAST.

FAILURE OF ZUNI DAM, N. MEX., SEPTEMBER 6, 1909.

No relation between the precipitation at these stations and the run-off is apparent, but probably some relation could be discovered were records available for a period of years.

The following table shows day by day the amount stored in the reservoir and gives the average run-off for the day, but no records were kept that will show the actual flood and its fluctuations:

Log of the Zuni reservoir, July 22 to Sept. 6, 1909.

Date.	Gage height (feet).	Total (acre-feet).	Gain (acre-feet).	Notes.
July 22.........	28.0	5,300	....	First rains.
23.........	30.7	6,460	1,160	
Aug. 1.........	31.4	6,780	320	
3.........	31.5	6,800	20	
4.........	32.1	7,080	280	
5.........	33.1	7,520	440	
6.........	33.8	7,900	380	
8.........	33.9	7,950	50	
9.........	34.1	8,010	60	Discharge through dam, 34 gallons a minute.
11.........	34.5	8,210	200	
14.........	34.6	8,260	50	
16.........	35.3	8,620	360	
17.........	35.5	8,700	80	Discharge, 40 gallons.
18.........	35.8	8,880	180	
19.........	37.5	9,700	820	Discharge clear, 66 gallons.
20.........	37.7	9,800	100	Discharge, 96 gallons a. m., 54 gallons p. m.
21.........	39.4	10,700	900	Discharge muddy, 138 gallons.
22.........	39.7	10,880	180	Muddy a. m., clear p. m., 66 gallons.
23.........	39.8	10,900	20	First noticed small crack on top of dam. Discharge, 66 gallons.
24.........	41.25	11,680	780	Muddy, discharge, 87 gallons.
25.........	41.4	11,760	80	Clear, discharge, 96 gallons. Slip on hillside of 5 or 6 feet, more cracks developed above tunnel.
26.........	....	....	....	Discharge, 96 gallons.
27.........	41.8	11,780	20	Discharge in morning, 87 gallons; 1 p. m. sink at north end and 15 feet of parapet wall fell out, discharge, 284 gallons, then fell to 187 gallons.
28.........	....	....	....	Discharge, 150 gallons, hole about 30 feet in diameter and 4 to 5 feet deep; 5 p. m. discharge, 96 gallons; 11 p. m. discharge, 96 gallons.
29.........	....	....	....	Discharge same. Slips at old quarry go deeper. Slide at tunnel mouth. Tunnel discharge, 96 gallons; had been 20.
30.........	....	....	....	Discharge, 134 gallons.
31.........	....	....	....	Discharge, 134 to 165 gallons.
Sept. 1.........	....	....	....	Discharge, 154 gallons.
2.........	41.9	12,040	200	Discharge, 114 gallons.
3.........	....	....	....	Discharge, 96 gallons.
5.........	42.9	12,580	540	Discharge, 124 gallons.
6, a. m...	45.6	14,020	1,440	Discharge, 467 gallons.
noon...	46.6	14,560	540	Discharge, about same.
4 p. m..	47.1	14,840	280	Several second-feet. Failure of Zuni reservoir began about 4 p. m. Bluewater dam went out at practically the same time.

On September 5, following several days of rain in the mountains, the run-off measured 540 acre-feet. Between dark that day and daylight the next morning the run-off was 1,440 acre-feet; by noon there was 540 additional feet accounted for, and between noon and 4 p. m., when the mesa failed, 280 acre-feet more, or a total of 2,260 acre-feet in about 21 hours.

This so-called failure of the Zuni dam has thus been described by James D. Schuyler, consulting engineer:

The extraordinary rupture and subsidence of the mesa adjacent to the Zuni dam on the south side, as the result of leakage through the basaltic cap rock, occurred September 6, 1909, after the reservoir had filled almost to overflowing, producing a complete wreck of the spillway, with its fine masonry side walls, and within a week had permitted to escape about one-half the contents of the reservoir. This accident has attracted widespread attention among engineers, as it is quite without precedent in the annals of hydraulic engineering. The foundations of dams sometimes give way underneath the superimposed structures, but it is quite a novel experience that a dam and its foundations should remain intact while the adjacent territory forming its abutment at highest level should prove so unstable as to sink from 4 to 9 feet and leak to the extent of 5,000 cubic feet per second around the end of the dam, when it had appeared to be particularly sound, solid, and water tight.

A fairly complete report of this failure was given in the Engineering News for December 2, 1909.

VIRGIN RIVER BASIN.

DESCRIPTION.[1]

Virgin River rises in the Colob Plateau, in the southwestern part of Utah, at an altitude ranging from 8,000 to 10,000 feet above sea level, flows in a general southerly course across the southwestern corner of Arizona into Nevada, where it turns and flows southward to its junction with the Colorado at Rioville, just above Boulder Canyon. The smaller creeks that drain the eastern portion of the plateau unite after descending to an altitude of 5,500 feet above the sea and form what is called the Parunuweap Fork of the Virgin. At and below the junction of these creeks the canyon valley in which they flow widens into what is known as Long Valley. Below Long Valley the East Fork enters Parunuweap Canyon and is simply a series of cascades for 15 miles, descending in this distance from 5,000 to 3,500 feet above sea level. Emerging from this canyon, it enters the valley of the Virgin. This valley is 44 miles long. Its upper portion is only an enlargement of the canyon; its lower portion is a broader valley, much broken by low basalt-covered mesas and sharp ridges of tilted sedimentary rocks. In the upper portion of the valley the river receives several tributaries, the principal ones being Little Zion, North Fork, La Verkin, and Ashe creeks. Midway of the valley two streams enter that come from Pine Valley Mountains, and near the foot Santa Clara River joins the Virgin, the united streams leaving the valley by a deep canyon cut through the Beaver Dam Mountains. The valley of the Virgin is at a lower altitude and has a warmer climate than any other portion of Utah. The soil of the irrigable lands is usually good, and wherever irrigation can be practiced it produces abundant crops.

VIRGIN RIVER AT VIRGIN, UTAH.

This station, which was established April 18, 1909, is located about half a mile east of and above the town of Virgin, Utah.

---

[1] Abstracted from report on the lands of the arid region of the United States, with a more detailed account of the lands of Utah, by J. W. Powell, 1878: Chapter 9, Irrigable lands of that portion of Utah drained by the Colorado River and its tributaries, by A. H. Thompson, pp. 151-153.

The station is about 1,000 feet below North Creek and is above Ashe and La Verkin creeks, which enter about 8 miles below. There are no diversions of any importance above.

The records are unaffected by ice or by artificial control above or below the station. The bed of the stream changes to a great extent during floods.

The first gage was used until August 7 when the section was changed by a flood causing the water to leave the gage. On August 31 the gage, cable, and bench marks were washed out and the section materially altered. On October 13 a new gage was installed at a different datum. Owing to the marked change in the channel, there can be no determined relation between the old and new gages.

Discharge measurements of Virgin River at Virgin, Utah, in 1909.

Date.	Hydrographer.	Width.	Area of section.	Gage height.	Discharge.
		Feet.	*Sq. ft.*	*Feet.*	*Sec.-ft.*
May 30	E. A. Porter..............................	89	139	2.65	457
July 13	.....do................................	84	64.1	1.65	88
23	Walter Spencer........................	90	151	2.8	557
Oct. 13a	E. A. Porter..............................	56	56.1	b 2.8	163

a Made after the change in channel conditions. b Gage height from new gage at different datum.

Daily gage height, in feet, of Virgin River at Virgin, Utah, for 1909.

[M. A. Hamilton, observer.]

Day.	Apr.	May.	June.	July.	Aug.	Oct.	Nov.	Dec.
1		4.0	2.6	1.3	1.5	....	2.95	2.8
2		3.8	2.6	1.3	1.6	....	2.9	2.9
3		3.8	2.6	1.2	1.7	....	2.8	2.9
4		3.8	2.6	1.3	1.6	....	2.75	3.1
5		3.6	2.7	1.3	1.5	....	2.9	3.0
6		3.4	2.7	1.4	1.5	....	2.85	2.8
7		3.4	2.6	1.4	....	....	3.0	2.7
8		3.7	2.6	1.5	....	....	3.0	2.8
9		3.5	2.5	1.6	....	....	2.95	2.9
10		3.4	2.5	1.6	....	....	2.9	2.7
11		3.4	2.5	1.7	....	....	3.0	2.9
12		3.2	2.4	1.7	....	....	3.1	2.8
13		3.2	2.4	1.7	....	2.8	3.1	2.8
14		3.3	2.4	1.7	....	2.85	3.4	· 2.8
15		3.1	2.4	1.7	....	2.75	3.3	2.85
16		3.1	2.4	1.7	....	2.8	3.35	2.8
17		3.0	2.4	1.7	....	2.65	3.1	2.7
18	3.2	3.0	2.4	1.7	....	2.7	3.0	2.6
19	3.5	3.0	2.3	1.7	....	3.2	2.95	....
20	3.9	3.0	2.3	1.7	....	2.65	3.1	....
21	3.7	2.9	2.1	1.9	....	2.8	3.0	....
22	3.3	2.9	2.1	2.1	....	2.7	2.9	2.6
23	3.0	2.9	2.1	2.8	....	2.75	2.95	2.7
24	3.2	2.8	2.0	2.3	....	2.8	2.8	2.7
25	3.0	2.8	2.0	....	....	2.85	2.8	2.6
26	3.2	2.8	2.0	....	....	2.9	2.9	2.6
27	4.4	2.7	1.9	....	....	2.75	2.85	2.8
28	4.4	2.7	1.7	....	....	2.9	2.8	2.7
29	4.9	2.7	1.7	....	....	2.8	2.85	2.9
30	4.6	2.6	1.6	....	....	2.95	2.75	3.1
31		2.6	....	....	....	3.0	....	5.1

NOTE.—On Oct. 13 new gage established and gage heights beginning this date are not comparable with those obtained earlier in the year.

Daily discharge, in second-feet, of Virgin River at Virgin, Utah, for 1909.

Day.	Apr.	May.	June.	July.	Aug.	Oct.	Nov.	Dec.
1	....	1,550	440	24	58	....	222	173
2	....	1,340	440	24	78	....	205	205
3	....	1,340	440	10	101	....	173	205
4	....	1,340	440	24	78	....	158	278
5	....	1,140	495	24	58	....	205	240
6	....	965	495	40	40	....	189	173
7	....	965	440	40	58	....	240	144
8	....	1,240	440	58	....	....	240	173
9	....	1,050	390	78	....	....	222	205
10	....	965	390	78	....	....	205	144
11	....	965	390	101	....	....	240	205
12	....	810	340	101	....	....	278	173
13	....	810	340	101	....	173	278	173
14	....	885	340	101	....	189	418	173
15	....	740	340	101	....	158	368	189
16	....	740	340	101	....	173	393	173
17	....	675	340	101	....	131	278	144
18	810	675	340	101	....	144	240	118
19	1,050	675	295	101	....	321	222	118
20	1,440	675	295	101	....	131	278	118
21	1,240	610	218	155	....	173	240	118
22	885	610	218	218	....	144	205	118
23	675	610	218	550	....	158	222	144
24	810	550	185	295	....	173	173	144
25	675	550	185	265	....	189	173	118
26	810	550	185	235	....	205	205	118
27	2,010	495	155	205	....	158	189	173
28	2,010	495	101	175	....	205	173	144
29	2,580	495	101	146	....	173	189	205
30	2,260	440	78	117	....	222	158	278
31	....	440	....	188	....	240	....	1,930

NOTE.—These discharges are based on rating curves applicable as follows: Apr. 18 to Aug. 7, fairly well defined below 600 second-feet; Oct. 13 to Dec. 31, not well defined.

Monthly discharge of Virgin River at Virgin, Utah, for 1909.

Month.	Discharge in second-feet.			Run-off (total in acre-feet).	Accuracy.
	Maximum.	Minimum.	Mean.		
Apr. 18-30	2,580	675	1,330	34,300	C.
May	1,550	440	819	50,400	C.
June	440	78	314	18,700	B.
July	550	24	128	7,870	B.
Aug. 1-7	101	40	67.3	934	B.
Oct. 13-31	321	131	182	6,860	B.
November	418	158	233	13,900	B.
December	1,930	118	226	13,900	C.

SANTA CLARA RIVER NEAR CENTRAL, UTAH.

This station, which was established April 21, 1909, is located about 1½ miles southeast of Central, Utah, the nearest post office, and about 6 miles west from the settlement of Pine Valley, Utah. It is about one-fourth mile from R. H. Hunt's ranch house in a small valley known as Eightmile Flat. The records show the total amount of water available for storage in the Pine Valley reservoir site, a few miles above the station.

The station is below all important tributaries except Mountain Meadows Creek, which enters about 10 miles below. A small canal whose maximum capacity is about 3.5 second-feet takes out water a short distance above the station.

The gage heights are not affected by ice. The bed of the stream is somewhat shifting, but fairly accurate results have been obtained. The datum of the staff gage on the left bank has remained unchanged since the station was established. Discharge measurements are made from a gaging bridge.

Discharge measurements of Santa Clara River near Central, Utah, in 1909.

Date.	Hydrographer.	Width.	Area of section.	Gage height.	Discharge.
		Feet.	*Sq. ft.*	*Feet.*	*Sec.-ft.*
Apr. 21a	E. A. Porter			3.80	102
June 1b	....do	30	38	3.60	60
July 15	....do	20	20	3.12	15
Oct. 6	....do	20	29	3.60	39
16	....do	23	30	3.50	30

a Made one-fourth mile above gage. Results not good.   b Made by wading.

Daily gage height, in feet, of Santa Clara River near Central, Utah, for 1909.

[Royal Hunt, observer.]

Day.	Apr.	May.	June.	July.	Aug.	Sept.	Oct.	Nov.	Dec.
1		4.0	3.7	3.25	3.1	7.0	3.6	3.4	3.6
2		3.9	3.8	3.25	3.1	4.4	3.6	3.45	3.6
3		4.0	3.7	3.6	3.1	4.0	3.7	3.4	3.55
4		4.1	3.8	3.4	3.1	4.0	3.7	3.35	3.5
5		4.1	3.8	3.3	3.0	4.0	3.6	3.35	3.5
6		4.2	3.8	3.3	3.0	4.0	3.6	3.35	3.5
7		4.2	3.8	3.3	3.1	3.9	3.6	3.35	3.5
8		4.2	3.7	3.3	3.1	3.8	3.6	3.35	3.5
9		4.2	3.6	3.25	3.1	3.7	3.6	3.4	3.7
10		4.2	3.6	3.25	3.1	3.6	3.6	3.4	3.7
11		4.0	3.5	3.2	3.1	3.6	3.55	3.4	3.6
12		4.0	3.5	3.2	3.0	3.6	3.55	3.4	3.5
13		3.8	3.4	3.2	3.05	3.55	3.5	3.4	3.5
14		3.7	3.4	3.1	3.0	3.55	3.5	3.4	3.45
15		3.7	3.4	3.1	3.0	3.5	3.5	3.4	3.45
16		3.8	3.4	3.1	3.2	3.5	3.55	3.4	3.45
17		3.7	3.4	3.1	3.3	3.5	3.5	3.4	3.4
18		3.7	3.4	3.1	3.3	3.5	3.5	3.4	3.4
19		3.7	3.3	3.1	3.3	3.4	3.5	3.4	3.4
20		3.7	3.3	3.1	3.3	3.4	3.45	3.4	3.45
21	3.8	3.8	3.3	3.1	3.3	3.4	3.45	3.45	3.45
22	3.7	3.8	3.2	3.0	3.3	3.4	3.45	3.45	3.5
23	3.7	3.7	3.2	3.1	3.2	3.4	3.45	3.45	3.5
24	3.6	3.6	3.2	3.1	3.2	3.4	3.45	3.45	3.4
25	3.6	3.6	3.2	3.1	3.2	3.4	3.45	3.45	3.4
26	3.7	3.7	3.2	3.1	3.15	3.3	3.45	3.5	3.4
27	3.9	3.7	3.1	3.1	3.2	3.7	3.45	3.5	3.4
28	4.1	3.6	3.1	3.1	3.2	4.0	3.45	3.55	3.4
29	4.2	3.7	3.2	3.1	3.3	3.8	3.45	3.55	3.4
30	3.8	3.7	3.2	3.1	3.3	3.7	3.45	3.55	3.4
31		3.7		3.1	3.6		3.4		6.5

Daily discharge, in second-feet, of Santa Clara River near Central, Utah, for 1909.

Day.	Apr.	May.	June.	July.	Aug.	Sept.	Oct.	Nov.	Dec.
1		117	69	27	15	475	37	27	37
2		106	77	27	15	100	37	30	37
3		117	69	46	15	64	43	27	34
4		124	75	32	15	64	43	25	32
5		124	75	27	12	64	37	25	32
6		136	75	27	12	64	37	25	32
7		136	75	27	15	56	37	25	32
8		136	66	27	15	49	37	25	32
9		132	56	23	15	43	37	27	43
10		132	56	23	15	37	37	27	43
11		110	50	21	15	37	34	27	37
12		110	50	21	12	37	34	27	32
13		90	44	21	14	34	32	27	32
14		77	41	15	12	34	32	27	30
15		77	41	15	12	32	32	27	30
16		86	41	15	19	32	34	27	30
17		77	41	15	23	32	32	27	27
18		77	41	15	23	32	32	27	27
19		74	34	15	23	27	32	27	27
20		74	34	15	23	27	30	27	30
21	102	84	34	15	23	27	30	30	30
22	92	84	29	12	23	27	30	30	32
23	92	74	29	15	19	27	30	30	32
24	79	63	26	15	19	27	30	30	27
25	79	63	26	15	19	27	30	30	27
26	89	72	26	15	17	23	30	32	27
27	109	72	22	15	19	43	30	32	27
28	132	63	22	15	19	64	30	34	27
29	140	69	24	15	23	49	30	34	27
30	96	69	24	15	23	43	30	34	27
31		69		15	37		27		393

NOTE.—These discharges are based on rating curves applicable as follows: Apr. 21 to July 15, indirect method for shifting channels used; July 15 to Dec. 31, fairly well defined below 50 second-feet, not well defined above.

Monthly discharge of Santa Clara River near Central, Utah, for 1909.

Month.	Discharge in second-feet.			Run-off (total in acre-feet).	Accuracy.
	Maximum.	Minimum.	Mean.		
Apr. 21–30	140	79	101	2,000	B.
May	136	63	93.4	5,740	B.
June	77	22	45.7	2,720	C.
July	46	12	19.9	1,220	C.
August	37	12	18.1	1,110	C.
September	475	23	56.6	3,370	C.
October	43	27	33.3	2,050	B.
November	34	25	28.3	1,680	B.
December	393	27	43	2,640	B.
The period				22,500	

SANTA CLARA RIVER NEAR ST. GEORGE, UTAH.

This station, which was established April 16, 1909, to determine the total unappropriated discharge of Santa Clara River, is located about 3 miles southwest of St. George, Utah, and about 2 miles above the mouth of the river.

The station is below all tributaries and diversions except two canals which head near the mouth of the river.

The gage heights are not affected by ice, but the bed of the stream shifts to a considerable extent. A fair record of run-off has, however, been obtained for 1909. The datum of the sloping gage on the left bank has remained unchanged since the station was established. The gage is underneath a footbridge from which the discharge measurements are made.

Discharge measurements of Santa Clara River near St. George, Utah, in 1909.

Date.	Hydrographer.	Width.	Area of section.	Gage height.	Discharge.
		Feet.	*Sq. ft.*	*Feet.*	*Sec.-ft.*
May 27	E. A. Porter.............................	18.5	13	2.02	35
31	....do.................................	20	14	2.10	45
July 14	....do.................................	5	1.2	1.35	(a)
Oct. 8	....do.................................	25.5	9.5	2.62	33
15	....do.................................	18	7.6	2.30	20

a Estimated eight-tenths second-foot.

Daily gage height, in feet, of Santa Clara River near St. George, Utah, for 1909.

[E. A. Everett, observer.]

Day.	Apr.	May.	June.	July.	Sept.	Oct.	Nov.	Dec.
1..	....	2.5	2.1	....	10.0	2.0	1.9	....
2..	....	2.4	2.1	....	4.0	2.0	1.9	....
3..	....	2.4	2.2	....	2.0	2.5	1.8	2.5
4..	....	2.5	2.2	....	1.5	2.6	1.8	2.4
5..	....	2.6	2.3	....	....	2.6	....	2.4
6..	....	2.7	....	....	4.0	2.6	1.8	2.4
7..	....	2.8	2.3	....	5.0	2.7	....	2.4
8..	....	2.8	2.3	....	1.5	2.6	....	2.4
9..	....	....	2.2	....	1.5	2.6	2.5	2.9
10..	....	2.8	2.1	....	1.5	....	2.7	2.7
11..	....	2.7	1.8	....	1.5	2.6	2.3	....
12..	....	2.5	1.7	....	1.5	2.5	2.2	....
13..	....	2.4	(a)	....	1.5	2.3	2.2	2.5
14..	....	2.4	....	1.35	1.5	2.3	....	....
15..	....	2.4	....	....	1.5	2.3	2.4	2.4
16..	3.0	2.4	....	....	1.5	2.3	....	....
17..	3.2	2.4	....	....	1.0	....	....	....
18..	....	2.3	....	....	1.0	2.2	2.2	2.3
19..	....	2.3	....	....	1.0	2.1	2.2	....
20..	2.6	2.3	....	....	1.0	2.1	2.2	2.4
21..	2.6	2.3	....	....	1.0	....	2.3	2.4
22..	2.5	2.4	....	....	1.0	2.1	2.3	2.4
23..	2.3	....	....	....	1.0	2.0	2.3	2.5
24..	2.2	2.3	....	....	1.0	....	....	2.5
25..	....	2.2	....	....	3.0	2.0	2.3	....
26..	2.4	2.1	....	....	4.0	2.0	....	....
27..	2.3	2.1	....	....	2.0	2.0	2.7	2.5
28..	2.4	2.1	....	....	2.0	2.0	....	2.4
29..	2.4	2.4	....	....	1.7	2.0	2.5	2.4
30..	2.3	....	....	....	1.7	....	2.5	2.4
31..	....	2.1	....	....	....	....	....	3.6

a See page 216.

NOTE.—Gage heights for September are unreliable.

Daily discharge, in second-feet, of Santa Clara River near St. George, Utah, for 1909.

Day.	Apr.	May.	June.	Oct.	Nov.	Dec.	Day.	Apr.	May.	June.	Oct.	Nov.	Dec.
1.......	......	79	42	3	1	28	16..........	145	69	.5	17	18	20
2.......	......	69	42	3	1	28	17..........	179	69	.5	14	15	18
3.......	......	69	50	28	......	28	18..........	150	59	.5	12	12	17
4.......	......	79	50	35	......	22	19..........	120	59	.5	7	12	20
5.......	......	90	59	35	......	22	20..........	90	59	.5	7	12	22
6.......	......	102	59	35	......	22	21..........	90	59	.5	7	14	22
7.......	......	115	59	43	9	22	22..........	79	69	.5	7	17	22
8.......	......	115	59	35	18	22	23..........	59	64	.5	3	17	28
9.......	......	115	50	35	28	61	24..........	50	59	.5	3	17	28
10.......	......	115	42	35	43	43	25..........	60	50	.5	3	17	28
11.......	......	102	20	35	17	38	26..........	69	42	.5	3	30	28
12.......	......	79	15	28	12	33	27..........	59	42	.5	3	43	28
13.......	......	69	.5	17	12	28	28..........	69	42	.5	3	36	22
14.......	......	69	.5	17	17	25	29..........	69	69	.5	3	28	22
15.......	......	69	.5	17	22	22	30..........	59	56	.5	3	28	22
							31..........	......	42	......	2	......	153

NOTE.—These discharges are based on rating curves applicable as follows: Apr. 16 to Aug. 31, fairly well defined below 60 second-feet; Sept. 1 to Dec. 31, not well defined; June 13 to Aug. 31, discharge due to seepage only and has been estimated.

Gage heights unreliable for September; no estimates made.

From June 13 to Aug. 31 the flow was due to seepage only.

Monthly discharge of Santa Clara River near St. George, Utah, for 1909.

Month.	Discharge in second-feet.			Run-off (total in acre-feet).	Accuracy.
	Maximum.	Minimum.	Mean.		
Apr. 16–30....................................	179	50	89.8	2,670	B.
May..	115	42	72.4	4,450	A.
June.......................................	....	....	a 18.5	1,100	B.
July.......................................	....	....	a .5	31	C.
August.....................................	....	....	a .5	31	C.
October....................................	43	2	16.1	990	B.
November..................................	43	....	16.5	982	B.
December..................................	153	17	30.5	1,880	B.

a From June 13 to Aug. 31 the flow was due to seepage only. Discharge for this period estimated on basis of measurement made July 14.

MUDDY RIVER NEAR MOAPA, NEV.

This station, which is located at the Narrows, about 7 miles from Moapa, Nev., the nearest railway point, was established January 1, 1904, to determine the amount of water available for storage in a reservoir site at this point.

No gage heights are available since 1906, but occasional measurements have been made since that time. Discharge measurements for 1909 were made in a flume which carries the entire flow of the river except during floods.

Measurements made in 1908 at the station were published in Water-Supply Paper 249, page 197, under "miscellaneous measurements in Virgin River drainage basin."

Discharge measurements of Muddy River near Moapa, Nev., in 1909.

Date.	Hydrographer.	Width.	Area of section.	Gage height.	Dis-charge.
		Feet.	*Sq. ft.*	*Feet.*	*Sec. ft.*
Jan. 4	E. C. La Rue	8	13.4	1.67	48
4	.....do...	8	14.8	1.85	55
4	.....do...	8	11.8	1.47	38
19	U. V. Perkins...	8	12.4	1.55	40
Mar. 17	A. L. F. McDermott...	8	4.0	.60	5.1
Apr. 27	U. V. Perkins...	8	10.4	1.30	46
May 14	.....do...	8	8.4	1.05	32
June 1	.....do...	8	4.7	.59	9.3
1	.....do...	8	6.4	.80	18
1	.....do...	8	7.4	.92	24
1	.....do...	8	8.8	1.10	35
July 2	E. C. La Rue...	8	2.9	.34	22
2	.....do...	8	8.7	1.05	30
2	.....do...	8	6.6	.80	15
2	.....do...	8	10.0	1.22	37
2	.....do...	8	4.2	.50	6.3
Nov. 27	.....do...	8	12.0	1.45	43

GILA RIVER BASIN.

DESCRIPTION.

Gila River rises in western and southwestern New Mexico, receiving its waters from mountains having an elevation of 7,000 to 8,000 feet. At the point where it crosses into Arizona it still has an elevation of 6,000 feet. From this place it flows between mountain ranges and falls rapidly, until at Florence, 180 miles away, it is about 1,500 feet above sea level. At a point about 15 miles above Florence the river emerges upon the plains, through which it winds for about 75 miles before receiving the waters of its principal tributary, the Salt. From the junction of the Salt the Gila continues west and southwest and enters the Colorado at Yuma, Ariz, near the southwest corner of the Territory.

The principal tributaries are San Pedro and Santa Cruz rivers from the south, and San Francisco, Salt,[1] Agua Fria, and Hassayampa rivers from the north.

San Francisco River rises in the southwestern part of Socorro County, N. Mex., and flows southwestward into Graham County, Ariz., where it unites with the Gila. The basin comprises about 2,600 square miles, of which 1,800 square miles are in New Mexico and 800 in Arizona.

San Pedro River rises in the northern part of the Mexican State of Sonora, flows northward for more than 100 miles, and empties into the Gila a few miles below the town of Dudleyville, 45 miles above Florence, Ariz. Rising in a country of very light snowfall, the river depends for the greater part of its water supply on the frequent showers of the rainy seasons. It flows over a sandy bed between high, steep banks, and during the dry season it shrinks to an insignifi-

---

[1] For description of Salt River, see p. 227.

cant stream of clear water, which rises and sinks in the sand with the varying depth of bed rock.

The floods of the upper Gila and its tributaries are commonly short and violent, occurring during the months of January and February. A period of high water occurs also usually during the late summer or early fall. The season of low water occurs in June and July.

The drainage basin of the Gila includes 7,000 square miles of merchantable timberland, 11,000 square miles of woodland, of which the San Francisco basin has 1,000 square miles of timberland, 45,000 square miles of land upon which there is no timber, 1,300 square miles of scattered timber, and 300 square miles of open land.

The average annual precipitation over the greater part of the contributary drainage area of Gila and San Francisco rivers in New Mexico is between 10 and 15 inches and in the high mountains of the headwater region it rises above 20 inches.

The winters are mild except in the mountainous sections, and very little ice forms on the rivers.

Irrigation in New Mexico has been confined chiefly to the bottom lands along the main streams and their tributaries, but the total area irrigated comprises only a very few thousand acres. Excellent opportunities for irrigation exist along both the Gila and the San Francisco. The United States Reclamation Service has made surveys for an irrigation project in the vicinity of Alma, N. Mex. Another promising district is that popularly known as the Lordsburg flat, which extends from Lordsburg, N. Mex., northward to Gila River, a distance of over 20 miles, and comprises over a quarter of a million acres of almost unbroken and very fertile land, at an elevation a little above 4,000 feet. This land could be irrigated by the stored water of Gila River, although the expense of reaching it would be considerable.

Good storage sites exist at various places along San Francisco and Gila rivers, among which may be mentioned the reservoir site on the San Francisco near Alma, and that on the Gila near Redrock, N. Mex.

Because of the torrential character of the Gila, water-power development is not feasible except where stored water is used. The San Francisco, being more of a mountain stream, presents better opportunities for the use of water power along its upper reaches. Most of the future water-power development along these streams will probably be in connection with irrigation projects. At present it is limited to one or two small plants on the San Francisco.

GILA RIVER NEAR REDROCK, N. MEX.

This station was originally established May 14, 1908, at the mouth of the Middle Box Canyon of the Gila (Pl. V, *B*), about 2 miles east of Redrock post office, N. Mex. On July 16, 1909, it was moved about one-eighth mile upstream in the canyon. The two nearest railroad points are Silver City, about 36 miles east of Redrock, and Lordsburg, about 30 miles south.

The records show the amount of water available for irrigation and power enterprises.

Mancos River, an intermittent stream, the first large tributary upstream from the station, joins the Gila about 12 miles above. A number of large washes come into the river above and below the station, and during flood stages the run-off from these tributaries is very great. The drainage area at the station is about 3,500 square miles A number of large irrigation ditches divert water above the station. Practically no power is developed in the headwaters of this basin, although opportunity for such development is good.

Except for fringe ice along the edges of the stream, ice does not interfere with the accuracy of the results.

The gage originally installed May 14, 1908, is bolted to a rock bluff on the left bank just a few feet above the mouth of the canyon. This gage was abandoned July 16, 1909, when a Friez automatic gage was installed on the left bank about one-eighth mile upstream and at a different datum. An auxiliary staff gage, which is necessary as a guide in setting the automatic gage, was not installed until August 18.

The accuracy of the base data obtained at this station is impaired by the adverse natural conditions. The station is isolated and the stream bed very shifting in character. Measurements by wading are more or less affected by quicksand.

Discharge measurements of Gila River near Redrock, N. Mex., in 1909.

Date.	Hydrographer.	Width.	Area of section.	Gage height.		Discharge.
				Automatic gage.	Staff gage.	
		Feet.	*Sq. ft.*	*Feet.*	*Feet.*	*Sec. ft.*
Feb. 1	J. B. Stewart	82.5	107		1.39	244
Mar. 1	....do	107	145		1.40	318
June 14	....do	57	42		1.00	97
July 11	....do	63	44		.82	67
13	....do	32.5	24		.67	42
16	....do	51.5	41	1.75	.95	85
Aug. 18	....do	154	142	2.80	1.65	398
Sept. 3	....do	41	51	2.20	.91	132
Oct. 14	W. B. Freeman	47.5	62	1.43		92

NOTE.—All measurements made by wading at various sections.

Daily gage height, in feet, of Gila River near Redrock, N. Mex., for 1909.

[J. G. Rutland and A. B. Conner, observers.]

Day.	Jan.	Feb.	Mar.	Apr.	May.	June.	July.	Aug.	Sept.	Oct.	Nov.	Dec.
1	0.75	1.4	1.2	1.6	1.35	1.1			2.35	1.45	1.4	1.5
2	.75	1.2	1.6	1.6	1.3	1.1			2.25	1.4	1.4	1.6
3	.7	1.0	1.6	1.5	1.3	1.1			2.25	1.5	1.4	1.7
4	.7	.8	1.6	1.5	1.2	1.0			2.3	1.5	1.4	1.65
5	.7	1.1	1.55	1.4	1.2	1.0			2.8	1.5	1.4	1.65
6	.7	1.0	1.7	1.2	1.3	1.0			4.9	1.45	1.4	1.65
7	.65	.95	1.6	1.2	1.2	.8			3.2	1.45	1.4	1.65
8	.65	1.3	1.6	1.3	1.2	.8			2.8	1.45	1.4	1.65
9	.65	1.0	1.6	1.4	1.2	.6			2.75	1.4	1.45	1.65
10	.6	1.6	1.8	1.4	1.2	.6			2.7	1.4	1.45	1.7
11	1.1	1.4	1.9	1.5	1.2	.6	0.8		3.3	1.4	1.45	1.5
12	.9	1.3	1.9	1.8	1.2	.7			3.4	1.4	1.45	1.5
13	.75	1.2	2.4	2.0	1.2	.8	.65		3.1	1.4	1.5	1.55
14	.7	1.0	2.6	1.6	1.2	.95		3.3	3.0	1.4	1.5	1.55
15	.7	1.0	1.9	1.5	1.3	.4	.65	3.6	2.9	1.4	1.5	1.55
16	.65	1.4	1.9	1.5	1.3	.4	.95	2.9	2.8	1.4	1.5	1.55
17	.8	1.3	2.2	1.4	1.3	.4		3.0	2.7	1.4	1.5	1.6
18	.8	1.2	2.0	1.4	1.3	.3		2.85	2.6	1.4	1.5	1.7
19	.7	1.6	1.9	1.3	1.2	.3		2.75		1.4	1.5	1.7
20	.7	1.4	2.4	1.3	1.2	.3		2.65		1.45	1.45	1.7
21	.95	1.4	2.2	1.6	1.3	.3		2.5		1.45	1.45	1.7
22	.85	1.2	1.9	1.6	1.3	.3		2.5		1.4	1.5	1.8
23	.8	1.1	2.1	1.5	1.3	.3		2.35		1.5	1.5	1.8
24	.8	1.0	2.0	1.4	1.2	.3		2.4		1.5	1.5	1.85
25	.75	1.2	1.9	1.4	1.2	.3		2.3	1.6	1.5	1.5	2.0
26	.7	1.3	2.3	1.3	1.2	.3		2.25	1.55	1.55	1.5	
27	.7	1.2	2.3	1.3	1.2	.3		2.15	1.5	1.5	1.55	
28	.7	1.2	2.3	1.3	1.2	.4		2.8	1.5	1.5	1.55	
29	.9		2.0	1.3	1.2	.4		2.3	1.5	1.55	1.5	
30	.8		2.4	1.4	1.2			2.2	1.45	1.4	1.5	
31	.7		2.6		1.2			2.3		1.4		

Daily discharge, in second-feet, of Gila River near Redrock, N. Mex., for 1908–9.

Day.	Nov.	Dec.	Day.	Nov.	Dec.	Day.	Nov.	Dec.
1908.			1908.			1908.		
1		80	11		51	21	53	52
2		70	12		76	22	53	52
3		58	13		67	23	106	43
4		58	14		67	24	80	43
5		50	15		49	25	74	83
6		97	16	105	49	26	65	55
7		90	17	96	83	27	395	47
8		80	18	75	64	28	330	48
9		62	19	75	52	29	132	40
10		54	20	66	52	30	80	42
						31		42

Daily discharge, in second-feet, of Gila River near Redrock, N. Mex., for 1908-9.

Day.	Jan.	Feb.	Mar.	Apr.	May.	June.	July.	Aug.	Sept.	Oct.	Nov.	Dec.
1909.												
1	47	248	174	415	288	142	25	185	173	72	90	103
2	47	160	390	415	253	142	30	185	147	68	90	116
3	42	98	390	360	253	142	30	185	143	78	90	181
4	42	59	395	360	205	100	35	185	158	90	90	123
5	42	126	365	299	205	100	35	220	280	90	90	123
6	42	98	462	195	253	100	40	210	1,120	84	90	123
7	37	86	395	195	205	60	45	210	410	84	90	123
8	37	203	395	243	205	60	50	260	278	84	90	123
9	37	98	395	299	205	36	55	320	282	86	95	123
10	33	370	536	299	205	36	60	370	268	86	95	181
11	116	250	610	360	205	36	65	420	470	86	95	103
12	70	210	610	565	205	46	52	470	505	88	95	103
13	47	166	1,030	730	193	58	40	530	395	88	103	109
14	42	102	1,210	435	193	85	40	590	378	90	103	109
15	42	102	620	370	240	24	40	710	345	90	103	109
16	38	258	620	370	240	24	85	435	312	90	103	109
17	56	210	860	308	240	24	58	473	284	90	103	116
18	56	166	698	308	240	16	105	416	255	90	103	181
19	44	380	620	250	193	16	233	375	230	90	103	181
20	44	258	1,040	250	193	16	220	333	205	95	95	181
21	84	258	860	435	240	16	220	280	180	95	95	181
22	65	174	620	435	240	16	220	273	155	90	103	149
23	56	137	775	376	240	16	220	226	125	103	103	149
24	56	107	707	313	178	16	220	232	100	103	103	159
25	49	174	629	313	178	16	233	202	82	103	103	193
26	44	217	962	255	178	16	233	284	75	109	103	190
27	44	174	962	255	178	16	220	158	72	109	109	190
28	44	174	962	255	178	24	185	315	72	109	109	190
29	74	....	708	255	178	24	185	180	78	109	103	190
30	56	....	1,050	313	178	24	185	152	72	90	103	190
31	44	....	1,230	....	178	....	185	167	....	90	....	190

NOTE.—These discharges for 1908-9 were obtained by the indirect method for shifting channels. Discharges estimated for days for which no gage heights are given.

Monthly discharge of Gila River near Redrock, N. Mex., for 1908-9.

Month.	Discharge in second-feet.			Run-off (total in acre-feet).	Accuracy.
	Maximum.	Minimum.	Mean.		
1908.					
Nov. 16–30	395	53	119	3,540	B.
December	97	40	59.9	3,680	B.
1909.					
January	116	33	50.9	3,130	B.
February	380	59	181	10,100	B.
March	1,230	174	686	42,200	B.
April	730	195	341	20,300	B.
May	288	178	212	13,000	B.
June	142	16	48.2	2,870	C.
July	233	25	118	7,260	D.
August	710	152	308	18,900	D.
September	1,120	72	255	15,200	D.
October	109	68	91.3	5,610	D.
November	109	90	98.3	5,850	D.
December	193	103	148	9,100	D.
The year	1,230	16	211	154,000	

NOTE.—Prior to Nov. 16, 1908, the data are too meager or unreliable on which to base estimates.

SAN FRANCISCO RIVER AT ALMA, N. MEX.

This station, which was established October 18, 1904, by the United States Reclamation Service, is located about half a mile southeast of Alma, N. Mex., and 85 miles northwest of Silver City, the most accessible railway point. It was discontinued by the Reclamation Service December 31, 1907, and was reestablished by the United States Geological Survey on January 1, 1909.

The station is a short distance below the mouth of Mineral Creek and about 5 miles above the mouth of Whitewater Creek. The waters of the San Francisco and its tributaries are used to some extent for mining and a few small ditches take out water above the station for irrigation of the bottom lands. The flow of this stream is very little affected by ice, though thin ice sometimes forms on the edges.

The rod gage established January 1, 1909, is at a different location and datum from those of the previous gage. It was washed out on September 6, 1909, and replaced October 10, 1909, by another sloping-rod gage 100 feet upstream at the same datum. Gage heights during the interval that the gages were out were read from a temporary gage installed by the observer and later reduced to the new gage datum. High-water measurements have been made from a cable 300 feet upstream from the location of the last gage.

Very good measurements can be obtained at this station, but they should be made frequently in order to obtain the best results, as the stream bed is shifting in character.

Discharge measurements of San Francisco River at Alma, N. Mex., in 1909.

Date.	Hydrographer.	Width.	Area of section.	Gage height.	Discharge.
		Feet.	*Sq. ft.*	*Feet.*	*Sec.-ft.*
Jan. 2	J. B. Stewart	38	19.6	0.74	37
Feb. 7	....do	39	19.0	1.00	52.0
Apr. 21	....do	61	60.0	1.55	214.0
June 6	....do	25	7.0	.45	8.7
6	....do	19	5.8	.45	8.5
July 4	....do	25.5	8.0	.58	15.9
Aug. 7	....do	32.5	13.2	.65	28.0
Oct. 10	W. B. Freeman	29.5	12.5	.75	15.8

Daily gage height, in feet, of San Francisco River at Alma, N. Mex., for 1909.

[Cora Jackson and Minnie Bowers. observers.]

Day.	Jan.	Feb.	Mar.	Apr.	May.	June.	July.	Aug.	Sept.	Oct.	Nov.	Dec.
1	0.7	1.0	1.2	2.0	1.0	0.5	0.3	0.6	0.6	1.0	0.5	0.95
2	.7	1.0	1.0	2.0	1.0	.5	.3	.6	.6	.9	.6	.85
3	.7	1.0	1.5	2.0	1.0	.5	.3	1.0	.9	.8	.5	.85
4	.7	1.0	2.0	2.5	1.0	.5	.5	.5	1.0	.8	.5	.85
5	.7	1.0	2.2	2.5	1.0	.5	.45	1.35	2.9	.8	.6	.95
6	.7	1.0	1.9	2.0	1.0	.45	.7	.7	1.0	.8	.6	.95
7	.7	1.0	2.0	2.0	1.0	.45	.5	1.0	1.0	.8	.6	1.0
8	.7	1.5	1.8	2.0	.85	.4	.5	1.0	1.0	.8	.55	1.0
9	.7	1.0	1.5	2.0	.85	.4	.5	1.2	1.0	.8	.55	1.0
10	.75	1.0	1.5	1.7	.85	.4	.5	2.2	.9	.75	.55	1.0
11	.75	1.0	1.5	1.7	.7	.4	.2	1.0	.9	.7	.6	.95
12	.75	1.0	1.4	1.7	.7	.4	.2	1.4	.9	.65	.6	.95
13	.8	1.8	1.2	1.7	.7	.4	.3	1.3	.8	.6	.6	.95
14	.8	1.5	1.2	1.7	.7	.4	.2	1.8	.8	.6	.7	.95
15	.8	1.5	1.5	1.7	.7	.4	.2	1.2	.8	.6	.7	.95
16	.8	1.3	1.5	1.7	.7	.4	.6	1.2	.8	.55	.7	.95
17	.8	1.3	2.3	2.0	.7	.4	.4	1.0	.8	.7	.7	.95
18	.8	1.0	3.0	1.7	.7	.4	1.2	1.3	.8	.65	.75	.95
19	.8	1.0	2.5	1.7	.7	.4	.6	.9	....	.6	.75	.95
20	.8	1.0	2.1	2.0	.7	.4	3.0	.8	....	.6	.75	.95
21	.8	1.0	2.2	1.6	.7	.35	.6	.8	....	.6	.7	.95
22	.8	1.0	2.0	1.4	.6	.35	.6	1.1	....	.5	.7	.95
23	.8	1.0	2.0	1.4	.6	.3	.6	.8	....	.5	.7	.95
24	.8	1.0	2.0	1.3	.6	.3	.8	1.1	....	.5	.7	.95
25	.8	1.0	1.6	1.15	.5	.3	.8	.7	....	.6	.75	.95
26	.8	1.1	1.8	1.1	.5	.3	1.0	.6	....	.7	.75	.95
27	2.0	1.1	2.0	1.1	.5	.3	.6	.5	....	.5	.7	.95
28	2.75	1.2	2.0	1.0	.5	.3	.6	.5	....	.4	.7	.95
29	2.0	....	2.0	1.0	.5	.3	.5	.5	....	.5	.8	.95
30	1.5	....	2.0	1.0	.5	.3	.6	.6	....	.5	.95	.95
31	1.0	....	2.0	....	.5	....	1.0	.6	....	.5	....	.95

NOTE.—Gage heights for 1909 are not referred to the same datum as that used by the United States Reclamation Service in 1907.

Daily discharge, in second-feet, of San Francisco River at Alma, N. Mex., for 1909.

Day.	Jan.	Feb.	Mar.	Apr.	May.	June.	July.	Aug.	Sept.	Oct.	Nov.	Dec.
1	33	52	88	485	52	11	4	24	16	39	6.	30
2	33	52	52	485	52	11	4	24	16	29	9	22
3	33	52	192	485	52	11	4	72	40	20	6	22
4	33	52	490	840	52	11	11	17	52	20	6	22
5	33	52	620	840	52	11	9	182	1,140	20	9	30
6	31	52	420	485	52	9	23	32	52	18	9	30
7	31	52	485	485	52	9	11	72	52	18	9	34
8	31	192	355	485	35	7	11	70	52	18	7	34
9	31	52	192	485	35	7	12	118	48	18	7	34
10	36	52	192	295	35	7	12	690	36	16	7	34
11	34	52	192	295	23	7	3	70	36	13	9	30
12	34	52	148	295	23	7	3	194	36	11	9	30
13	39	355	88	295	23	7	6	143	28	9	9	30
14	39	192	88	295	23	7	3	410	28	9	13	30
15	39	192	192	295	23	7	3	116	28	9	13	30
16	37	114	192	295	23	7	19	105	28	7	13	30
17	37	114	690	485	23	7	10	62	26	13	13	30
18	37	52	1,200	295	23	7	249	137	26	11	16	30
19	37	52	840	295	23	7	20	48	30	9	16	30
20	37	52	550	485	23	7	1,200	37	30	9	16	30
21	35	52	620	240	23	5.5	22	37	30	9	13	30
22	35	52	485	148	16	5.5	22	80	30	6	13	30
23	35	52	485	148	16	4	22	37	30	6	13	30
24	35	52	485	114	16	4	38	73	30	6	13	30
25	35	52	240	78	11	4	40	25	30	9	16	30
26	33	68	355	68	11	4	68	18	30	13	16	30
27	510	68	485	68	11	4	23	12	30	6	13	30
28	1,050	88	485	52	11	4	23	12	30	3	13	30
29	510	....	485	52	11	4	17	12	30	6	18	30
30	208	....	485	52	11	4	24	18	30	6	30	30
31	56	....	485	....	11	....	70	18	....	6	....	30

NOTE.—These discharges have been obtained by the indirect method for shifting channels.

Monthly discharge of San Francisco River at Alma, N. Mex., for 1909.

Month.	Discharge in second-feet.			Run-off (total in acre-feet).	Accu racy.
	Maximum.	Minimum.	Mean.		
January...............................	1,050	31	104	6,400	C.
February...............................	355	52	84.7	4,700	B.
March...............................	1,200	52	399	24,500	C.
April...............................	840	52	324	19,300	B.
May...............................	52	11	27.3	1,680	B.
June...............................	11	4	6.90	411	C.
July...............................	1,200	3	64.1	3,940	C.
August...............................	690	12	95.6	5,880	C.
September...............................	1,140	16	70.0	4,170	D.
October....................·....	39	3	12.6	775	D.
November...............................	30	6	12.0	714	D.
December...............................	34	22	29.7	1,830	D.
The year...............................	1,200	3	102	74,300	

SANTA CRUZ RIVER NEAR NOGALES, ARIZ.

This station, which was established March 22, 1907, was located about 5 miles from Nogales, near Yerba Buena ranch. Conditions of flow are subject to change. It was discontinued November 30, 1907, and reestablished April 1, 1909. No discharge measurements were made in 1909.

This station is maintained and the data have been furnished for publication by the United States Reclamation Service.

Daily gage height, in feet, of Santa Cruz River near Nogales, Ariz., for 1909.

[Mr. Harrison, observer.]

Day.	Apr.	May.	June.	July.	Aug.	Sept.	Oct.	Nov.	Dec.
1.................	2.8	3.4	2.9	2.65	3.8	3.4	3.3	3.2	3.4
2.................	2.8	2.8	2.9	2.65	3.75	3.4	3.3	3.2	3.4
3.................	2.8	2.8	2.8	3.92	3.75	3.4	3.3		3.4
4.................	2.85	3.3	2.8	3.25	3.7	3.4	3.3	3.3	3.4
5.................	2.85	....	2.8	3.22	3.65	3.4	....	3.3	3.4
6.................	2.8	....	2.8	3.35	3.68	3.4	3.3	3.3	3.4
7.................	2.7	3.3	2.8	2.8	3.6	3.4	3.3	3.3	3.4
8.................	2.8	3.3	2.8	2.7	3.9	3.4	....	3.25	3.4
9.................	2.8	3.3	2.8	2.7	3.8	3.35	3.3	3.2	3.4
10.................	2.7	3.3	2.8	2.7	3.7	3.3	3.3		3.4
11.................	2.7	3.3	2.75	2.7	3.6	3.3	3.3	3.2	3.4
12.................	2.8	3.3	2.75	2.7	3.6	3.3	3.3	3.4	3.4
13.................	2.8	2.9	2.75	2.7	3.95	....	3.3	3.35	3.4
14.................	2.8	3.2	2.75	2.7	3.6	3.3	3.3	3.35	3.35
15-.................	2.8	3.2	2.75	2.7	3.9	3.3	3.3	3.3	3.35
16.................	2.8	3.2	2.7	3.3	5.2	....	3.3	3.3	3.35
17.................	2.8	3.2	2.7	4.02	5.75	3.3	3.3	3.3	3.3
18.................	....	3.2	2.7	5.3	4.9	3.3	3.3	3.3	3.3
19.................	2.8	3.2	2.7	3.25	3.9	3.3	3.3	3.3	3.35
20.................	2.8	3.2	2.7	3.25	3.8	3.3	3.3	3.3	3.3
21.................	2.8	....	2.7	3.72	3.8	3.3	3.3	3.3	3.3
22.................	2.75	3.6	2.7	4.58	3.8	3.3	3.3	3.3	3.3
23.................	2.8	3.6	2.7	3.75	4.4	3.3	3.3	3.3	3.35
24.................		3.6	2.7	3.62	3.7	3.3	3.3	3.3	3.35
25.................	2.8	3.6	2.7	3.2	3.6	3.3	3.3	3.3	3.35
26.................	3.4	3.6	2.7	3.08	3.6	3.35	3.3	3.3	3.35
27.................	3.4	3.5	2.65	3.0	3.6	3.3	3.3	3.35	3.35
28.................	3.4	3.5	2.65	2.9	3.6	3.3	3.3	3.6	3.4
29.................	3.4	3.5	2.65	2.9	3.5	3.3	3.3	3.5	3.4
30.................	3.4	3.0	2.65	2.85	3.4	3.3	3.25	3.4	3.4
31.................	....	3.0	....	2.82	3.4	....	3.2	....	3.4

SANTA CRUZ RIVER AND DITCHES AT TUCSON, ARIZ.

This station was established October 15, 1905, at Congress Street Bridge, Tucson, Ariz. The gage height records were discontinued November 12, 1907, but discharge measurements have been made since then by engineers of the United States Reclamation Service, who maintain the station.

Manning and Farmers ditches divert practically the entire flow during the low period of Santa Cruz River. These ditches are taken out just above the gaging station, and their flow is determined by current-meter or weir measurements, supplemented by daily records, kept by the ditch managers, of the amount of water contained in each. On April 16 and 17, 1908, a permanent Cippoletti weir was established on Manning ditch 3 miles below the head gate. This water is used to irrigate lands on the north and south sides of Santa Cruz River in and about the vicinity of Tucson.

Conditions of flow are changeable.

The results published herewith were furnished by the United States Reclamation Service.

Discharge measurements of Santa Cruz River at Tucson, Ariz., in 1908–9.

[By engineers of the United States Reclamation Service.]

Date.	Width.	Area of section.	Gage height.	Dis-charge.	Date.	Width.	Area of section.	Gage height.	Dis-charge.
1908.	*Feet.*	*Sq. ft.*	*Feet.*	*Sec.-ft.*	*1909.*	*Feet.*	*Sq. ft.*	*Feet.*	*Sec.-ft.*
Feb. 5.....	53	53	0.92	213	July 19...	46	28	0.22	85
July 28....	38	21	.16	61	19...	94	274	3.20	....
Aug. 18....	42	42	.38	163	19...	91	170	2.06	....
Sept. 9....	89	172	2.00	1,080	20 ...	36	20	.50	32
9....	58	93	1.07	616	23...	55	79	.82	372
					24...	93	292	3.40	....
1909.					26...	84	79	.89	432
Mar. 16...	9.0	1.5	....	2.4	Aug. 6...	72	36	.23	91
24...	6.5	1.4	....	2.3	15...	78	89	.90	445
Apr. 5.....	7.5	6.2	.03	14	16...	86	147	2.02	974
July 17....	30	27	.14	83	17...	82	58	1.10	287

Discharge measurements of Manning ditch near Tucson, Ariz., in 1908–9.

[By engineers of the United States Reclamation Service.]

Date.	Dis-charge.	Date.	Dis-charge.
1908.	*Sec.-ft.*	**1909.**	*Sec.-ft.*
Mar. 2................................	11.2	Jan. 15................................	6.2
Apr. 11................................	9.9	22................................	8.3
18................................	7.3	Feb. 19................................	9.6
21................................	8.6	Mar. 30................................	7.6
21................................	10.2	Apr. 8................................	9.1
May 9................................	8.7	May 29................................	6.1
30................................	5.3	June 7................................	5.0
June 1................................	5.3	11................................	5.5
21................................	4.4	17................................	5.0
July 14................................	4.3	23................................	3.9
25................................	8.2	July 7................................	5.9
30................................	11.7	8................................	4.8
31................................	10.4	14................................	3.1
Aug. 8................................	8.8	14................................	4.8
11................................	8.5	15................................	3.4
25................................	8.2	Aug. 2................................	7.1
Sept. 7................................	11.2	27................................	6.6
Nov. 3................................	7.6	Sept. 3................................	9.5
11................................	5.5		

NOTE.—1908. Measurements Mar. 2, Apr. 11, 21 (second), 1908, and July 14 (second), 1909, were made by current meter at head gate. All others were made by weirs 3 miles below head gate.
From Feb. 2 to 22, 1908, ditch was dry. Diversion dam washed out 4 p. m. Mar. 22 to noon Mar. 23. Apr. 14 to 18, ditch dry for cleaning. Apr. 16–17, permanent Cippoletti weir established 3 miles below head gate.
Measurement July 25, 30, and 31 contain Farmers ditch water.
1909. .Measurement Mar. 30, 1909, made before cleaning ditch; that of Apr. 8 made after cleaning. May 17–18, water developed at head. May 19–22, ditch cleaned. Diversion dam washed out July 4 and again July 6, 1909.

Monthly discharge of Santa Cruz River and ditches at Tucson, Ariz., for 1908–9.

Month.	River discharge in second-feet.			Run-off (total in acre-feet).			Total.
	Maximum.	Minimum.	Mean.	River.	Manning ditch.	Farmers ditch.	
1908.							
January......................	16	....	2.3	130	600	240	970
February.....................	213	....	30.3	1,750	240	120	2,110
March........................	19	....	2.1	130	660	380	1,170
April........................	19	....	1.6	90	420	300	810
May..........................	....	....	....	....	530	300	830
June.........................					300	250	500
July.........................	6,780	....	132	8,100	460	....	8,560
August.......................	220	....	37	2,280	225	40	2,540
September....................	1,080	....	31.5	1,870	430	210	2,510
October......................	....	....	....	....	490	250	740
November.....................	....	....	....	....	470	240	710
December.....................	25	0.5	12.6	780	180	70	1,030
The year..............	6,780	....	20.8	15,100	5,000	2,400	22,500
1909.							
January......................	4	.4	1.9	120	430	280	830
February.....................	4	1.0	2.5	140	540	260	940
March........................	4	1.0	2.2	130	510	250	890
April........................	14	....	3.8	230	400	240	870
May..........................	6	....	.8	50	300	200	550
June.........................					240	180	420
July.........................	1,740	....	94.0	5,820	120	120	6,060
August.......................	1,610	....	121	7,510	130	....	7,640
September....................	940	....	21.0	1,260	500	....	1,760
October......................	....	....	....	200	....	....	....
November.....................	....	....	....	160	....	....	....
December.....................	....	....	....	200	....	....	....
The period............	....	....	....	15.800	....	....	....

NOTE.—The river carried water from Manning ditch 12½ days in October, 10 days in November, and 13½ days in December.

SALT RIVER DRAINAGE BASIN.

DESCRIPTION.

Salt River, though considered a tributary of the Gila, is in fact larger both in catchment area and in discharge. It receives the drainage from central Arizona, its principal tributary, the Verde, flowing southeasterly and south from the mountains and table-lands south of Colorado River. The Verde Valley is situated in Yavapai County, Ariz., on the headwaters of the stream, and extends from a canyon above Camp Verde to a point about 10 miles below the fort. About a mile below the junction of the Verde and 30 miles above Phoenix the Salt enters upon the plains of the Gila Valley.

The Salt River project involves the construction of a large storage reservoir controlled by the Roosevelt dam, on Salt River, at Roosevelt, Ariz., about 70 miles northeast of Phoenix, and the Granite Reef dam, on the same stream, about 40 miles below the Roosevelt dam, diverting water into the old Arizona canal on the right side of the river and into the highland canal on the left side of the river; the enlargement of these two canals, and the consolidation of the canal systems in the Salt River valley, in the vicinity of Phoenix and Mesa, Ariz., into two systems receiving water from these two canals. A power plant is being constructed at the storage dam for generating power from stored water in the reservoir and from water diverted from a power canal heading at a diversion dam on the Salt River about $18\frac{1}{2}$ miles above the storage dam. This power will be partly sold for industrial uses and partly used for pumping water from underground sources onto high lands in the Gila Indian reservation and in the Salt River valley. The power canal diversion dam, the power canal, the Roosevelt dam, and the Granite Reef dam are completed; the power plant, the improvements of the Arizona canal system, and the wells for underground pumping are under construction.[1]

SALT RIVER AT McDOWELL, ARIZ.

This station, which was established April 20, 1897, to determine the amount of water available for irrigation, is located one-third mile above the junction of Salt and Verde rivers, 30 miles northeast of Phoenix, 15 miles northeast of Mesa, and $1\frac{3}{4}$ miles above the Arizona canal diversion dam.

The bed of the river at this point is sandy and shifting, and frequent measurements are required to properly determine the daily discharge.

The station is maintained by the United States Reclamation Service.

---

[1] Description of the Salt River project taken from Seventh Ann. Rept. U. S. Recl. Service, 1907-8, pp. 50-51.

Discharge measurements of Salt River at McDowell, Ariz., in 1909.

[By W. Richins.]

Date.	Gage height.	Dis- charge.	Date.	Gage height.	Dis- charge.
	Feet.	*Sec.-ft.*		*Feet.*	*Sec.-ft.*
Jan. 4	2.80	137	July 16	4.15	363
8	2.30	48	20	4.10	333
12*a*	2.20	50	23	4.25	411
18*a*	2.15	*b* 56	26	4.35	511
23*a*	4.25	1,080	28	4.60	732
26*a*	6.30	3,770	30	4.65	746
29*a*	7.65	6,410	Aug. 3	4.50	619
Feb. 2*a*	5.95	4,500	6	4.40	571
5*a*	5.95	3,910	10	4.85	864
11	6.10	3,420	13	4.95	1,030
12	6.25	4,110	17	5.65	1,690
16	6.60	4,080	20	6.00	1,900
19	6.45	3,540	24	5.75	1,660
23	6.60	3,760	27	5.65	1,560
26	5.95	2,680	31	5.40	1,450
Mar. 2	6.55	3,520	Sept. 3	5.20	1,200
5	5.50	*c* 1,800	7	5.40	1,510
9	6.50	3,200	8	8.15	4,790
12	6.00	2,360	10	8.35	6,000
16	6.40	3,320	14	7.45	3,910
19	3.50	*ϩ* 194	17	6.80	2,940
23	6.75	3,350	21	6.05	1,920
26	6.75	3,420	23	5.80	1,660
30	7.15	4,100	24	5.70	1,470
Apr. 2	7.35	4,520	29	5.25	1,060
6	7.40	4,260	Oct. 1	5.15	910
9	7.20	3,720	5	4.85	670
13	4.90	*ϩ* 852	8	4.75	548
16	7.55	4,510	12	4.60	516
20	7.95	5,820	15	4.50	446
23	7.85	5,290	17	4.45	404
27	7.85	5,370	22	4.40	383
30	5.90	*c* 1,870	26	4.30	328
May 4	3.45	118	29	4.30	334
7	6.55	3,120	Nov. 2	4.25	314
7	6.55	3,760	5	4.25	328
8	5.80	*c* 1,710	8	5.20	998
10	3.80	153	9	5.30	1,150
12	3.65	112	10	5.80	1,530
14	3.50	98	11	5.85	1,560
17	3.45	109	12	5.85	1,600
19	3.85	270	16	5.75	1,430
21	4.25	449	17	5.75	1,420
25	4.75	738	19	5.75	1,450
28	4.95	886	23	5.70	1,400
June 1	4.95	884	26	5.65	1,350
4	4.95	887	28	5.50	1,230
8	4.90	732	30	5.50	1,250
11	4.85	719	Dec. 3	5.05	986
15	4.80	720	7	4.50	560
18	4.70	675	9	4.30	457
22	4.60	559	10	3.85	*c* 237
25	4.60	557	14	3.45	95
29	4.50	506	17	3.40	78
July 2	4.40	444	20	3.90	*d* 294
6	4.35	438	23	3.95	322
9	4.35	428	28	3.35	*c* 106
13	4.25	413	31	3.30	108

a Made by J. C. Leaming.
b Gage height affected by backwater in Verde River.
c Low discharge is caused by closing gate, at Roosevelt dam.
d Water turned back in river.

Daily gage height, in feet, of Salt River at McDowell, Ariz., for 1909.

[W. Richins, observer.]

Day.	Jan.	Feb.	Mar.	Apr.	May.	June.	July.	Aug.	Sept.	Oct.	Nov.	Dec.
1	5.80	6.15	6.70	7.38	4.42	4.95	4.42	4.55	5.32	5.12	4.25	5.28
2	5.45	5.95	6.52	7.32	3.75	4.95	4.40	4.50	5.28	5.08	4.25	5.25
3	3.50	5.85	6.50	4.35	3.58	4.95	4.40	4.48	5.20	5.00	4.25	5.08
4	2.82	5.95	6.65	5.65	3.42	4.92	4.40	4.42	5.15	4.92	4.25	5.05
5	2.65	5.98	6.25	6.90	3.38	4.92	4.35	4.40	5.20	4.85	4.25	4.80
6	2.50	5.92	6.42	7.40	3.35	4.90	4.35	4.40	5.25	4.80	4.20	4.60
7	2.38	5.95	6.40	7.25	6.52	4.90	4.35	4.40	6.02	4.78	4.20	4.52
8	2.30	5.95	6.45	7.20	5.80	4.90	4.35	4.40	8.22	4.72	5.20	4.40
9	2.30	5.85	6.48	7.20	4.00	4.90	4.32	4.78	8.48	4.70	5.32	4.28
10	2.22	6.00	6.05	7.25	3.82	4.88	4.30	4.85	8.30	4.65	5.78	3.98
11	2.20	6.22	6.00	7.40	3.68	4.85	4.30	4.80	8.10	4.60	5.82	3.70
12	2.20	6.38	6.00	7.55	3.62	4.85	4.25	4.80	7.80	4.60	5.85	3.60
13	2.20	6.20	6.15	4.55	3.58	4.85	4.25	4.92	7.58	4.55	5.85	3.50
14	2.20	6.60	6.50	3.75	3.50	4.80	4.20	4.95	7.40	4.52	5.80	3.45
15	3.62	6.60	6.62	7.45	3.45	4.80	4.18	4.95	7.15	4.50	5.80	3.45
16	3.32	6.60	6.42	7.55	3.45	4.78	4.15	5.50	6.95	4.50	5.75	3.42
17	2.30	6.55	6.00	7.80	3.45	4.75	4.10	5.70	6.75	4.45	5.75	3.40
18	2.15	6.40	3.85	4.55	3.65	4.70	4.10	5.92	6.55	4.45	5.45	3.40
19	2.10	6.38	3.70	7.80	3.90	4.70	4.10	6.00	6.35	4.45	5.75	3.40
20	2.10	6.55	6.75	7.92	4.05	4.65	4.10	5.98	6.15	4.40	5.75	3.88
21	2.10	6.50	6.70	7.32	4.28	4.60	4.10	5.92	6.05	4.40	5.75	3.90
22	2.10	6.58	6.70	7.90	4.42	4.60	4.28	5.80	5.98	4.40	5.70	3.95
23	4.30	6.60	6.72	7.85	4.55	4.60	4.25	5.75	5.82	4.35	5.70	3.95
24	5.42	6.38	6.68	7.85	4.62	4.60	4.25	5.72	5.68	4.35	5.70	3.62
25	5.40	5.95	6.78	7.85	4.78	4.58	4.35	5.62	5.58	4.30	5.65	3.40
26	6.32	5.95	6.72	7.85	4.88	4.50	4.38	5.62	5.45	4.30	5.62	3.35
27	4.28	6.00	6.10	7.85	4.95	4.50	4.50	5.62	5.35	4.30	5.70	3.35
28	6.65	6.10	6.20	7.62	4.95	4.50	4.60	5.58	5.30	4.28	5.50	3.35
29	7.55		7.20	7.32	5.00	4.48	4.65	5.50	5.22	4.30	5.40	3.35
30	6.90		7.18	5.85	4.95	4.45	4.62	5.45	5.12	4.30	5.48	3.35
31	6.55		7.40		4.95		4.60	5.70		4.30		3.35

NOTE.—Rise Jan. 15–17 due to backwater from Verde River. No change in amount turned out at Roosevelt.

Gage heights are affected by regulation at Roosevelt dam.

Maximum recorded gage heights occurred as follows: Jan. 29, 7.70; Feb. 1, 6.15; May 7, 6.55; Aug. 31, 6.00; Sept. 9, 8.50; Oct. 1, 5.15; and Dec. 1, 5.30 feet.

Minimum recorded gage heights occurred as follows: Mar. 19, 3.50; April 14, 3.70; Sept. 30, 5.10; and Oct. 28, 4.25 feet.

Daily discharge, in second-feet, of Salt River at McDowell, Ariz., for 1909.

Day.	Jan.	Feb.	Mar.	Apr.	May.	June.	July.	Aug.	Sept.	Oct.	Nov.	Dec.
1	2,800	4,700	3,920	4,700	920	944	450	670	1,340	880	320	1,090
2	2,230	4,500	3,480	4,500	200	920	450	630	1,250	840	314	1,060
3	440	4,100	3,450	460	140	900	450	610	1,150	780	320	1,000
4	145	4,000	3,570	1,500	110	870	450	590	1,120	760	325	940
5	120	4,000	1,550	3,100	90	830	450	580	1,200	670	328	790
6	80	3,700	3,280	4,280	70	800	448	571	1,280	620	330	660
7	60	3,600	3,140	3,900	3,100	750	440	560	2,440	590	330	610
8	60	3,500	3,170	3,720	1,710	734	440	560	5,600	530	998	530
9	60	3,000	3,140	3,720	240	730	430	820	6,400	530	1,180	450
10	55	3,100	2,380	3,900	160	720	430	864	5,800	530	1,500	300
11	50	3,400	2,360	5,300	150	719	430	840	5,200	520	1,560	180
12	50	3,600	2,360	6,000	130	720	420	860	4,550	516	1,600	150
13	50	3,300	2,650	640	120	720	413	990	4,120	490	1,560	110
14	50	4,600	3,480	240	98	720	390	1,030	3,850	470	1,500	95
15	50	4,400	3,750	4,200	100	723	380	1,000	3,450	452	1,500	90
16	50	4,000	4,510	100	710	363	1,540	3,150	450	1,430	80	
17	50	3,900	2,450	5,300	109	690	350	1,620	2,840	440	1,420	78
18	56	3,700	310	320	180	654	340	1,970	2,550	430	1,100	80
19	55	3,600	260	5,300	290	660	340	2,010	2,250	412	1,450	80
20	55	3,700	3,900	5,600	360	600	333	1,880	2,040	390	1,450	290
21	55	3,600	3,600	3,800	470	559	330	1,810	1,920	375	1,440	300
22	55	3,740	3,500	5,500	550	560	420	1,680	1,820	361	1,410	320
23	1,110	3,760	3,450	5,300	610	560	411	1,640	1,700	350	1,400	324
24	2,300	3,300	3,350	5,320	660	560	420	1,640	1,440	345	1,400	190
25	2,200	2,600	3,500	5,330	750	558	520	1,500	1,350	335	1,350	110
26	3,800	2,590	3,370	5,350	860	530	550	1,520	1,240	327	1,300	110
27	800	2,600	2,200	5,360	880	520	732	1,520	1,130	330	1,420	110
28	3,760	2,750	2,300	5,100	884	510	740	1,520	1,100	330	1,230	105
29	6,000		4,400	4,600	930	500	750	1,490	1,050	334	1,120	110
30	5,500		4,200	1,800	940	470	740	1,480	960	330	1,220	110
31	5,150		4,820		940		710	1,830		325		110

NOTE.—These discharges were obtained by the indirect method for shifting channels.

Monthly discharge of Salt River at McDowell, Ariz., for 1909.

[Drainage area, 6,260 square miles.]

Month.	Discharge in second-feet.				Run-off.	
	Maximum.	Minimum.	Mean.	Per square mile.	Depth in inches on drainage area.	Total in acre-feet.
January........................	6,000	50	1,200	0.192	0.22	73,800
February.......................	4,600	2,590	3,620	.578	.60	201,000
March.........................	4,820	260	3,050	.487	.56	188,000
April..........................	6,000	240	3,940	.629	.70	234,000
May...........................	3,100	70	544	.087	.10	33,400
June..........................	944	470	681	.109	.12	40,500
July..........................	750	330	468	.075	.09	28,800
August........................	2,010	560	1,220	.195	.22	75,000
September.....................	6,400	960	2,510	.401	.45	149,000
October.......................	880	325	485	.077	.09	29,800
November......................	1,600	314	1,130	.180	.20	67,200
December......................	1,090	78	341	.054	.06	21,000
The year.....................	6,400	50	1,580	.252	3.41	1,140,000

NOTE.—These estimates were recomputed to conform to the computation rules used by the U. S. Geological Survey. The values differ slightly from those computed by the Reclamation Service.

VERDE RIVER AT McDOWELL, ARIZ.

This station, which was established April 20, 1897, is located 30 miles northeast of Phoenix, 15 miles northeast of Mesa, 2¼ miles above the Arizona canal diversion dam, and three-fourths of a mile above the mouth of the river.

As the bed of the stream at this point is sandy and shifting, frequent measurements are required to properly determine the daily discharge.

This station is maintained by the United States Reclamation Service.

Discharge measurements of Verde River at McDowell, Ariz., in 1909.

[By W. Richins.]

Date.	Gage height.	Discharge.	Date.	Gage height.	Discharge.
	Feet.	*Sec.-feet.*		*Feet.*	*Sec.-feet.*
Jan. 4......................	4.30	519	Apr. 2......................	7.20	2,710
8......................	4.20	453	6......................	7.45	2,860
12a......................	4.05	481	9......................	6.65	1,260
17a......................	6.50	3,650	13......................	6.65	1,390
20a......................	5.50	1,450	16......................	6.55	1,240
22a......................	5.10	1,000	20......................	6.45	1,050
26a......................	6.20	2,720	23......................	6.15	663
30a......................	5.55	1,500	27......................	5.80	424
Feb. 3a......................	5.20	954	30......................	5.60	333
9a......................	5.70	2,240	May 4......................	5.45	248
12......................	5.50	1,180	7......................	5.35	248
16......................	6.20	2,240	10......................	5.25	185
19......................	5.90	1,430	12......................	5.25	205
23......................	6.20	1,820	14......................	5.20	181
26......................	5.80	1,240	17......................	5.20	191
Mar. 2......................	6.35	1,840	19......................	5.20	184
5......................	6.50	1,960	21......................	5.20	191
9......................	6.60	2,000	25......................	5.15	162
12......................	5.95	1,180	28......................	5.15	165
16......................	6.35	1,910	June 1......................	5.10	164
19......................	6.45	1,760	4......................	5.10	162
23......................	6.35	1,500	8......................	5.05	148
26......................	6.35	1,470	11......................	4.95	132
30......................	7.90	5,700	15......................	4.90	129

a Made by J. C. Leaming.

Discharge measurements of Verde River at McDowell, Ariz., in 1909—Continued.

Date.	Gage height.	Discharge.	Date.	Gage height.	Discharge.
	Feet.	*Sec.-feet.*		*Feet.*	*Sec.-feet.*
June 18	4.90	132	Oct. 1	5.60	196
22	4.90	126	5	5.50	161
25	4.90	124	8	5.50	159
29	4.90	114	12	5.50	175
July 2	4.85	119	15	5.50	162
6a	5.45	416	19	5.50	162
9	5.10	233	22	5.45	144
13	4.95	160	26	5.45	155
16	4.85	147	29	5.45	156
July 20	5.10	188	Nov. 2	5.45	157
23	5.88	651	5	5.50	227
26	6.80	1,760	8	5.45	181
28	6.75	1,396	9	5.45	194
30	6.05	607	10	5.45	195
Aug. 3	5.60	328	11	5.45	192
6	5.55	310	12	5.50	201
10	6.10	623	16	5.50	245
13	7.30	2,300	17	5.55	267
17	8.08	4,020	19	5.55	254
20	6.90	1,340	23	5.55	241
24	6.50	801	26	5.50	213
27	6.65	1,050	28	5.55	260
31	6.10	492	30	5.60	297
Sept. 3	5.95	398	Dec. 3	5.60	288
7	7.10	1,800	7	5.60	290
8	7.10	1,760	9	5.60	283
10	6.65	828	10	5.65	326
14	6.05	374	14	5.85	472
17	5.85	275	17	5.70	330
21	5.75	247	20	5.65	322
23	5.70	211	23	6.00	503
24	5.65	207	28	5.90	382
29	5.55	180	31	5.80	337

a Made from cable.

NOTE.—From June 11 to July 20, excepting July 6, measurements were made about 500 feet above cable by wading. Water too shallow at cable to make an accurate measurement. Beginning July 23, measurements were resumed from the cable.

Daily gage height, in feet, of Verde River at McDowell, Ariz., for 1909.

Day.	Jan.	Feb.	Mar.	Apr.	May.	June.	July.	Aug.	Sept.	Oct.	Nov.	Dec.
1	4.40	5.50	6.48	7.18	5.55	5.10	4.85	5.70	6.08	5.60	5.45	5.60
2	4.40	5.32	6.40	7.08	5.50	5.0	4.85	5.62	5.98	5.55	5.45	5.60
3	4.35	5.20	6.40	6.82	5.45	5.08	4.85	5.68	5.92	5.50	5.45	5.60
4	4.30	5.20	6.40	7.15	5.45	5.08	4.85	5.68	5.85	5.50	5.48	5.55
5	4.30	5.20	6.60	7.45	5.40	5.0	4.85	5.62	6.00	5.50	5.50	5.55
6	4.28	5.20	6.82	7.32	5.38	5.07	5.38	5.55	6.78	5.50	5.50	5.62
7	4.22	5.20	6.95	7.05	5.35	5.05	5.30	6.65	7.18	5.50	5.45	5.60
8	4.20	5.20	6.95	6.80	5.30	5.05	5.18	6.05	7.05	5.50	5.45	5.60
9	4.15	5.75	6.58	6.58	5.30	5.00	5.08	6.20	6.78	5.50	5.45	5.60
10	4.15	5.90	6.28	6.50	5.25	4.98	5.02	6.10	6.60	5.50	5.45	5.62
11	4.10	5.70	6.02	6.50	5.25	4.95	5.00	6.22	6.42	5.50	5.45	5.65
12	4.05	5.45	5.95	6.55	5.25	4.95	4.95	6.18	6.20	5.50	5.50	5.70
13	4.05	5.40	5.78	6.65	5.22	4.90	4.92	6.72	6.12	5.50	5.50	5.75
14	4.05	5.80	5.65	6.50	5.20	4.90	4.90	6.72	6.02	5.50	5.50	5.82
15	7.98	6.20	5.60	6.48	5.20	4.90	4.85	7.75	5.92	5.50	5.50	5.78
16	7.50	6.18	6.35	6.65	5.20	4.90	4.82	8.22	5.88	5.50	5.52	5.70
17	6.42	5.92	6.58	6.60	5.20	4.90	4.80	7.92	5.85	5.50	5.55	5.70
18	6.00	5.75	6.40	6.55	5.20	4.90	6.20	7.52	5.80	5.50	5.55	5.70
19	5.75	5.95	6.42	6.52	5.20	4.90	5.22	7.18	5.75	5.50	5.55	5.65
20	5.50	6.08	6.45	6.45	5.20	4.90	5.08	7.12	5.75	5.48	5.55	5.65
21	5.30	6.10	6.40	6.38	5.20	4.90	5.20	7.10	5.75	5.45	5.55	5.68
22	5.10	6.32	6.38	6.25	5.20	4.90	5.52	5.70	5.70	5.45	5.55	5.88
23	5.00	6.15	6.32	6.12	5.15	4.90	5.78	6.48	5.70	5.45	5.55	6.05
24	8.00	5.88	6.15	6.02	5.15	4.90	5.55	6.62	5.65	5.45	5.52	6.05
25	6.90	5.72	6.20	5.95	5.15	4.90	5.42	6.45	5.65	5.45	5.50	6.00
26	6.30	5.82	6.50	5.88	5.15	4.90	6.35	6.70	5.60	5.45	5.50	6.00
27	6.02	5.88	6.95	5.80	5.15	4.90	6.05	6.60	5.60	5.45	5.50	5.92
28	5.85	5.40	6.95	5.75	5.15	4.90	6.70	6.45	5.55	5.45	5.55	5.90
29	5.98	......	8.00	5.68	5.15	4.90	6.70	6.20	5.55	5.45	5.60	5.88
30	5.75	......	7.82	5.00	5.10	4.90	6.00	6.12	5.58	5.45	5.60	5.82
31	5.58	......	7.40	......	5.10	......	5.82	6.08	......	5.45	......	5.80

NOTE.—Gage heights are affected by regulation at Roosevelt dam. Maximum recorded gage heights occurred as follows: Jan. 15, 8.90; Mar. 29, 8.30; July 28, 6.75; Aug. 16, 8.80; Sept. 7, 7.25; Dec. 23, 6.10.

Daily discharge, in second-feet, of Verde River at McDowell, Ariz., for 1909.

Day.	Jan.	Feb.	Mar.	Apr.	May.	June.	July.	Aug.	Sept.	Oct.	Nov.	Dec.
1	510	1,360	2,130	2,540	300	164	120	380	480	197	155	290
2	520	1,100	1,920	2,400	270	164	120	340	400	180	157	290
3	520	954	1,860	1,780	250	164	120	360	370	170	160	273
4	519	1,000	1,830	2,370	248	162	120	360	320	160	190	270
5	510	1,050	2,120	2,950	245	150	120	340	420	161	227	270
6	490	1,100	2,550	2,580	245	150	300	320	1,200	160	220	300
7	470	1,150	2,800	1,900	248	150	320	1,400	1,920	160	200	291
8	453	1,200	2,750	1,500	220	149	270	600	1,650	159	181	300
9	450	2,050	1,960	1,160	210	140	230	730	1,080	160	194	299
10	470	2,100	1,490	1,100	187	135	190	626	770	165	195	327
11	480	1,600	1,170	1,120	195	132	180	710	630	165	192	340
12	481	1,250	1,100	1,210	203	130	170	610	470	175	201	350
13	480	1,070	1,000	1,390	190	130	160	2,020	420	165	205	300
14	480	1,550	900	1,130	181	130	155	1,600	380	165	205	472
15	6,800	2,240	900	1,100	190	130	150	3,400	340	162	205	300
16	4,650	2,200	1,910	1,380	190	130	145	4,900	320	162	245	340
17	3,400	1,640	2,030	1,280	191	130	140	3,850	272	162	267	330
18	2,300	1,300	1,790	1,180	190	130	960	2,670	200	162	260	330
19	1,820	1,490	1,720	1,140	185	130	300	1,900	250	162	254	320
20	1,460	1,650	1,720	1,050	190	130	190	1,700	250	155	250	318
21	1,200	1,680	1,610	940	191	130	280	1,820	247	150	250	330
22	1,000	2,500	1,580	790	185	126	440	1,000	220	143	250	400
23	900	1,740	1,500	650	170	125	600	780	211	150	241	500
24	7,000	1,320	1,250	570	170	125	460	970	208	150	230	500
25	4,300	1,160	1,300	520	162	125	390	760	208	150	220	490
26	3,000	1,250	1,690	480	165	120	1,120	1,120	200	151	213	450
27	2,350	1,290	2,550	424	165	120	800	1,150	200	150	220	410
28	1,990	850	2,600	400	166	120	1,350	830	190	150	260	384
29	2,200	......	5,050	370	165	116	810	600	180	154	290	370
30	1,820	......	4,680	333	165	120	560	550	190	155	297	350
31	1,550	......	3,450	......	165	......	420	500	......	155	......	337

NOTE.—These discharges were obtained by the indirect method for shifting channels.

Monthly discharge of Verde River at McDowell, Ariz., for 1909.

[Drainage area, 6,000 square miles.]

Month.	Discharge in second-feet.				Depth in inches on drainage area.	Run-off (total in acre-feet).
	Maximum.	Minimum.	Mean.	Per square mile.		
January	6,800	450	1,760	0.293	0.34	108,000
February	2,500	850	1,460	.243	.25	81,000
March	5,050	900	2,030	.338	.39	125,000
April	2,950	333	1,260	.210	.23	74,900
May	300	162	200	.033	.04	12,300
June	164	116	135	.022	.02	8,050
July	1,350	120	379	.063	.07	23,300
August	4,900	320	1,260	.210	.24	77,100
September	1,920	180	475	.079	.09	28,300
October	197	143	160	.027	.03	9,840
November	297	155	221	.037	.04	13,200
December	500	270	354	.059	.07	21,800
The year	6,800	116	805	.134	1.81	583,000

MISCELLANEOUS MEASUREMENTS IN COLORADO RIVER DRAINAGE BASIN.

The following miscellaneous discharge measurements were made in Colorado River drainage basin during 1909. They are arranged by drainage basins in downstream order.

Miscellaneous measurements in Colorado River drainage basin in 1909.

Green River basin.

Date.	Stream.	Tributary to—	Locality.	Gage height.	Discharge.
				Feet.	*Sec.-ft.*
Nov. 13	Lake Fork	Duchesne River	Lower Canyon, Utah		156
June 14	Great Western canal	Cottonwood Creek	Orangeville, Utah		17.3
July 28	....do	....do	....do		12.1
June 14	Clipper canal	....do	....do	3.2	20.3
July 28	....do	....do	....do	3.5	28.4
June 14	Blue Cut canal	....do	....do	3.2	54.2
July 27	....do	....do	....do	2.76	34.8
Aug. 23	....do	....do	....do	2.25	4.5
June 15	Mammoth canal	....do	....do	3.65	30.5
July 27	....do	....do	....do	4.13	43.0
Apr. 28	North canal	Ferron Creek	Ferron, Utah		24.9
June 12	....do	....do	....do	2.65	90.9
July 26	....do	....do	....do	2.50	61.5
Aug. 23	....do	....do	....do	2.00	23.5
Apr. 28	South canal	....do	....do		21.7
June 12	....do	....do	....do		15.6
July 26	....do	....do	....do		30.5
May 4	Huntington Creek	San Rafael River	At county bridge, near Huntington, Utah.	3.25	08.1
June 18	....do	....do	....do	3.8	696
July 29	....do	....do	....do	1.7	10.1
June 17	Cleveland canal	Huntington Creek	Huntington, Utah	4.0	111
July 29	....do	....do	....do	4.15	129
Aug. 21	....do	....do	....do	2.6	1.5
June 17	North canal	....do	....do	3.05	51.8
Aug. 21	....do	....do	....do	2.24	15.9
Sept. 21	....do	....do	....do	2.05	2.5
June 17	Huntington canal	....do	....do	1.7	63.4
Aug. 21	....do	....do	....do	1.0	17.9
Sept. 21	....do	....do	....do	1.0	18.3

Grand River basin.

Date.	Stream.	Tributary to—	Locality.	Gage height.	Discharge.
Aug. 10	East Inlet to Grand Lake	Grand Lake	Grand Lake, Colo	−4.60	48
June 27	South Fork of Grand River.	Grand River	At Grigg's ranch, near Lehman, Colo.		718
Aug. 9	....do	....do	Lehman's ranch, Colo	4.80	112
Do...	Stillwater Creek	....do	King's ranch, near - Granby, Colo.		*a* 3
Aug. 11	....do	....do	....do		*a* 3
June 17.	Willow Creek	....do	One-half mile above mouth, near Dexter, Colo.	6.10	910
May 4	....do	....do	At mouth, near Spitzer..	3.63	85
June 17	....do	....do	....do	6.10	910
Jan. 20	Blue River	....do	Governor mine, Colo		1.6
Mar. 23	....do	....do	Above Breckenridge, Colo.		1.5
Jan. 20	....do	....do	One-fourth mile above Spruce Creek Plant, near Breckenridge, Colo.		2.1
May 12	....do	....do	At Breckenridge, Colo		45
Mar 22	....do	....do	At Dillon, Colo		60
May 12	Ten Mile Creek	Blue River	....do		133
July 14	Gypsum Creek	Eagle River	One-fourth mile north of Gypsum, Colo.		79
Jan. 17	Roaring Fork	Grand River	Aspen, Colo		29
Mar. 20	....do	....do	....do		35
Apr. 24	....do	....do	....do		68
Jan. 17	Hunters Creek	Roaring Fork	....do		5.6
Mar. 20	....do	....do	....do		5.0
Apr. 24	....do	....do	....do		15
Mar. 20	Castle Creek	....do	....do	.9	50
Apr. 24	....do	....do	....do	.95	51
Jan. 17	....do	....do	At mouth, near Aspen, Colo.		61
Apr. 24	Maroon Creek	....do	Near Aspen, Colo		17
Jan. 17	....do	....do	At mouth, near Aspen Colo.		11
Mar. 20	....do	....do	....do		12
Jan. 18	Frying Pan River	....do	Thomasville, Colo		34
Mar. 21	....do	....do	....do		48
Do...	Lime Creek	Frying Pan River	At mouth, Thomasville, Colo.		9.2
Jan. 14	Crystal River	Roaring Fork	Redstone, Colo	2.40	58

a Estimated.

Miscellaneous measurements in Colorado River drainage basin in 1909—Continued.

Grand River basin—Continued.

Date.	Stream.	Tributary to—	Locality.	Gage height.	Discharge.
				Feet.	*Sec.-ft.*
Mar. 17	Crystal River............	Roaring Fork..........	Redstone, Colo.........	2.40	56
Jan. 15	.....do..................	.....do................	Sewell, Colo............	1.85	111
Do...	.....do..................	.....do................	Hot Springs, Colo......		76
Mar. 18	.....do..................	.....do................	....do....:...........		62
Jan. 19	Spruce and Crystal canal.	Crystal River..........	One-fourth mile below headgate.		.5
Jan. 15	Coal Creek...............	.....do................	Redstone, Colo.........		6.2
Mar. 17	.....do..................	.....do................	....do.................		11
Jan. 15	Avalanche Creek........	.....do................	At mouth, Redstone, Colo.		20
Mar. 18	.....do..................	.....do................	....do.................		20
Do...	Thompson Creek........	.....do................	Sewell, Colo............		*a* 9
Apr. 22	.....do..................	.....do................	....do.................	1.12	40
Aug. 26	Whitewater Creek......	Gunnison River........	One-half mile above head gate of ditch, T. 12 S., R. 98 W., Colorado.		2.3
Oct. 10	.....do..................	.....do................	....do.................		2.5
Mar. 4	.....do..................	.....do................	One-fourth mile above head gate.		1.3
4	.....do..................	.....do................	....do.................		1.4
3	.....do..................	.....do................	At head gate of ditch in N. E. ¼ sec. 8, T. 12 S., R. 98 W., Colorado.		1.2
4	.....do..................	.....do................	....do.................		1.3
4	.....do..................	.....do................	....do.................		1.3
Aug. 25	.....do..................	.....do................	....do.................		2.2
26	.....do..................	.....do................	....do.................		2.6

Fremont River basin.

Date.	Stream.	Tributary to—	Locality.	Gage height.	Discharge.
Aug. 29	Upper Fremont Canyon.	Fremont River........	In Rabbit Valley, Utah.		2.5
Do...	Right Fork of Fremont River.	.....do...............	....do.................		6.0
Aug. 26	Ivy Creek...............	.....do................	....do.................		2.0
Aug. 28	Solomon Creek..........	.....do................	....do.................		2.0
Aug. 30	Bulberry Creek.........	.....do................	....do.................		.6
Do...	Donkey Creek...........	.....do................	....do.................		2.0
Aug. 11	Sand Creek..............	.....do................	....do.................		2.0
Do...	Fish Creek..............	.....do................	....do.................		1.5
Aug. 12	Grover Creek...........	.....do................	....do.................		1.8
June 30	Torry canal.............	.....do................	In Rabbit Valley, Utah, (near Loa).		11
Aug. 30	.....do.................	.....do................	In Rabbit Valley, Utah.		3.0
Aug. 13	Oak Creek..............	.....do................	....do.................		3.0
Aug. 26	Queatah-up-pah canal..	.....do................	....do.................		1.8
July 26	Muddy Creek...........	.....do................	At county bridge, near Emery, Utah.		5.6
June 13	Independent canal......	Muddy Creek..........	Emery, Utah..........	2.15	7.9
Do...	Emery canal...........	.....do................	....do.................	3.90	60.4
Aug. 25	.....do.................	.....do................	....do.................	3.43	13.7

San Juan River basin.

Date.	Stream.	Tributary to—	Locality.	Gage height.	Discharge.
Sept. 30	San Juan River........	Colorado River........	Above mouth of Piedras River, near Arboles, Colo.		*a* 200
Do...	Navajo Creek..........	San Juan River........	At mouth, near Pajosa Junction, Colo.		*a* 25
Do...	Piedras River..........	.....do................	At Arboles, Colo.......		*a* 100
Do...	Los Pinos River........	.....do................	At railroad bridge at La Boca, Colo.		*a* 60
Do...	Spring Creek...........	Los Pinos River.......	At La Boca, Colo......		*a* 10
Do...	Rio Florida...........	Animas River..........	At Denver and Rio Grande railroad crossing, south of Durango, Colo.		*a* 20

Little Colorado River basin.

Date.	Stream.	Tributary to—	Locality.	Gage height.	Discharge.
Apr. 7	Manuelito Creek........	Little Colorado River...	Manuelito, N. Mex.....	*b* − .41	8.4
May 24	.....do.................	.....do................	....do.................	*b* − .35	11.7
June 26	.....do.................	.....do................	....do.................	*b* −1.03	.2

a Estimated.   *b* Distance down from reference mark to water surface.

Miscellaneous measurements in Colorado River drainage basin in 1909—Continued.

Virgin River basin.

Date	Stream.	Tributary to—	Locality.	Gage height.	Discharge.
				Feet.	*Sec.-ft.*
Oct. 13	Virgin River............	Colorado River........	Above Zion Creek, near Springdale, Utah.	....	48.5
Do...	Zion Creek..............	Virgin River............	Just above mouth, near Springdale, Utah.	....	62.5
July 12	Hurricane canal........	.....do.................	Near Virgin, Utah......	....	13.7
Do...	La Verkin canal........	.....do.................	.....do.................	....	11.9
July 14	South side Santa Clara canal.	Santa Clara River.......	Near Santa Clara, Utah..	....	2.6
Do...	North side Santa Clara canal.	.....do.................	.....do.................		4.9
July 15	Six Mile Flat canal.....	.....do.................	Near Central, Utah.....		2.8
July 14	South side St. George canal.	.....do.................	Near St. George, Utah...	....	4.1
Do...	North side St. George canal.	.....do.................	.....do.................		3.1
Do...	Seep canal	.....do.................	.....do.................		4.8
Oct. 18	Pinto Creek............	....	Near Pinto, Utah......	....	4.3

Gila River basin.

Date	Stream.	Tributary to—	Locality.	Gage height.	Discharge.
Apr. 25	Gila River..............	Colorado River........	7 miles above Gila Farm, N. Mex.	1.40	256
Do...	Mogollon River.........	Gila River.............	1 mile above mouth, above Gila, N. Mex.	....	48
Apr. 19	Duck Creek.............	.....do.................	Cliff, N. Mex..........	5.10	3
Oct. 9	.....do.................	.....do.................	.....do.................		a 3
Do...	Mangus Creek..........	.....do	Mangus, N. Mex........		a 1
Feb. 7	Whitewater Creek.......	San Francisco River....	Glenwood, N. Mex......	b — .38	8.3
Apr. 21	.....do.................	.....do.................	.....do.................	b — .14	84
June 6	.....do.................	.....do.................	.....do.................	b — .42	25
July 4	.....do.................	.....do.................	.....do.................	b — .60	9.8
Oct. 9	.....do.................	.....do.................	.....do.................	....	a 6.5

a Estimated.

b Distance down from reference point to water surface. Reference point is a staple driven into root of large cottonwood tree about 20 feet downstream from tree which is the second one on the right bank above the ford.

INDEX.